Nonlinear Continuum Mechanics of Solids

Springer
Berlin
Heidelberg
New York
Barcelona
Hong Kong
London
Milano
Paris
Singapore
Tokyo

Yavuz Başar · Dieter Weichert

Nonlinear Continuum Mechanics of Solids

Fundamental Mathematical and Physical Concepts

With 35 figures and 5 tables

Springer

Prof. Dr.-Ing. Yavuz Başar
Ruhr-Universität Bochum
Institut für Statik und Dynamik
Universitätsstraße 150
44780 Bochum
Germany

Prof. Dr.-Ing. Dieter Weichert
RWTH Aachen
Institut für Allgemeine Mechanik
Templergraben 64
52056 Aachen
Germany

ISBN 3-540-66601-X Springer Verlag Berlin Heidelberg New York

Cataloging-in-Publication Data applied for

Başar, Yavuz:
Nonlinear continuum mechanics of solids : fundamental concepts and perspectives /
Yavuz Başar ; Dieter Weichert. – Berlin ; Heidelberg ; New York ; Barcelona ;
Hong Kong ; London ; Milano ; Paris ; Singapore ; Tokyo : Springer, 2000

ISBN 3-540-66601-X

Springer-Verlag is a company in the specialist publishing group BertelsmannSpringer

Printed in Germany

Coverdesign: Medio, Berlin
Typesetting: Camera-ready copy by author
SPIN: 10701373 68/3020 – 5 4 3 2 1 0 – Printed on acid-free paper

Preface

After the fast development of computational methods in the last decades, attention in today's research in continuum and computational mechanics is focusing increasingly on the refinement of theoretical formulations. Nowadays, the problem ist not so much to transform existing theoretical models into numerical codes, but to improve the accuracy of these models in various fields (material modelling, large strain analysis, damage mechanics, optimization of nonlinear structures, etc.) in order to ensure a more realistic simulation especially of nonlinear phenomena. Evidently, research on such complex topics requires a profound understanding of nonlinear continuum mechanics and expertise in tensor analysis.

The content of this book reflects essentially the lectures given by the authors at Ruhr-University Bochum, Technical University Aachen and Technical University Lille for graduate students of engineering and material sciences, applied mathematicians and research engineers wishing to be brought rapidly within reach of their specific research area. Hence, fundamental concepts which are of general interest and not special topics have been found to be relevant for the presentation. The authors are aware of the gap between *"what is taught in classical engineering education and what is required by research on current topics of nonlinear structural and continuum mechanics"*. This book attempts to bridge this gap.

In this book equations are developed in absolute tensor notation which is used almost exclusively in modern literature of continuum mechanics. This notation provides a general and elegant formulation of the theoretical background particularly for nonlinear problems and is, moreover, of considerable help in transforming the theory into numerical codes. In contrast to many other works on tensor calculus, here the aim is to present the fundamentals of continuum mechanics of solids together with the mathematical background in an unified description. Accordingly, the mathematical tools are presented so as to enable the reader to study the book without permanent reference to other works.

The first chapter presents the basic rules of tensor calculus in absolute notation and introduces the special tensors relevant for continuum mechanics. It also deals extensively with the eigenvalue problems of second-order tensors, the orthogonal and rotation tensors and the differentiation rules of tensors. Thus it involves all basic mathematical concepts needed in the sequel.

The second chapter is devoted to a detailed description of deformations of solids under systematical consideration of geometrical nonlinearities. Here, various deformation and strain measures are defined, their mechanical interpretation is given through the corresponding eigenvalue problems and a systematical classification of the strain tensors is presented. This section involves also further relevant topics such as pull-back and push-forward operations, and the definition of the rate of deformation tensor as well as isotropic tensor functions the last being of special significance for material modelling. The detailed derivations in this part should enable the reader to get experienced with the tensor calculus. An in-depth study of this chapter together with the foregoing one is recommended for an easy understanding of the book. Formulae and definitions of tensor algebra in index notation, which are prerequisites for chapter 1, are summarized in appendix 1.

Chapter 3 starts with the definition of the CAUCHY stress tensor where emphasis is placed on its mechanical interpretation. Subsequently, various stress tensors are defined by purely mathematical transformations and then shown to be energy conjugate to the strain tensors from the previous chapter through the rate of internal energy. A precise definition of the internal energy is, however, given in chapter 5 in connection with the law of conservation of energy.

The notion of the material time derivative is explained in chapter 4 and then applied to define the velocity and the acceleration vector. The material time derivatives of some geometrical variables such as volume, surface and line elements are also given in this chapter.

Chapter 5 presents in a systematic way the balance laws: conservation of mass, balance of momentum, balance of moment of momentum, balance of kinetic energy and conservation of energy. Equations of motion are obtained as local formulation of balance of momentum. Similarly, the symmetry of the CAUCHY stress tensor introduced in chapter 3 as a postulate is proved through the local formulation of balance of moment of momentum. This chapter closes with the derivation of the principle of virtual work as weak formulation of the equations of motion and the dynamic boundary conditions.

Material modelling at large elastic strains is extensively discussed in chapter 6. The discussion starts with the general principles to be considered in formulating material laws and the definition of objective tensors. Hyperelastic materials are defined first in a general form and then particular attention is paid to isotropic materials. In this context many practically important material models are presented. Finally, some useful connections between them are established through linearization.

Each chapter includes a number of applications in order to help the reader to get experienced with the theory. Some of them present also important results needed in the subsequent derivations.

It is a pleasure to thank Mrs U. Hollstegge for typing all the manuscript and to Mrs B. Trimborn for preparing the figures. Thanks are also due to Dr.-Ing. A. Eckstein, Dipl.-Ing. D. Lürding, Dipl.-Ing. O. Kintzel for help in proof-reading and Dr.-Ing. U. Hanskötter and Dr.-Ing. M. Itskov for many helpful suggestions.

The first author also wishes to register a note of sincere appreciation to the German National Science Foundation (DFG) for the support of many research projects in the field of computational and continuum mechanics which have been a real motivation for this book.

Thanks are also due to Springer Verlag for the care in getting the book into print.

August 1999 Y. Başar and D. Weichert

Contents

1 Mathematical fundamentals

This section defines tensors as invariant quantities and introduces tensorial operations in absolute notation. Emphasis is given to the definition of some special tensors playing an important role in continuum mechanics. In addition, some useful results such as the definition of the gradient and the divergence of tensors are presented. This section aims to present the mathematical background for an easy understanding of the following sections.

1.1 Simple tensors

First-order tensors. In this section we deal with simple tensors whose definition is based on physical or geometrical (invariant) vectors of the 3D-Euclidean space E3. We use curvilinear coordinates Θ^i to define points of E3, $\mathbf{g}_i$ and $\mathbf{g}^i$ denoting the associated covariant and contravariant base vectors. Vectors will in general be denoted by bold lower-case letters, e.g. by **a**, **b**, ... Their components with respect to a new coordinate system $\bar{\Theta}^i$ will be presented by $(\bar{...})$ and those referring to the initial coordinate system Θ^i without any mark, thus

$$\mathbf{a} = a_i\,\mathbf{g}^i = a^i\,\mathbf{g}_i = \bar{a}_i\,\bar{\mathbf{g}}^i = \bar{a}^i\,\bar{\mathbf{g}}_i \ . \tag{1.1.1}$$

The variables a_i, a^i are first-order tensor components, while the vector **a** itself being independent of any coordinate system is said to form a *first-order tensor.* Before extending this concept to the definition of higher-order tensors we first recall the usual vectorial operations with their appropriate notations:

scalar product: $$\mathbf{a}\cdot\mathbf{b} = \mathbf{b}\cdot\mathbf{a} = (a_i\,\mathbf{g}^i)\cdot(b^j\,\mathbf{g}_j) = a_i\,b^i = a^i\,b_i \ , \tag{1.1.2}$$

vectorial product: $$\mathbf{c} = \mathbf{a}\times\mathbf{b} = -\mathbf{b}\times\mathbf{a} = \varepsilon_{ijk}\,a^i\,b^j\,\mathbf{g}^k = \varepsilon^{ijk}\,a_i\,b_j\,\mathbf{g}_k \ , \tag{1.1.3}$$

mixed product: $$[\mathbf{a}\,\mathbf{b}\,\mathbf{c}] = \mathbf{a}\cdot(\mathbf{b}\times\mathbf{c}) = \varepsilon_{ijk}\,a^i\,b^j\,c^k = \varepsilon^{ijk}\,a_i\,b_j\,c_k \ . \tag{1.1.4}$$

We see that the first and the last operation lead to invariant scalar-valued quantities, *zero-order tensors*, while the result of the vectorial product is a new vector **c** in E3.

Tensorial product. We introduce a new operation for arbitrary vectors, the so-called *tensorial (dyadic) product.* If this operation (notation with $\otimes$) is applied to invariant vectors **a** and **b** the result will be a new invariant quantity **S** called a *second-order simple tensor* or *dyad.* Thus

$$\mathbf{S} = \mathbf{a} \otimes \mathbf{b} \ . \tag{1.1.5}$$

Similarly, simple tensors of arbitrary order *(polyads)* can be constructed, e.g. a third-order tensor:

$$\mathbf{R} = \mathbf{c} \otimes \mathbf{d} \otimes \mathbf{e} \ . \tag{1.1.6}$$

By definition the tensorial product is *not commutative*

$$\mathbf{a} \otimes \mathbf{b} \neq \mathbf{b} \otimes \mathbf{a} \ , \tag{1.1.7}$$

but it is supposed to satisfy the following requirements:

the distributive rule: $$\mathbf{a} \otimes (\mathbf{b} + \mathbf{c}) = \mathbf{a} \otimes \mathbf{b} + \mathbf{a} \otimes \mathbf{c} \ , \tag{1.1.8}$$

the associative rule: $$(\alpha\, \mathbf{a}) \otimes \mathbf{b} = \mathbf{a} \otimes (\alpha\, \mathbf{b}) = \alpha\, (\mathbf{a} \otimes \mathbf{b}) \ , \tag{1.1.9}$$

α being a scalar.

In terms of vector components defined according to (1.1.1) we obtain from (1.1.5) with the bases $\mathbf{g}_i$ and $\mathbf{g}^i$

$$\begin{aligned}\mathbf{S} &= a^i\, b^j\, \mathbf{g}_i \otimes \mathbf{g}_j = a^i\, b_j\, \mathbf{g}_i \otimes \mathbf{g}^j = a_i\, b^j\, \mathbf{g}^i \otimes \mathbf{g}_j = a_i\, b_j\, \mathbf{g}^i \otimes \mathbf{g}^j \\ &= S^{ij}\ \mathbf{g}_i \otimes \mathbf{g}_j = S^{i}_{\cdot j}\ \mathbf{g}_i \otimes \mathbf{g}^j = S_{i\cdot}^{\ j}\ \mathbf{g}^i \otimes \mathbf{g}_j = S_{ij}\ \mathbf{g}^i \otimes \mathbf{g}^j \end{aligned} \tag{1.1.10}$$

indicating that a second order tensor can be equivalently represented by four different sets of components S^{ij}, $S^{i}_{\cdot j}$, $S_{i\cdot}^{\ j}$, S_{ij}.

Simple contraction. Tensors of arbitrary orders, e.g. $\mathbf{S}$ (1.1.5) and $\mathbf{R}$ (1.1.6), may be related by the *simple contraction*, denoted as $\mathbf{S\,R}$ (without any mark). This operation provides the scalar multiplication of adjacent vectors of contributing tensors such that

$$\mathbf{S\,R} = (\mathbf{a} \otimes \mathbf{b})\,(\mathbf{c} \otimes \mathbf{d} \otimes \mathbf{e}) = (\mathbf{b} \cdot \mathbf{c})\,(\mathbf{a} \otimes \mathbf{d} \otimes \mathbf{e}) \ . \tag{1.1.11}$$

Thus, it leads to a tensor $\mathbf{S\,R}$ the order of which is twice less than the sum of the orders of the participants $\mathbf{S}$ and $\mathbf{R}$. The simple contraction obeys by virtue of (1.1.8), (1.1.9) and (1.1.11) the following rules

the distributive rule: $$\mathbf{R}\,(\mathbf{S} + \mathbf{T}) = \mathbf{R\,S} + \mathbf{R\,T} \ , \tag{1.1.12}$$

$$(\mathbf{S} + \mathbf{T})\,\mathbf{R} = \mathbf{S\,R} + \mathbf{T\,R} \ , \tag{1.1.13}$$

the associative rule: $$(\mathbf{S\,T})\,\mathbf{R} = \mathbf{S}\,(\mathbf{T\,R}) \ , \tag{1.1.14}$$

$$\alpha\,(\mathbf{S\,T}) = (\alpha\,\mathbf{S})\,\mathbf{T} = \mathbf{S}\,(\alpha\,\mathbf{T}) \ , \tag{1.1.15}$$

where α is a scalar. Evidently, in equations (1.1.12) and (1.1.13) the tensors $\mathbf{S}$ and $\mathbf{T}$ are supposed to be of the same order. In general the simple contraction is *not commutative*:

$$\begin{aligned}\mathbf{R\,S} &= (\mathbf{c} \otimes \mathbf{d} \otimes \mathbf{e})\,(\mathbf{a} \otimes \mathbf{b}) = (\mathbf{e} \cdot \mathbf{a})\,(\mathbf{c} \otimes \mathbf{d} \otimes \mathbf{b}) \\ &\neq \mathbf{S\,R} = (\mathbf{a} \otimes \mathbf{b})\,(\mathbf{c} \otimes \mathbf{d} \otimes \mathbf{e}) = (\mathbf{b} \cdot \mathbf{c})\,(\mathbf{a} \otimes \mathbf{d} \otimes \mathbf{e}) \ .\end{aligned}$$

According to the above mentioned rules the following relations hold:

$$(\mathbf{a} \otimes \mathbf{b})\,\mathbf{u} = (\mathbf{b} \cdot \mathbf{u})\,\mathbf{a} = (\mathbf{u} \cdot \mathbf{b})\,\mathbf{a} \;,$$

$$(\mathbf{a} \otimes \mathbf{b})\,(\mathbf{u} + \mathbf{v}) = [\mathbf{b} \cdot (\mathbf{u} + \mathbf{v})]\,\mathbf{a} = (\mathbf{b} \cdot \mathbf{u})\,\mathbf{a} + (\mathbf{b} \cdot \mathbf{v})\,\mathbf{a} \;. \qquad (1.1.16)$$

If **S** and **R** are first-order tensors, the simple contraction (1.1.11) corresponds to the scalar-product of vectors and is only in this case commutative.

Double contraction. A further important operation applicable to higher-order tensors **S** and **R** is the *double contraction*, denoted by **S**:**R**. In this case two scalar products are to be carried out in the form

$$\mathbf{S} : \mathbf{R} = (\mathbf{a} \otimes \mathbf{b}) : (\mathbf{c} \otimes \mathbf{d} \otimes \mathbf{e}) = (\mathbf{a} \cdot \mathbf{c})\,(\mathbf{b} \cdot \mathbf{d})\,\mathbf{e} \qquad (1.1.17)$$

leading to a tensor **S**:**R** whose order is four times less than the sum of the orders of the participants **S** and **R**.

The double contraction satisfies the following rules:

the distributive rule: $\mathbf{R} : (\mathbf{S} + \mathbf{T}) = \mathbf{R} : \mathbf{S} + \mathbf{R} : \mathbf{T}\;,$ (1.1.18)

the associative rule: $(\alpha\,\mathbf{R}) : \mathbf{S} = \mathbf{R} : (\alpha\,\mathbf{S}) = \alpha\,(\mathbf{R} : \mathbf{S})\;.$ (1.1.19)

Generally the double contraction is *not commutative*:

$$\mathbf{R} : \mathbf{S} = (\mathbf{c} \otimes \mathbf{d} \otimes \mathbf{e}) : (\mathbf{a} \otimes \mathbf{b}) = (\mathbf{e} \cdot \mathbf{b})\,(\mathbf{d} \cdot \mathbf{a})\,\mathbf{c} \neq \mathbf{S} : \mathbf{R} = (\mathbf{a} \cdot \mathbf{c})\,(\mathbf{b} \cdot \mathbf{d})\,\mathbf{e}\;.$$

An exception is however the case where the participant tensors **S** and **R** are both of second-order:

$$\mathbf{S} : \mathbf{R} = (\mathbf{a} \otimes \mathbf{b}) : (\mathbf{c} \otimes \mathbf{d}) = (\mathbf{a} \cdot \mathbf{c})\,(\mathbf{b} \cdot \mathbf{d})$$

$$= (\mathbf{c} \cdot \mathbf{a})\,(\mathbf{d} \cdot \mathbf{b}) = (\mathbf{c} \otimes \mathbf{d}) : (\mathbf{a} \otimes \mathbf{b}) = \mathbf{R} : \mathbf{S}\;. \qquad (1.1.20)$$

Further we may write for an arbitrary second-order tensor **S**

$$\mathbf{S} : (\mathbf{c} \otimes \mathbf{d}) = \mathbf{c} \cdot (\mathbf{S}\,\mathbf{d}) = (\mathbf{c}\,\mathbf{S}) \cdot \mathbf{d} = \mathbf{c}\,\mathbf{S}\,\mathbf{d} \;, \qquad (1.1.21)$$

which can easily be proved by using for **S** the expression (1.1.5). Thus,

$$\mathbf{S} : (\mathbf{c} \otimes \mathbf{d}) = (\mathbf{a} \otimes \mathbf{b}) : (\mathbf{c} \otimes \mathbf{d}) = (\mathbf{a} \cdot \mathbf{c})\,(\mathbf{b} \cdot \mathbf{d}) \;, \qquad (1.1.22)$$

$$\mathbf{c} \cdot (\mathbf{S}\,\mathbf{d}) = \mathbf{c} \cdot [(\mathbf{a} \otimes \mathbf{b})\,\mathbf{d}] = [(\mathbf{b} \cdot \mathbf{d})\,\mathbf{a}] \cdot \mathbf{c} = (\mathbf{a} \cdot \mathbf{c})\,(\mathbf{b} \cdot \mathbf{d}) \;. \qquad (1.1.23)$$

1.2 General tensors

Tensors as invariant quantities. We now consider arbitrary tensor components

$$a_i \,,\; S_{ij} \,,\; T_{ijk}$$

defined with respect to a curvilinear coordinate system Θ^i. The values of the above components in connection with new coordinates $\bar{\Theta}^i$ will be presented by $(\bar{...})$. We recall that both kinds of components S_{ij} and $\bar{S}_{ij}$ are related by

$$\bar{S}_{ij} = \frac{\partial\Theta^k}{\partial\bar{\Theta}^i}\frac{\partial\Theta^l}{\partial\bar{\Theta}^j} S_{kl} , \quad S_{ij} = \frac{\partial\bar{\Theta}^k}{\partial\Theta^i}\frac{\partial\bar{\Theta}^l}{\partial\Theta^j}\bar{S}_{kl} . \quad \text{(summation over k and l)} \tag{1.2.1}$$

Here and in the sequel it is assumed that there exists a sufficiently smooth, one to one mapping between the system of coordinates Θ^i and $\bar{\Theta}^i$.

In (1.1.10) we have already observed that second-order tensor components $S^{ij} = a^i b^j$ form by using the basis $\mathbf{g}_i \otimes \mathbf{g}_j$ an invariant quantity **S** called tensor. Using suitable bases

$$\mathbf{g}_i \otimes \mathbf{g}_j \otimes \mathbf{g}_k \otimes \ldots \tag{1.2.2}$$

as tensorial product of the base vectors $\mathbf{g}_i$, $\mathbf{g}_j$, ... the above idea can readily be extended to associate arbitrary sets of tensor components with an invariant quantity called tensor. Here we recall that expressions of the form (1.2.2) obey both of the rules (1.1.8) and (1.1.9) valid for tensorial products ($\otimes$). The property (1.1.9) has already been considered in the derivation of (1.1.10).

We now refer to S_{ij} to construct – using the basis $\mathbf{g}^i \otimes \mathbf{g}^j$ – the *second-order tensor*

$$\mathbf{S} = S_{ij}\,\mathbf{g}^i \otimes \mathbf{g}^j = S_{i\cdot}^{\cdot j}\,\mathbf{g}^i \otimes \mathbf{g}_j = S_{\cdot j}^{i\cdot}\,\mathbf{g}_i \otimes \mathbf{g}^j = S^{ij}\,\mathbf{g}_i \otimes \mathbf{g}_j , \tag{1.2.3}$$

where the summation rule over repeated indices applies. The relation (1.2.3) demonstrates **S** to be a general representation of four possible sets of components S_{ij}, $S_{i\cdot}^{\cdot j}$, ... Since for any coordinate transformation $\Theta^i \rightarrow \bar{\Theta}^i$ the equality

$$S_{ij}\,\mathbf{g}^i \otimes \mathbf{g}^j = \bar{S}_{ij}\,\bar{\mathbf{g}}^i \otimes \bar{\mathbf{g}}^j = \mathbf{S} , \tag{1.2.4}$$

holds we furthermore see that **S** is independent of any special coordinate system, therefore an *invariant* quantity. Accordingly, relations to be established in terms of such invariant variables will hold for arbitrary coordinate systems: a significant advantage of the *symbolic notation*. As a further example we form with T_{ijk} a *third-order tensor*

$$\mathbf{T} = T_{ijk}\,\mathbf{g}^i \otimes \mathbf{g}^j \otimes \mathbf{g}^k = T^{i}_{\cdot jk}\,\mathbf{g}_i \otimes \mathbf{g}^j \otimes \mathbf{g}^k = \ldots . \tag{1.2.5}$$

Our next goal is the generalisation of the operations introduced in the previous section for *polyads (simple tensors)* to the general tensors of the form (1.2.3) and (1.2.5).

Simple contraction. The application of the rule (1.1.11) to the tensors **S** and **R**

$$\begin{aligned}\mathbf{S}\,\mathbf{R} &= (S_{ij}\,\mathbf{g}^i \otimes \mathbf{g}^j)\,(R_{rst}\,\mathbf{g}^r \otimes \mathbf{g}^s \otimes \mathbf{g}^t)\\ &= S_{ij}\,(\mathbf{g}^j \cdot \mathbf{g}^r)\,R_{rst}\,\mathbf{g}^i \otimes \mathbf{g}^s \otimes \mathbf{g}^t\end{aligned}$$

$$= S_{ij}\ g^{jr}\ R_{rst}\,\mathbf{g}^i \otimes \mathbf{g}^s \otimes \mathbf{g}^t$$

$$= S_{i\cdot}^{\ \cdot r}\ R_{rst}\,\mathbf{g}^i \otimes \mathbf{g}^s \otimes \mathbf{g}^t = S_{ir}\ R^{r}_{\cdot st}\,\mathbf{g}^i \otimes \mathbf{g}^s \otimes \mathbf{g}^t = \ldots \tag{1.2.6}$$

produces a *contraction* with respect to the last index of the first component and the first index of the second component. Thus it leads to a tensor **S R** the order of which is twice less than the order of the tensorial product **S** ⊗ **R**.

It is possible to contract any higher-order tensor from the left and right side by means of simple contractions, e.g. in the form

$$\mathbf{T\,S\,u} = (T_{ij}\,\mathbf{g}^i \otimes \mathbf{g}^j)\,(S^{st}\,\mathbf{g}_s \otimes \mathbf{g}_t)\,(u_m\,\mathbf{g}^m) = (\mathbf{g}^j \cdot \mathbf{g}_s)\,(\mathbf{g}_t \cdot \mathbf{g}^m)\,T_{ij}\,S^{st}\,u_m\,\mathbf{g}^i$$

$$= \delta^j_s\,\delta^m_t\,T_{ij}\,S^{st}\,u_m\,\mathbf{g}^i = T_{ij}\,S^{jm}\,u_m\,\mathbf{g}^i\ . \tag{1.2.7}$$

Evidently, a vector **u** can not be contracted from both sides. So, an expression of the form **T u S** is nonsense.

According to the above definitions, the validity of the following rules for simple contractions can easily be proved:

the associative rule: $(\mathbf{T\,S})\,\mathbf{R} = \mathbf{T}\,(\mathbf{S\,R})\ ,$ (1.2.8)

the distributive rule: $\mathbf{T}\,(\mathbf{R}+\mathbf{S}) = \mathbf{T\,R}+\mathbf{T\,S}\ ,$ (1.2.9)

$(\mathbf{R}+\mathbf{S})\,\mathbf{T} = \mathbf{R\,T}+\mathbf{S\,T}\ ,$ (1.2.10)

where in the last two relations **R** and **S** are supposed to be of the same order. Generally the simple contraction is *not commutative*.

Powers of second-order tensors. The simple contraction allows to define powers of a second-order tensor **S** in the form:

$$\mathbf{S}^0 = \mathbf{I}\ ,\quad \mathbf{S}^1 = \mathbf{S}\ ,\quad \mathbf{S}^2 = \mathbf{S\,S}\ ,\quad \ldots \tag{1.2.11}$$

where by definition $\mathbf{S}^0$ is identical with the so-called identity tensor $\mathbf{I} = \mathbf{g}_i \otimes \mathbf{g}^i$. In accordance with (1.2.11) we also note that

$$\mathbf{S}^m\,\mathbf{S}^n = \mathbf{S}^{m+n} = \mathbf{S}^n\,\mathbf{S}^m\ ,\quad (\alpha\,\mathbf{S})^m = \alpha^m\,\mathbf{S}^m\ ,\quad (\mathbf{S}^m)^n = \mathbf{S}^{mn}\ . \tag{1.2.12}$$

Evaluation of tensor components. In (1.2.3) the tensor **S** is expressed in terms of its components S_{ij} . Conversely, it is possible to express S_{ij} in terms of **S** and the associated base vectors. By using the rule (1.2.6) we find

$$\mathbf{g}_i\,\mathbf{S}\,\mathbf{g}_j = \mathbf{g}_i\,(S_{mn}\,\mathbf{g}^m \otimes \mathbf{g}^n)\,\mathbf{g}_j = (\mathbf{g}_i \cdot \mathbf{g}^m)\,(\mathbf{g}^n \cdot \mathbf{g}_j)\,S_{mn} = \delta^m_i\,\delta^n_j\,S_{mn}$$

leading as final result to:

$$S_{ij} = \mathbf{g}_i\,\mathbf{S}\,\mathbf{g}_j\ ,\quad S^{i}_{\cdot j} = \mathbf{g}^i\,\mathbf{S}\,\mathbf{g}_j\ ,\quad S_{i\cdot}^{\ j} = \mathbf{g}_i\,\mathbf{S}\,\mathbf{g}^j\ ,\quad S^{ij} = \mathbf{g}^i\,\mathbf{S}\,\mathbf{g}^j\ . \tag{1.2.13}$$

The above rule will often serve to evaluate tensor components if the tensor itself is given as an invariant quantity.

Application. The expression $\mathbf{S}\,\mathbf{u}$ with a second-order tensor $\mathbf{S}$ and a vector $\mathbf{u}$ represents a vector. Thus its scalar product with the vector $\mathbf{v}$ can be formed as follows:

$$(\mathbf{S}\,\mathbf{u}) \cdot \mathbf{v} = \mathbf{v} \cdot (\mathbf{S}\,\mathbf{u}) = [(S^{ij}\,\mathbf{g}_i \otimes \mathbf{g}_j)\,(u_k\,\mathbf{g}^k)] \cdot v_n\,\mathbf{g}^n = S^{ij}\,u_j\,v_i \ .$$

The same result can also be obtained from the expression $\mathbf{v}\,\mathbf{S}\,\mathbf{u}$ so that

$$\mathbf{v}\,\mathbf{S}\,\mathbf{u} = (\mathbf{S}\,\mathbf{u}) \cdot \mathbf{v} = \mathbf{v} \cdot (\mathbf{S}\,\mathbf{u}) \ . \tag{1.2.14}$$

Application. Possible component representations of the expression $\mathbf{A}\,\mathbf{B}\,\mathbf{C}\,\mathbf{v}$ with second-order tensors $\mathbf{A}$, $\mathbf{B}$, $\mathbf{C}$ and the vector $\mathbf{v}$ are:

$$A_{ij}\,B^{j}_{\cdot k}\,C^{k}_{\cdot m}\,v^m = A_i^{\cdot j}\,B_{jk}\,C^{k}_{\cdot m}\,v^m = A_i^{\cdot j}\,B_j^{\cdot k}\,C_{km}\,v^m = \ldots \ .$$

Double contraction. The corresponding rule has already been introduced in (1.1.17) for simple tensors. Its application to arbitrary tensors $\mathbf{T}$ and $\mathbf{R}$

$$\begin{aligned}\mathbf{T} : \mathbf{R} &= (T^{lmn}\,\mathbf{g}_l \otimes \mathbf{g}_m \otimes \mathbf{g}_n) : (R^{ijk}\,\mathbf{g}_i \otimes \mathbf{g}_j \otimes \mathbf{g}_k) \\ &= T^{lmn}\,(\mathbf{g}_m \cdot \mathbf{g}_i)\,(\mathbf{g}_n \cdot \mathbf{g}_j)\,R^{ijk}\,(\mathbf{g}_l \otimes \mathbf{g}_k) \\ &= T^{lmn}\,g_{mi}\,g_{nj}\,R^{ijk}\,\mathbf{g}_l \otimes \mathbf{g}_k \end{aligned} \tag{1.2.15}$$

delivers

$$\mathbf{T} : \mathbf{R} = T^{l}_{\cdot ij}\,R^{ijk}\,\mathbf{g}_l \otimes \mathbf{g}_k = T^{lij}\,R_{ij}^{\cdot\cdot k}\,\mathbf{g}_l \otimes \mathbf{g}_k = \ldots \ . \tag{1.2.16}$$

This operation is called *double contraction* since it reduces the order of the tensorial product $\mathbf{T} \otimes \mathbf{R}$ by four due to the two scalar products to be performed.

Generally the double contraction is *not commutative*. But if the participant tensors $\mathbf{A}$ and $\mathbf{B}$ are both of second order this property holds:

$$\mathbf{A} : \mathbf{B} = A_{ij}\,B^{ij} = B^{ij}\,A_{ij} = \mathbf{B} : \mathbf{A} \ . \tag{1.2.17}$$

From (1.2.16) the following rules can easily be deduced for the double contraction

the distributive rule: $\quad \mathbf{R} : (\mathbf{S} + \mathbf{T}) = \mathbf{R} : \mathbf{S} + \mathbf{R} : \mathbf{T} \ ,$ (1.2.18)

the associative rule: $\quad (\alpha\,\mathbf{T}) : \mathbf{S} = \mathbf{T} : (\alpha\,\mathbf{S}) = \alpha\,(\mathbf{T} : \mathbf{S}) \ ,$ (1.2.19)

where in the distributive rule $\mathbf{S}$ and $\mathbf{T}$ are supposed to be tensors of the same order. In component form relation (1.2.18) reads as

$$R^{ijk}\,(S_{jk} + T_{jk}) = R^{ijk}\,S_{jk} + R^{ijk}\,T_{jk} \tag{1.2.20}$$

confirming its validity.

Application. The scalar-valued function

$$\rho = H^{ijkl} \alpha_{ij} \alpha_{kl}$$

has in symbolic notation the form

$$\rho = \boldsymbol{\alpha} : \mathbf{H} : \boldsymbol{\alpha} \quad \text{with} \quad \boldsymbol{\alpha} = \alpha_{ij} \mathbf{g}^i \otimes \mathbf{g}^j \;, \quad \mathbf{H} = H^{klmn} \mathbf{g}_k \otimes \mathbf{g}_l \otimes \mathbf{g}_m \otimes \mathbf{g}_n \;.$$

Partial derivatives of second-order tensors. To close this section we want to form the partial derivative of a second-order tensor $\mathbf{S} = S^{mn} \mathbf{g}_m \otimes \mathbf{g}_n$ with respect to the coordinate Θ^i. The corresponding result will be denoted as usual by $\mathbf{S}_{,i}$. We first recall that the covariant derivative of the tensor components S^{mn} is defined by

$$S^{mn}|_i = S^{mn}{}_{,i} + \Gamma^m_{ir} S^{rn} + \Gamma^n_{ir} S^{mr} \,, \tag{1.2.21}$$

in terms of the CHRISTOFFEL symbol $\Gamma^r_{ij} = \Gamma^r_{ji}$ of the second type, while the relation

$$\mathbf{g}_{i,j} = \frac{\partial \mathbf{g}_i}{\partial \Theta^j} = \Gamma^r_{ij} \mathbf{g}_r \tag{1.2.22}$$

holds for the partial derivatives of the basis $\mathbf{g}_i$. If we now construct the partial derivative $\mathbf{S}_{,i}$ we receive

$$\begin{aligned} \mathbf{S}_{,i} = \frac{\partial \mathbf{S}}{\partial \Theta^i} &= S^{mn}{}_{,i} \mathbf{g}_m \otimes \mathbf{g}_n + S^{mn} \mathbf{g}_{m,i} \otimes \mathbf{g}_n + S^{mn} \mathbf{g}_m \otimes \mathbf{g}_{n,i} \\ &= (S^{mn}{}_{,i} + \Gamma^m_{ir} S^{rn} + \Gamma^n_{ir} S^{mr}) \, \mathbf{g}_m \otimes \mathbf{g}_n \\ &= S^{mn}|_i \, \mathbf{g}_m \otimes \mathbf{g}_n \;. \end{aligned} \tag{1.2.23}$$

Thus the covariant derivatives $S^{mn}|_i$ turn out to be the components of $\mathbf{S}_{,i}$ with respect to the basis $\mathbf{g}_m \otimes \mathbf{g}_n$. If we define the covariant derivative of $\mathbf{g}_i$, similar to that of tensor components of type A_i, by

$$\mathbf{g}_i|_j = \mathbf{g}_{i,j} - \Gamma^r_{ij} \mathbf{g}_r = \mathbf{0} \tag{1.2.24}$$

we see from (1.2.22) that the result is zero. Accordingly, covariant and partial derivatives of $\mathbf{S}$ are identical operations:

$$\mathbf{S}_{,i} = \mathbf{S}|_i = S^{mn}|_i \, \mathbf{g}_m \otimes \mathbf{g}_n = S_{mn}|_i \, \mathbf{g}^m \otimes \mathbf{g}^n \,, \tag{1.2.25}$$

this result being valid for tensors of arbitrary order.

1.3 Special tensors

Tensors and vectors. In the following we shall introduce a number of special tensors playing an important role in tensor analysis and continuum mechanics. Most of the definitions will refer to second-order tensors which will be denoted by upper-case letters:

$$\mathbf{S} = S_{ij} \mathbf{g}^i \otimes \mathbf{g}^j \;, \quad \mathbf{T} = T_{ij} \mathbf{g}^i \otimes \mathbf{g}^j \;, \tag{1.3.1}$$

while vectors will be denoted as usual by lower-case letters, e.g.

$$\mathbf{u} = u_i \, \mathbf{g}^i \; , \qquad \mathbf{w} = w_i \, \mathbf{g}^i \; . \tag{1.3.2}$$

Identity tensor. The *identity tensor* **I** is closely related to the base vectors $\mathbf{g}_i$ and is defined by

$$\mathbf{I} = \mathbf{g}_i \otimes \mathbf{g}^i = \mathbf{g}^i \otimes \mathbf{g}_i \; , \tag{1.3.3}$$

where, as usual, the summation rule is to be applied for repeated indices. The tensor **I** possesses according to (1.2.13) the following components

$$\mathbf{I} = g_{ij} \, \mathbf{g}^i \otimes \mathbf{g}^j = \delta_i^j \, \mathbf{g}^i \otimes \mathbf{g}_j = \delta_j^i \, \mathbf{g}_i \otimes \mathbf{g}^j = g^{ij} \, \mathbf{g}_i \otimes \mathbf{g}_j \; , \tag{1.3.4}$$

indicating **I** to be an invariant quantity associated with the metric tensor components $g_{ij} = \mathbf{g}_i \cdot \mathbf{g}_j$. Accordingly, **I** may be also called *metric tensor*. By means of the above definition (1.3.3) it can be shown that

$$\mathbf{I}\,\mathbf{T} = \mathbf{T}\,\mathbf{I} = \mathbf{T} \; . \tag{1.3.5}$$

This relation holds for arbitrary tensors **T**, particularly also for vectors **u**. In view of this property and (1.1.21) we receive:

$$\mathbf{I} : (\mathbf{c} \otimes \mathbf{d}) = \mathbf{c}\,\mathbf{I}\,\mathbf{d} = \mathbf{c} \cdot (\mathbf{I}\,\mathbf{d}) = \mathbf{c} \cdot \mathbf{d} = c^i \, d_i \tag{1.3.6}$$

showing that **I** permits to express the scalar product of the vectors **c** and **d** in form of a double contraction.

If the basis $\mathbf{g}_i$ is changed into a new one $\bar{\mathbf{g}}_i$ associated with a new set of coordinates $\bar{\Theta}^i$

$$\mathbf{g}_i = \frac{\partial \bar{\Theta}^j}{\partial \Theta^i} \, \bar{\mathbf{g}}_j \quad , \qquad \mathbf{g}^i = \frac{\partial \Theta^i}{\partial \bar{\Theta}^k} \, \bar{\mathbf{g}}^k$$

then we receive from (1.3.3)

$$\mathbf{I} = \mathbf{g}_i \otimes \mathbf{g}^i = \frac{\partial \bar{\Theta}^j}{\partial \Theta^i} \frac{\partial \Theta^i}{\partial \bar{\Theta}^k} \, \bar{\mathbf{g}}_j \otimes \bar{\mathbf{g}}^k = \delta_k^j \, \bar{\mathbf{g}}_j \otimes \bar{\mathbf{g}}^k \tag{1.3.7}$$

which yields:

$$\mathbf{I} = \mathbf{g}_i \otimes \mathbf{g}^i = \bar{\mathbf{g}}_i \otimes \bar{\mathbf{g}}^i = \mathbf{i}_i \otimes \mathbf{i}^i \; . \tag{1.3.8}$$

Accordingly, the identity tensor **I** may be constructed from an arbitrary system of base vectors, particularly from those associated with orthogonal Cartesian coordinates, $\mathbf{i}_i$.

For later applications it is suitable to extend the definition (1.2.3) to an identity tensor of fourth-order

$$\overset{4}{\mathbf{I}} = \mathbf{I} \otimes \mathbf{I} = \mathbf{g}_i \otimes \mathbf{g}^i \otimes \mathbf{g}_k \otimes \mathbf{g}^k = g^{ij} \, g^{kl} \, \mathbf{g}_i \otimes \mathbf{g}_j \otimes \mathbf{g}_k \otimes \mathbf{g}_l \tag{1.3.9}$$

which, by virtue of (1.2.15), satisfies the relations

$$\overset{4}{\mathbf{I}} : \mathbf{S} = \mathbf{S} : \overset{4}{\mathbf{I}} = (\mathrm{tr}\,\mathbf{S})\,\mathbf{I}\ , \quad \mathbf{S} : \overset{4}{\mathbf{I}} : \mathbf{S} = S_{i\cdot}^{\cdot i}\, S_{j\cdot}^{\cdot j} = (\mathrm{tr}\,\mathbf{S})^2\ , \tag{1.3.10}$$

where $\mathbf{S}$ is an arbitrary second-order tensor and tr $\mathbf{S}$ is an abbreviation for the double contraction $\mathrm{tr}\,\mathbf{S} = \mathbf{S} : \mathbf{I} = S_{i\cdot}^{\cdot i}$.

Inverse tensors. The *inverse tensor* $\mathbf{S}^{-1}$ of any second-order tensor $\mathbf{S}$ is defined by the equalities

$$\mathbf{S}\,\mathbf{S}^{-1} = \mathbf{S}^{-1}\,\mathbf{S} = \mathbf{I}\ , \tag{1.3.11}$$

$\mathbf{I}$ being the identity tensor which according to (1.3.5) satisfies the relation

$$\mathbf{I} = \mathbf{I}^{-1}\ . \tag{1.3.12}$$

The inverse tensor $\mathbf{S}^{-1}$ permits to solve any equation of the form

$$\mathbf{w} = \mathbf{S}\,\mathbf{v} \tag{1.3.13}$$

for $\mathbf{v}$ as

$$\mathbf{v} = \mathbf{S}^{-1}\,\mathbf{w}\ . \tag{1.3.14}$$

By means of the definition (1.3.11), the following identities can be derived for inverse tensors:

$$(\mathbf{S}^{-1})^{-1} = \mathbf{S}\ , \quad (\alpha\,\mathbf{S})^{-1} = \alpha^{-1}\,\mathbf{S}^{-1}\ , \quad (\mathbf{S}\,\mathbf{T})^{-1} = \mathbf{T}^{-1}\,\mathbf{S}^{-1}\ . \tag{1.3.15}$$

Application. As an example the last identity in (1.3.15) will be proved. For this purpose we start according to (1.3.11) from

$$(\mathbf{S}\,\mathbf{T})\,(\mathbf{S}\,\mathbf{T})^{-1} = \mathbf{I}\ ,$$

which by multiplication with $\mathbf{T}^{-1}\,\mathbf{S}^{-1}$ from the left side delivers

$$\mathbf{T}^{-1}\,\mathbf{S}^{-1}\,\mathbf{S}\,\mathbf{T}\,(\mathbf{S}\,\mathbf{T})^{-1} = \mathbf{T}^{-1}\,\mathbf{S}^{-1}\ .$$

In view of (1.3.5) and (1.3.11) we then deduce that

$$(\mathbf{S}\,\mathbf{T})^{-1} = \mathbf{T}^{-1}\,\mathbf{S}^{-1}\ ,$$

in accordance with (1.3.15).

Transposed tensors. A transposed tensor will be denoted by $(\ldots)^T$. By definition, any vector $\mathbf{u}$ is identical with its transposed $\mathbf{u}^T$. The same convention is also valid for a scalar α such that

$$\mathbf{u}^T = \mathbf{u}\ , \quad \alpha^T = \alpha\ . \tag{1.3.16}$$

The transposed tensor of a simple tensor

$$\mathbf{A} = \mathbf{a} \otimes \mathbf{b}$$

is defined as a tensor obtained from $\mathbf{A}$ by interchanging the vectors $\mathbf{a}$ and $\mathbf{b}$:

$$\mathbf{A}^{\mathrm{T}} = \mathbf{b} \otimes \mathbf{a} \ . \tag{1.3.17}$$

This definition can be readily extended to a second-order tensor

$$\mathbf{S} = S_{ij}\, \mathbf{g}^i \otimes \mathbf{g}^j \ .$$

If we replace the basis $\mathbf{g}^i \otimes \mathbf{g}^j$ according to (1.3.17) by its transposed form $\mathbf{g}^j \otimes \mathbf{g}^i$ the result

$$\mathbf{S}^{\mathrm{T}} = (S^{\mathrm{T}})_{ij}\, \mathbf{g}^i \otimes \mathbf{g}^j = S_{ij}\, \mathbf{g}^j \otimes \mathbf{g}^i = S_{ji}\, \mathbf{g}^i \otimes \mathbf{g}^j \tag{1.3.18}$$

is called the *transposed tensor* of $\mathbf{S}$ denoted by $\mathbf{S}^{\mathrm{T}}$. From (1.3.18) it follows that

$$(S^{\mathrm{T}})_{ij} = S_{ji} \ , \quad (S^{\mathrm{T}})^{i}_{\cdot j} = S_{j}^{\cdot i} \ , \quad (S^{\mathrm{T}})_{i}^{\cdot j} = S^{j}_{\cdot i} \ , \quad (S^{\mathrm{T}})^{ij} = S^{ji} \ . \tag{1.3.19}$$

The definition (1.3.18) of the transposed tensor $\mathbf{S}^{\mathrm{T}}$ may be equivalently replaced by the relation

$$\mathbf{u} \cdot (\mathbf{S}\, \mathbf{v}) = \mathbf{v} \cdot (\mathbf{S}^{\mathrm{T}}\, \mathbf{u}) = \mathbf{u}\, \mathbf{S}\, \mathbf{v} = \mathbf{v}\, \mathbf{S}^{\mathrm{T}}\, \mathbf{u} \tag{1.3.20}$$

holding for arbitrary vectors $\mathbf{u}$ and $\mathbf{v}$. This can be easily proved by making use of component relations to obtain

$$\begin{aligned}
\mathbf{u} \cdot (\mathbf{S}\, \mathbf{v}) &= (u_i\, \mathbf{g}^i) \cdot [(S_{jk}\, \mathbf{g}^j \otimes \mathbf{g}^k)\, (v_l\, \mathbf{g}^l)] \\
&= (u_i\, \mathbf{g}^i) \cdot (S_{jk}\, g^{kl}\, v_l\, \mathbf{g}^j) = u_i\, g^{ij}\, S_{jk}\, g^{kl}\, v_l \\
&= u_i\, S^{il}\, v_l = \mathbf{u}\, \mathbf{S}\, \mathbf{v} \ ,
\end{aligned}$$

$$\begin{aligned}
\mathbf{v} \cdot (\mathbf{S}^{\mathrm{T}}\, \mathbf{u}) &= (v_l\, \mathbf{g}^l) \cdot [(S^{\mathrm{T}})_{jk}\, \mathbf{g}^j \otimes \mathbf{g}^k)\, (u_i\, \mathbf{g}^i)] \\
&= (v_l\, \mathbf{g}^l) \cdot (S^{\mathrm{T}})_{jk}\, (g^{ki}\, u_i\, \mathbf{g}^j) = v_l\, g^{lj}\, (S^{\mathrm{T}})_{jk}\, g^{ki}\, u_i \\
&= v_l\, (S^{\mathrm{T}})^{li}\, u_i = u_i\, S^{il}\, v_l = \mathbf{v}\, \mathbf{S}^{\mathrm{T}}\, \mathbf{u} \ ,
\end{aligned}$$

which confirms in view of (1.3.19) the equality (1.3.20). By a similar procedure it can also be shown that

$$(\mathbf{T} + \mathbf{S})^{\mathrm{T}} = \mathbf{T}^{\mathrm{T}} + \mathbf{S}^{\mathrm{T}} \ , \tag{1.3.21}$$

$$(\mathbf{T}\, \mathbf{S})^{\mathrm{T}} = \mathbf{S}^{\mathrm{T}}\, \mathbf{T}^{\mathrm{T}} \ , \quad (\alpha\, \mathbf{S})^{\mathrm{T}} = \alpha\, \mathbf{S}^{\mathrm{T}} \ . \tag{1.3.22}$$

In view of (1.3.16) and (1.3.22) we have for any vector given as simple contraction $\mathbf{a} = \mathbf{S}\, \mathbf{u}$

$$\mathbf{a} = \mathbf{S}\, \mathbf{u} = \mathbf{a}^{\mathrm{T}} = \mathbf{u}\, \mathbf{S}^{\mathrm{T}} \ . \tag{1.3.23}$$

A further important identity is

$$\mathbf{A} : (\mathbf{B}\, \mathbf{C}) = (\mathbf{B}^{\mathrm{T}}\, \mathbf{A}) : \mathbf{C} = (\mathbf{A}\, \mathbf{C}^{\mathrm{T}}) : \mathbf{B} \ , \tag{1.3.24}$$

whose validity can be confirmed by transforming the above expressions into component form

$$A_{ij}\, B^{i}_{\cdot k}\, C^{kj} = (B^{T})_{k\cdot}^{\cdot i}\, A_{ij}\, C^{kj} = A_{ij}\, (C^{T})^{jk}\, B^{i}_{\cdot k} \tag{1.3.25}$$

and by considering (1.3.19). Similarly, it can be proved that

$$\mathbf{S} : \mathbf{T} = \mathbf{S}^{T} : \mathbf{T}^{T} \ . \tag{1.3.26}$$

To derive a further important identity we set according to (1.3.11)

$$\mathbf{S}\,\mathbf{S}^{-1} = \mathbf{I} \ . \tag{1.3.27}$$

In view of (1.3.22) and the equality (1.3.3)

$$\mathbf{I} = \mathbf{I}^{T} = \mathbf{g}_{i} \otimes \mathbf{g}^{i} = \mathbf{g}^{i} \otimes \mathbf{g}_{i} \tag{1.3.28}$$

equation (1.3.27) gives

$$(\mathbf{S}\,\mathbf{S}^{-1})^{T} = (\mathbf{S}^{-1})^{T}\,\mathbf{S}^{T} = \mathbf{I}^{T} = \mathbf{I} \tag{1.3.29}$$

indicating $(\mathbf{S}^{-1})^{T}$ to be the inverse of the tensor $\mathbf{S}^{T}$. Thus

$$(\mathbf{S}^{-1})^{T} = (\mathbf{S}^{T})^{-1} = \mathbf{S}^{-T} \ . \tag{1.3.30}$$

Accordingly, the order of the application of the operations $(...)^{T}$ and $(...)^{-1}$ to a tensor does not affect the result.

Application. Show that $(\mathbf{ABCD})^{T} = \mathbf{D}^{T}\,\mathbf{C}^{T}\,\mathbf{B}^{T}\,\mathbf{A}^{T}$. Proof according to (1.3.22):

$$(\mathbf{A}\,(\mathbf{BCD}))^{T} = (\mathbf{BCD})^{T}\,\mathbf{A}^{T} = \mathbf{D}^{T}\,(\mathbf{BC})^{T}\,\mathbf{A}^{T} = \mathbf{D}^{T}\,\mathbf{C}^{T}\,\mathbf{B}^{T}\mathbf{A}^{T} \ . \tag{1.3.31}$$

Symmetric tensors. A second-order tensor $\mathbf{S}$ is said to be *symmetric* if it is identical with the transposed tensor $\mathbf{S}^{T}$. According to (1.3.19) we then may write

$$\mathbf{S} = S_{ij}\,\mathbf{g}^{i} \otimes \mathbf{g}^{j} = \mathbf{S}^{T} = (\mathbf{S}^{T})_{ij}\,\mathbf{g}^{i} \otimes \mathbf{g}^{j} = S_{ji}\,\mathbf{g}^{i} \otimes \mathbf{g}^{j} \ , \tag{1.3.32}$$

showing in view of the arbitrariness of the basis $\mathbf{g}^{i} \otimes \mathbf{g}^{j}$ that

$$S_{ij} = S_{ji} \ , \quad S_{i\cdot}^{\cdot j} = S^{j}_{\cdot i} = S^{j}_{i} \ , \quad S^{ij} = S^{ji} \ . \tag{1.3.33}$$

The above property permits to omit the dot in the mixed components S^{j}_{i}. Note that the simple contraction $\mathbf{S}\,\mathbf{S}^{T}$ in terms of any second-order tensor $\mathbf{S}$ forms a symmetric tensor since by virtue of (1.3.22):

$$(\mathbf{S}\,\mathbf{S}^{T})^{T} = (\mathbf{S}^{T})^{T}\,\mathbf{S}^{T} = \mathbf{S}\,\mathbf{S}^{T} \ . \tag{1.3.34}$$

Application. Show that $\mathbf{S} = \mathbf{F}^{-1}\,\mathbf{T}\,\mathbf{F}^{-T}$ is symmetric, if $\mathbf{T}$ is symmetric. Proof by considering (1.3.30) and (1.3.31):

$$\mathbf{S}^{T} = (\mathbf{F}^{-1}\,\mathbf{T}\,\mathbf{F}^{-T})^{T} = \mathbf{F}^{-1}\,\mathbf{T}^{T}\,\mathbf{F}^{-T} = \mathbf{F}^{-1}\,\mathbf{T}\,\mathbf{F}^{-T} = \mathbf{S} \ .$$

Application. Show that $\mathbf{A}:\mathbf{B} = \mathbf{A}^T:\mathbf{B}$ if $\mathbf{B}$ is symmetric. Proof according to (1.3.26) and (1.3.32):

$$\mathbf{A}:\mathbf{B} = \mathbf{A}^T:\mathbf{B}^T = \mathbf{A}^T:\mathbf{B} \ .$$

Skew-symmetric tensors. A second-order tensor $\mathbf{T}$ is said to be *skew-symmetric* if it is equal to the negative value of the transposed tensor $\mathbf{T}^T$. Thus we have

$$\mathbf{T} = T_{ij}\,\mathbf{g}^i \otimes \mathbf{g}^j = -\mathbf{T}^T = -T_{ji}\,\mathbf{g}^i \otimes \mathbf{g}^j \tag{1.3.35}$$

leading to

$$T_{ij} = -T_{ji} \quad \text{for} \quad i \neq j \ , \qquad T_{ij} = 0 \quad \text{for} \quad i = j \tag{1.3.36}$$

and

$$T_{i\cdot}^{\cdot j} = -T^{j}_{\cdot i} \tag{1.3.37}$$

for arbitrary indices. Note that in general $T_{1\cdot}^{\cdot 1} \neq 0$ and $T_{2\cdot}^{\cdot 2} \neq 0$.

Application. If $\mathbf{T}$ is a skew-symmetric tensor and $\mathbf{a}$ an arbitrary vector, then

$$\mathbf{a}\,\mathbf{T}\,\mathbf{a} = 0 \ .$$

To prove this we use the property $\mathbf{T} = -\mathbf{T}^T$ and (1.3.18). Thus we see that

$$\mathbf{a}\,\mathbf{T}\,\mathbf{a} = \mathbf{a}\,\mathbf{T}^T\,\mathbf{a} = -\mathbf{a}\,\mathbf{T}\,\mathbf{a} = 0 \ .$$

Splitting of second-order tensors. Any second-order tensor $\mathbf{A}$ can be expressed as

$$\begin{aligned} \mathbf{A} = A_{ij}\,\mathbf{g}^i \otimes \mathbf{g}^j &= \tfrac{1}{2}(A_{ij} + A_{ji})\,\mathbf{g}^i \otimes \mathbf{g}^j + \tfrac{1}{2}(A_{ij} - A_{ji})\,\mathbf{g}^i \otimes \mathbf{g}^j \\ &= \tfrac{1}{2}(\mathbf{A} + \mathbf{A}^T) + \tfrac{1}{2}(\mathbf{A} - \mathbf{A}^T) \ . \end{aligned} \tag{1.3.38}$$

Introducing the abbreviations

$$\begin{aligned} \operatorname{sym}\mathbf{A} &= \tfrac{1}{2}(A_{ij} + A_{ji})\,\mathbf{g}^i \otimes \mathbf{g}^j = \tfrac{1}{2}(\mathbf{A} + \mathbf{A}^T) \ , \\ \operatorname{skew}\mathbf{A} &= \tfrac{1}{2}(A_{ij} - A_{ji})\,\mathbf{g}^i \otimes \mathbf{g}^j = \tfrac{1}{2}(\mathbf{A} - \mathbf{A}^T) \end{aligned} \tag{1.3.39}$$

we may then write

$$\mathbf{A} = \operatorname{sym}\mathbf{A} + \operatorname{skew}\mathbf{A} \ . \tag{1.3.40}$$

This shows that any second-order tensor $\mathbf{A}$ may be split up into a symmetric tensor sym $\mathbf{A}$ and a skew-symmetric one skew $\mathbf{A}$.

Permutation tensor. In (1.1.3) we have introduced the set of components ε_{ijk} permitting to express the vector product $\mathbf{g}_i \times \mathbf{g}_j$ in tensorial form as

$$\mathbf{g}_i \times \mathbf{g}_j = \varepsilon_{ijk}\,\mathbf{g}^k \ . \tag{1.3.41}$$

If we now introduce the third-order tensor

$$\mathbf{E} = \varepsilon_{ijk}\, \mathbf{g}^i \otimes \mathbf{g}^j \otimes \mathbf{g}^k \tag{1.3.42}$$

and remember that ε_{ijk} possesses the symmetry properties

$$\varepsilon_{ijk} = -\varepsilon_{jik} = -\varepsilon_{kji} = -\varepsilon_{ikj} = \varepsilon_{kij} = \varepsilon_{jki} \tag{1.3.43}$$

we find by considering (1.2.15)

$$\mathbf{E} : (\mathbf{g}_i \otimes \mathbf{g}_j) = \varepsilon_{ijk}\, \mathbf{g}^k = -\varepsilon_{ikj}\, \mathbf{g}^k = -\mathbf{g}_i\, \mathbf{E}\, \mathbf{g}_j \ . \tag{1.3.44}$$

Accordingly, equation (1.3.41) takes the form

$$\mathbf{g}_i \times \mathbf{g}_j = \mathbf{E} : (\mathbf{g}_i \otimes \mathbf{g}_j) = -\mathbf{g}_i\, \mathbf{E}\, \mathbf{g}_j \tag{1.3.45}$$

showing that the *permutation tensor* **E** permits to express a vectorial product in symbolic notation. By multiplying the above result by $u^i\, v^j$ we receive a similar formulation

$$\mathbf{u} \times \mathbf{v} = \mathbf{E} : (\mathbf{u} \otimes \mathbf{v}) = -\mathbf{u}\, \mathbf{E}\, \mathbf{v} \ , \tag{1.3.46}$$

for arbitrary vectors $\mathbf{u} = u^i\, \mathbf{g}_i$ and $\mathbf{v} = v^i\, \mathbf{g}_i$.

Trace of second-order tensors. The *trace* of an arbitrary dyadic $\mathbf{a} \otimes \mathbf{b}$ is an invariant scalar defined by

$$\mathrm{tr}\,(\mathbf{a} \otimes \mathbf{b}) = \mathbf{I} : (\mathbf{a} \otimes \mathbf{b}) = \mathbf{a}\, \mathbf{I}\, \mathbf{b} = \mathbf{a} \cdot \mathbf{b} \tag{1.3.47}$$

with **I** as the identity tensor (1.3.4). Thus, in terms of the vector components

$$\mathbf{a} = a_i\, \mathbf{g}^i \ , \quad \mathbf{b} = b_j\, \mathbf{g}^j \tag{1.3.48}$$

equation (1.3.47) reads as

$$\mathrm{tr}\,(\mathbf{a} \otimes \mathbf{b}) = a_i\, b^i = a^i\, b_i \ . \tag{1.3.49}$$

The *trace of a second-order tensor* **S** is defined similarly by

$$\mathrm{tr}\, \mathbf{S} = \mathbf{I} : \mathbf{S} = (\mathbf{g}_i \otimes \mathbf{g}^i) : (S_{mn}\, \mathbf{g}^m \otimes \mathbf{g}^n) = S_{i\cdot}^{\cdot i} = S_{\cdot i}^{i\cdot} \ . \tag{1.3.50}$$

In view of this definition we can easily verify that the equations

$$\mathrm{tr}\, \mathbf{S}^T = \mathrm{tr}\, \mathbf{S} \ ,$$

$$\mathrm{tr}\,(\mathbf{S} + \mathbf{T}) = \mathrm{tr}\, \mathbf{S} + \mathrm{tr}\, \mathbf{T} \ ,$$

$$\mathrm{tr}\,(\mathbf{S}\, \mathbf{T}) = \mathrm{tr}\,(\mathbf{T}\, \mathbf{S}) \ ,$$

$$\mathrm{tr}\,(\mathbf{S}\, \mathbf{T}^T) = \mathrm{tr}\,(\mathbf{T}\, \mathbf{S}^T) = \mathbf{T} : \mathbf{S} = \mathrm{tr}\,(\mathbf{S}^T\, \mathbf{T}) = \mathrm{tr}\,(\mathbf{T}^T\, \mathbf{S}) = \mathbf{T}^T : \mathbf{S}^T \tag{1.3.51}$$

hold for arbitrary second-order tensors **S** and **T**.

Application. By virtue of the definition (1.3.50) the following identity can be derived for any symmetric tensor **C**:

$$C^r_r C^s_s - C^r_s C^s_r = (\mathrm{tr}\,\mathbf{C})^2 - \mathrm{tr}\,\mathbf{C}^2 \ . \tag{1.3.52}$$

Axial vectors. The definition of an axial vector is closely related to skew-symmetric tensors. We start from a second-order, skew-symmetric tensor

$$\mathbf{T} = \frac{1}{2}(\mathbf{a}\otimes\mathbf{b} - \mathbf{b}\otimes\mathbf{a}) = \frac{1}{2}(a_i\, b_j\, \mathbf{g}^i\otimes\mathbf{g}^j - b_i\, a_j\, \mathbf{g}^i\otimes\mathbf{g}^j) \ . \tag{1.3.53}$$

The simple contraction of this relation from the right side by an arbitrary vector **u** gives:

$$\mathbf{T\,u} = \frac{1}{2}[(\mathbf{b}\cdot\mathbf{u})\,\mathbf{a} - (\mathbf{a}\cdot\mathbf{u})\,\mathbf{b}] = \frac{1}{2}\mathbf{u}\times(\mathbf{a}\times\mathbf{b}) = \frac{1}{2}(\mathbf{b}\times\mathbf{a})\times\mathbf{u} \ . \tag{1.3.54}$$

Introducing under consideration of (1.3.46) the *axial vector*

$$\mathbf{t} = \frac{1}{2}(\mathbf{b}\times\mathbf{a}) = \frac{1}{2}\mathbf{E} : (\mathbf{b}\otimes\mathbf{a}) \tag{1.3.55}$$

we finally obtain

$$\mathbf{T\,u} = \mathbf{t}\times\mathbf{u} = \hat{\mathbf{t}}\,\mathbf{u} \ , \tag{1.3.56}$$

for arbitrary vectors **u**. Thus, the axial vector **t** turns out to be a quantity which permits to express the contraction **T u** of arbitrary vectors **u** by a given skew-symmetric tensor **T** in form of a vector product $\mathbf{t}\times\mathbf{u}$. Note that, in accordance with the number of independent components of **T**, the vector **t** is determined by three components. The notation

$$\hat{\mathbf{t}} = \mathbf{t}\times \quad \rightarrow \quad \hat{\mathbf{t}}\,\mathbf{u} = \mathbf{t}\times\mathbf{u} \tag{1.3.57}$$

introduced in (1.3.56) is sometimes useful to omit the symbol $\times$.

If the skew-symmetric tensor **T** is to be applied only to a single vector, e.g. $\mathbf{u} = \mathbf{g}^3$ then equation (1.3.56) is satisfied by an axial vector $\mathbf{t} = t_\alpha\, \mathbf{g}^\alpha$ $(\alpha = 1,2)$ possessing only two components since in this case $t_3\, \mathbf{g}^3 \times \mathbf{g}^3 = \mathbf{0}$.

Norm of second-order tensors. By considering the last equation in (1.3.51) the *absolute value* or *the norm* of a second-order tensor **S** is defined by

$$\|\mathbf{S}\| = \sqrt{\mathbf{S}:\mathbf{S}} = \sqrt{S^{ij}\, S_{ij}} = \sqrt{\mathrm{tr}\,(\mathbf{S}\,\mathbf{S}^T)} = \sqrt{\mathrm{tr}\,(\mathbf{S}^T\,\mathbf{S})} \ . \tag{1.3.58}$$

Thus, the SCHWARZ inequality may be expressed for second-order tensors **T** and **S** as:

$$\|\mathbf{T\,S}\| \le \|\mathbf{T}\|\,\|\mathbf{S}\| = \sqrt{\mathbf{T}:\mathbf{T}}\,\sqrt{\mathbf{S}:\mathbf{S}} = \sqrt{\mathrm{tr}\,(\mathbf{T}\,\mathbf{T}^T)}\,\sqrt{\mathrm{tr}\,(\mathbf{S}\,\mathbf{S}^T)} \ . \tag{1.3.59}$$

Determinants. The *determinant* of a second-order tensor $\mathbf{S} = S_{ij}\, \mathbf{g}^i\otimes\mathbf{g}^j = S^i_{\cdot j}\, \mathbf{g}_i\otimes\mathbf{g}^j = \dots$ may be essentially constructed by different component types S_{ij}, $S^i_{\cdot j}$ ... and will generally lead to different values. In this book the determinant of **S** (notation: det **S**) will be exclusively understood to be the value $|S^i_{\cdot j}|$ constructed by the mixed components $S^i_{\cdot j}$. By using the permutation tensor e_{ijk} associated with orthogonal Cartesian coordinates the corresponding result reads as

$$\det \mathbf{S} = | S^{i}_{\cdot j} | = e_{ijk}\, S^{i}_{\cdot 1}\, S^{j}_{\cdot 2}\, S^{k}_{\cdot 3} = III_{S} \; . \tag{1.3.60}$$

In section 1.8 we shall see that det **S** is an invariant which can be alternatively denoted by III_S , also called the *third invariant* of **S**.

1.4 Orthogonal tensors

Orthogonal tensors play a very important role in continuum mechanics, e.g. the rotation tensor to be introduced in section 1.10 is an orthogonal tensor. A second-order tensor **Q** is said to be *orthogonal* if the transposed tensor $\mathbf{Q}^T$ and the inverse tensor $\mathbf{Q}^{-1}$ are identical

$$\mathbf{Q}^{T} = \mathbf{Q}^{-1} \; . \tag{1.4.1}$$

By virtue of (1.3.11) this implies the equalities

$$\mathbf{Q}\,\mathbf{Q}^{T} = \mathbf{Q}\,\mathbf{Q}^{-1} = \mathbf{Q}^{-1}\,\mathbf{Q} = \mathbf{Q}^{T}\,\mathbf{Q} = \mathbf{I} \tag{1.4.2}$$

characterizing any orthogonal tensor **Q**. Since $\mathbf{Q}\,\mathbf{Q}^T$ is by virtue of (1.4.2) a symmetric tensor, relation (1.4.2) involves six independent component relations. Accordingly, an orthogonal tensor **Q** possesses only three independent components.

If the tensor **Q** is applied to two arbitrary vectors **a** and **b**

$$\tilde{\mathbf{a}} = \mathbf{Q}\,\mathbf{a} \; , \quad \tilde{\mathbf{b}} = \mathbf{Q}\,\mathbf{b} \tag{1.4.3}$$

the scalar product of the resulting vectors $\tilde{\mathbf{a}}$ and $\tilde{\mathbf{b}}$ satisfies by virtue of (1.4.2) the relation

$$\tilde{\mathbf{a}} \cdot \tilde{\mathbf{b}} = (\mathbf{Q}\,\mathbf{a}) \cdot (\mathbf{Q}\,\mathbf{b}) = \mathbf{a} \cdot (\mathbf{Q}^{T}\,\mathbf{Q}\,\mathbf{b}) = \mathbf{a} \cdot (\mathbf{I}\,\mathbf{b}) = \mathbf{a} \cdot \mathbf{b} \tag{1.4.4}$$

which is also valid for **b** = **a**. This indicates the main characteristic of an orthogonal tensor which upon operating on a set of vectors preserves the length of each vector as well as the angle between two arbitrary vectors. In other words, an orthogonal tensor may describe pure rotational movements of given vector sets in E3*.

In continuum mechanics orthogonal tensors will be mainly used to describe the rotation of the set of base vectors $\mathbf{g}_i$ (i = 1, 2, 3) and, accordingly, the rotation of arbitrary vectors defined with respect to $\mathbf{g}_i^{\,*}$. Let $\tilde{\mathbf{g}}_i$ be a set of vectors obtained from the base vectors $\mathbf{g}_i$ by a pure rotation. This can be described according to (1.4.3) by

$$\tilde{\mathbf{g}}_i = \mathbf{Q}\,\mathbf{g}_i \tag{1.4.5}$$

in terms of an orthogonal tensor **Q**. It can easily be verified that the above relation is automatically satisfied if

---

* An orthogonal tensor **Q** describing the rigid-body rotation of vectors is characterized by det **Q** = +1. For det **Q** = −1 the tensor **Q** describes the reflection of vectors with respect to a plane.

$$\mathbf{Q} = \tilde{\mathbf{g}}_i \otimes \mathbf{g}^i \ . \tag{1.4.6}$$

In fact, we then receive

$$\mathbf{Q}\,\mathbf{g}_i = (\tilde{\mathbf{g}}_j \otimes \mathbf{g}^j)\,\mathbf{g}_i = \tilde{\mathbf{g}}_j\,\delta_i^j = \tilde{\mathbf{g}}_i \ .$$

Using (1.4.5), we obtain again in view of the orthogonality of **Q**:

$$\tilde{\mathbf{g}}_i \cdot \tilde{\mathbf{g}}_j = (\mathbf{Q}\,\mathbf{g}_i) \cdot (\mathbf{Q}\,\mathbf{g}_j) = \mathbf{g}_i \cdot (\mathbf{Q}^T\,\mathbf{Q})\,\mathbf{g}_j = \mathbf{g}_i \cdot (\mathbf{I}\,\mathbf{g}_j)$$

and hence:

$$\tilde{\mathbf{g}}_i \cdot \tilde{\mathbf{g}}_j = \tilde{g}_{ij} = \mathbf{g}_i \cdot \mathbf{g}_j = g_{ij} \tag{1.4.7}$$

showing the equality of the metric tensors $\tilde{g}_{ij}$ and g_{ij} related to $\tilde{\mathbf{g}}_i$ and $\mathbf{g}_i$.

Finally, we shall show an important statement connected with the definition of orthogonal tensors: Any second-order tensor **F** can be split up multiplicatively into an arbitrary orthogonal tensor **R** and a second-order tensor **U** or **V** in the form:

$$\mathbf{F} = \mathbf{R}\,\mathbf{U} = \mathbf{V}\,\mathbf{R} \ . \tag{1.4.8}$$

This can easily be confirmed by making use of the property $\mathbf{R}\,\mathbf{R}^T = \mathbf{I}$. It then follows that

$$\mathbf{F} = \mathbf{I}\,\mathbf{F} = \mathbf{R}\,\mathbf{R}^T\,\mathbf{F} = \mathbf{R}\,\mathbf{U} \ , \qquad \mathbf{F} = \mathbf{F}\,\mathbf{I} = \mathbf{F}\,\mathbf{R}^T\,\mathbf{R} = \mathbf{V}\,\mathbf{R} \tag{1.4.9}$$

with

$$\mathbf{U} = \mathbf{R}^T\,\mathbf{F} \quad \text{and} \quad \mathbf{V} = \mathbf{F}\,\mathbf{R}^T \ , \tag{1.4.10}$$

in accordance with the above statement.

Due to the orthogonality of **R** the tensors **U** and **R** possess together twelve independent components which have to satisfy only nine scalar equations involved in (1.4.8). Therefore the decomposition (1.4.8) is in general not unique. In the applications of continuum mechanics the tensor **R** will be selected so as to produce a symmetric tensor **U** or **V** corresponding to three constraints.

1.5 Spherical tensor, deviatoric tensor

Any second-order tensor $\mathbf{T} = T^{ij}\,\mathbf{g}_i \otimes \mathbf{g}_j$ can be split up according to

$$\mathbf{T} = \operatorname{dev}\mathbf{T} + \operatorname{sph}\mathbf{T} \tag{1.5.1}$$

into a *spherical tensor* sph **T** and a *deviatoric tensor* dev **T**, the definitions of which are given, in view of (1.3.50), by

$$\operatorname{sph}\mathbf{T} = \frac{1}{3}(\mathbf{T} : \mathbf{I})\,\mathbf{I} = \frac{1}{3}(\operatorname{tr}\mathbf{T})\,\mathbf{I} = \frac{1}{3}T^{k}_{\cdot k}\,g^{ij}\,\mathbf{g}_i \otimes \mathbf{g}_j \ , \tag{1.5.2}$$

$$\operatorname{dev}\mathbf{T} = \mathbf{T} - \operatorname{sph}\mathbf{T} = \mathbf{T} - \frac{1}{3}(\operatorname{tr}\mathbf{T})\,\mathbf{I} = (T^{ij} - \frac{1}{3}T^{k}_{\cdot k}\,g^{ij})\,\mathbf{g}_i \otimes \mathbf{g}_j \ . \tag{1.5.3}$$

We see that sph **T** is always symmetric while this is the case for dev **T** only if the initial tensor **T** has this property. The definitions (1.5.2) and (1.5.3) imply that the spherical part of a deviatoric tensor vanishes:

$$\begin{aligned}\text{sph}\,(\text{dev}\,\mathbf{T}) &= \frac{1}{3}\,\text{tr}\,(\text{dev}\,\mathbf{T})\,\mathbf{I} = \frac{1}{3}\,\text{tr}\left(\mathbf{T} - \frac{1}{3}\,(\text{tr}\,\mathbf{T})\,\mathbf{I}\right)\mathbf{I} \\ &= \frac{1}{3}\left(\text{tr}\,\mathbf{T} - \frac{1}{3}\,\text{tr}\,\mathbf{T}\,\text{tr}\,\mathbf{I}\right)\mathbf{I} = \mathbf{0}\ , \end{aligned} \qquad (1.5.4)$$

since $\text{tr}\,\mathbf{I} = 3$. Similarly, it can also be verified that the double contraction of a spherical tensor and a deviatoric one is zero, thus:

$$\begin{aligned}\text{dev}\,\mathbf{C} : \text{sph}\,\mathbf{D} &= \left(\mathbf{C} - \frac{1}{3}\,(\text{tr}\,\mathbf{C})\,\mathbf{I}\right) : \frac{1}{3}\,(\text{tr}\,\mathbf{D})\,\mathbf{I} \\ &= \frac{1}{3}\,\text{tr}\,\mathbf{C}\,\text{tr}\,\mathbf{D} - \frac{1}{9}\,\text{tr}\,\mathbf{C}\,\text{tr}\,\mathbf{D}\,\text{tr}\,\mathbf{I} = 0\ . \end{aligned} \qquad (1.5.5)$$

Consequently, we may write

$$\text{dev}\,\mathbf{C} : \mathbf{D} = \text{dev}\,\mathbf{C} : \text{dev}\,\mathbf{D} = \mathbf{C} : \text{dev}\,\mathbf{D}\ , \qquad (1.5.6)$$

$$\text{sph}\,\mathbf{C} : \mathbf{D} = \text{sph}\,\mathbf{C} : \text{sph}\,\mathbf{D} = \mathbf{C} : \text{sph}\,\mathbf{D}\ , \qquad (1.5.7)$$

$$\mathbf{C} : \mathbf{D} = \text{dev}\,\mathbf{C} : \text{dev}\,\mathbf{D} + \text{sph}\,\mathbf{C} : \text{sph}\,\mathbf{D}\ . \qquad (1.5.8)$$

It should be noticed that the tensors introduced in this section play an important role in continuum mechanics, e.g. in modelling incompressible material behaviour.

1.6 Differential operators

The Nabla-Operator. We first introduce the so-called *Nabla-Operator*

$$\nabla = \mathbf{g}^k \frac{\partial}{\partial \Theta^k}\ , \qquad (1.6.1)$$

which defines an operation, where, after multiplication by $\mathbf{g}^k$, a function is to be differentiated with respect to Θ^k. We illustrate this by examples:

$$\nabla\,\Phi = \frac{\partial \Phi}{\partial \Theta^k}\,\mathbf{g}^k = \Phi_{,k}\,\mathbf{g}^k\ , \qquad (1.6.2)$$

$$\nabla \cdot \mathbf{u} = \mathbf{g}^k \cdot \frac{\partial \mathbf{u}}{\partial \Theta^k} = \mathbf{u}_{,k} \cdot \mathbf{g}^k\ , \qquad (1.6.3)$$

$$\nabla \times \mathbf{u} = \mathbf{g}^k \frac{\partial}{\partial \Theta^k} \times \mathbf{u} = \mathbf{g}^k \times \mathbf{u}_{,k} = \text{rot}\,\mathbf{u} \qquad (1.6.4)$$

where the abbreviation

$$\operatorname{rot}\mathbf{u} = \mathbf{g}^k \times \mathbf{u}_{,k} = \mathbf{g}^k \times \mathbf{u}|_k = \mathbf{g}^i \times (u_j\, \mathbf{g}^j)|_i = \varepsilon^{ijk}\, u_j|_i\, \mathbf{g}_k \tag{1.6.5}$$

has been introduced forming a vector. Note that $(\ldots)|_k$ denotes covariant derivatives with respect to Θ^k and the equality $\mathbf{u}_{,k} = \mathbf{u}|_k$ holds for arbitrary vectors $\mathbf{u}$. We also recall that $\mathbf{g}_i|_j = \mathbf{0}$ by equation (1.2.24).

The Gradient. The *gradient* of a scalar-valued function Φ is a vector defined by

$$\operatorname{grad}\Phi = \frac{\partial \Phi}{\partial \Theta^k}\, \mathbf{g}^k = \Phi_{,k}\, \mathbf{g}^k = \nabla\, \Phi\ . \tag{1.6.6}$$

Thus we see that grad and ∇ are, in the case of application to scalar valued functions, identical operations. The gradient of a vector $\mathbf{u}$ is a second-order tensor

$$\operatorname{grad}\mathbf{u} = \frac{\partial \mathbf{u}}{\partial \Theta^k} \otimes \mathbf{g}^k = \mathbf{u}_{,k} \otimes \mathbf{g}^k = u_i|_k\, \mathbf{g}^i \otimes \mathbf{g}^k \tag{1.6.7}$$

defined as the tensorial product of the derivative $\mathbf{u}_{,k}$ with the contravariant base vector $\mathbf{g}^k$. Since $\mathbf{u}_{,k}$ transforms according to the covariant rule, grad $\mathbf{u}$ is invariant. Similarly, the *gradient* of a second-order tensor $\mathbf{A} = A_{ij}\, \mathbf{g}^i \otimes \mathbf{g}^j$ can be defined. This leads by virtue of (1.2.25) to a third-order tensor of the form

$$\operatorname{grad}\mathbf{A} = \frac{\partial \mathbf{A}}{\partial \Theta^k} \otimes \mathbf{g}^k = \mathbf{A}_{,k} \otimes \mathbf{g}^k = A_{ij}|_k\, \mathbf{g}^i \otimes \mathbf{g}^j \otimes \mathbf{g}^k\ . \tag{1.6.8}$$

Consequently the gradient describes a rule which increases the order of the initial tensor by one. For the gradient of a product $\Phi\, \mathbf{u}$ we find according to (1.6.6) and (1.6.7)

$$\operatorname{grad}(\Phi\, \mathbf{u}) = (\Phi\, \mathbf{u})_{,k} \otimes \mathbf{g}^k = \mathbf{u} \otimes \operatorname{grad}\Phi + \Phi \operatorname{grad}\mathbf{u}\ . \tag{1.6.9}$$

The Divergence. The *divergence* of a vector $\mathbf{u}$ is a scalar-valued invariant defined by the rule

$$\operatorname{div}\mathbf{u} = \operatorname{grad}\mathbf{u} : \mathbf{I} = u^k|_k = \mathbf{u}_{,k} \cdot \mathbf{g}^k = \nabla \cdot \mathbf{u}\ , \tag{1.6.10}$$

while for a second-order tensor $\mathbf{A}$ the definition reads by considering (1.6.8) as

$$\operatorname{div}\mathbf{A} = \operatorname{grad}\mathbf{A} : \mathbf{I} = \operatorname{grad}\mathbf{A} : \mathbf{g}_k \otimes \mathbf{g}^k = A^{ij}|_j\, \mathbf{g}_i = \mathbf{A}_{,k}\, \mathbf{g}^k\ . \tag{1.6.11}$$

Accordingly, the divergence describes a rule which reduces the order of a tensor by one. Because of the property $\mathbf{g}_i|_j = \mathbf{0}$, div $\mathbf{A}$ can be given alternatively by

$$\operatorname{div}\mathbf{A} = A^{ij}|_j\, \mathbf{g}_i = (A^{ij}\, \mathbf{g}_i)|_j = \mathbf{A}^j|_j \tag{1.6.12}$$

in terms of the contravariant vector $\mathbf{A}^j = A^{ij}\, \mathbf{g}_i$. For later derivations we also need the divergence of a simple contraction $\mathbf{u}\, \mathbf{A}$, which can be evaluated according to (1.6.10) as

$$\operatorname{div}(\mathbf{u}\,\mathbf{A}) = (\mathbf{u}\,\mathbf{A})_{,k} \cdot \mathbf{g}^k = (\mathbf{u}_{,k}\,\mathbf{A}) \cdot \mathbf{g}^k + (\mathbf{u}\,\mathbf{A}_{,k}) \cdot \mathbf{g}^k$$

$$= \mathbf{A} : (\mathbf{u}_{,k} \otimes \mathbf{g}^k) + \mathbf{u} \cdot (\mathbf{A}_{,k}\,\mathbf{g}^k) \ . \qquad (1.6.13)$$

In view of (1.6.7) (1.6.11) the final result is

$$\operatorname{div}(\mathbf{u}\,\mathbf{A}) = \mathbf{A} : \operatorname{grad}\mathbf{u} + \mathbf{u} \cdot \operatorname{div}\mathbf{A} \ . \qquad (1.6.14)$$

The GAUSS-GREEN theorem. This theorem permits to transform the volume integral of a divergence into a surface integral or vice versa. If **u** is a vector field defined on a volume V and $\mathbf{n} = n_i\,\mathbf{g}^i$ is the field of unit outward normals to the boundary surface A of V, the GAUSS-GREEN theorem reads as

$$\iiint_V \operatorname{div}\mathbf{u}\, dV = \iint_A \mathbf{u}\cdot\mathbf{n}\, dA \ \rightarrow \ \iiint_V u^j|_j\, dV = \iint_A u^j\, n_j\, dA \ . \qquad (1.6.15)$$

For a second-order tensor field $\mathbf{A} = A^{ij}\,\mathbf{g}_i \otimes \mathbf{g}_j$ defined again throughout V the above relation takes the form

$$\iiint_V \operatorname{div}\mathbf{A}\, dV = \iint_A \mathbf{A}\,\mathbf{n}\, dA \ \rightarrow \ \iiint_V A^{ij}|_j\, dV = \iint_A A^{ij}\, n_j\, dA \qquad (1.6.16)$$

and can be also given by means of (1.6.12) alternatively by

$$\iiint_V \mathbf{A}^i|_i\, dV = \iint_A \mathbf{A}^i\, n_i\, dA \qquad (1.6.17)$$

in terms of the contravariant vector $\mathbf{A}^i = A^{ji}\,\mathbf{g}_j$.

1.7 Differentiation rules

For different purposes we have to calculate the partial derivatives of tensors with respect to the coordinates Θ^i or the derivatives of tensors with respect to other tensors. The corresponding differentiation rules are introduced in this section.

Partial derivatives with respect to coordinates. We first recall that the partial derivative of an arbitrary tensor **S** with respect to Θ^i is, in view of the identity $\mathbf{g}_i|_j = \mathbf{0}$, identical with its covariant derivative $\mathbf{S}|_i$ and given by (1.2.25):

$$\mathbf{S}_{,i} = \frac{\partial \mathbf{S}}{\partial \Theta^i} = \mathbf{S}|_i = S_{mn}|_i\ \mathbf{g}^m \otimes \mathbf{g}^n \qquad (1.7.1)$$

for a second order tensor $\mathbf{S} = S_{mn}\,\mathbf{g}^m \otimes \mathbf{g}^n$. Note that the function $\mathbf{S}_{,i}$ obeys the covariant transformation rule. By considering again $\mathbf{g}_i|_j = \mathbf{0}$ the second covariant derivative $\mathbf{S}|_{ij}$ can be defined as follows:

$$\mathbf{S}|_{ij} = (S_{mn}|_i\, \mathbf{g}^m \otimes \mathbf{g}^n)|_j = S_{mn}|_{ij}\, \mathbf{g}^m \otimes \mathbf{g}^n \,, \tag{1.7.2}$$

where the covariant derivatives $S_{mn}|_{ij} = (S_{mn}|_i)|_j$ are now to be constructed considering $S_{mn}|_i$ as third-order tensor components.

Partial derivative with respect to a tensor. Let us consider a scalar-valued function

$$\Pi = \Pi\,(\mathbf{A}) \tag{1.7.3}$$

depending upon the second-order tensor $\mathbf{A} = A_{ij}\, \mathbf{g}^i \otimes \mathbf{g}^j$. The partial derivative of Π with respect to $\mathbf{A}$, denoted by $\Pi,_{\mathbf{A}}$ is defined by

$$\Pi,_{\mathbf{A}} = \frac{\partial \Pi}{\partial \mathbf{A}} = \frac{\partial \Pi}{\partial A_{ij}}\, \mathbf{g}_i \otimes \mathbf{g}_j = B^{ij}\, \mathbf{g}_i \otimes \mathbf{g}_j = \frac{\partial \Pi}{\partial A_{i\cdot}^{\cdot j}}\, \mathbf{g}_i \otimes \mathbf{g}^j = B^{i}_{\cdot j}\, \mathbf{g}_i \otimes \mathbf{g}^j = \ldots \tag{1.7.4}$$

and is therefore a second-order tensor. In view of this definition the choice of the component type, i.e. A_{ij}, $A_{i\cdot}^{\cdot j}$, ..., in forming the partial derivative $\Pi,_{\mathbf{A}}$ is immaterial since the result is independently of this selection always the same tensor. The second partial derivative of Π with respect to $\mathbf{A}$ is defined similarly by

$$\Pi,_{\mathbf{A}\otimes\mathbf{A}} = \frac{\partial^2 \Pi}{\partial \mathbf{A}\, \partial \mathbf{A}} = \frac{\partial^2 \Pi}{\partial A_{ij}\, \partial A_{mn}}\, \mathbf{g}_i \otimes \mathbf{g}_j \otimes \mathbf{g}_m \otimes \mathbf{g}_n = C^{ijmn}\, \mathbf{g}_i \otimes \mathbf{g}_j \otimes \mathbf{g}_m \otimes \mathbf{g}_n \,, \tag{1.7.5}$$

the result being in this case a fourth-order tensor.

The definitions introduced above for a zero-order tensor Π can readily be extended to tensors of arbitrary orders. In view of their importance we consider here second-order tensors $\mathbf{B} = B_{ij}\, \mathbf{g}^i \otimes \mathbf{g}^j$ and suppose the components B_{ij} to be functions of A_{ij} similar to Π in the previous example. The partial derivative of $\mathbf{B}$ with respect to $\mathbf{A}$ denoted by $\mathbf{B},_{\mathbf{A}}$ is then defined by

$$\mathbf{B},_{\mathbf{A}} = \frac{\partial \mathbf{B}}{\partial \mathbf{A}} = \frac{\partial B_{ij}}{\partial A_{kl}}\, \mathbf{g}^i \otimes \mathbf{g}^j \otimes \mathbf{g}_k \otimes \mathbf{g}_l = C_{ij\cdot\cdot}^{\cdot\cdot kl}\, \mathbf{g}^i \otimes \mathbf{g}^j \otimes \mathbf{g}_k \otimes \mathbf{g}_l \,, \tag{1.7.6}$$

and forms therefore a fourth-order tensor $\mathbf{C}$. We see that the partial differentiation with respect to a second-order tensor increases the order of the original tensor by two. In view of the definition (1.7.6) the partial derivative $\mathbf{A},_{\mathbf{A}}$ leads to

$$\mathbf{A},_{\mathbf{A}} = \frac{\partial \mathbf{A}}{\partial \mathbf{A}} = \frac{\partial A_{ij}}{\partial A_{kl}}\, \mathbf{g}^i \otimes \mathbf{g}^j \otimes \mathbf{g}_k \otimes \mathbf{g}_l = \delta_i^k\, \delta_j^l\, \mathbf{g}^i \otimes \mathbf{g}^j \otimes \mathbf{g}_k \otimes \mathbf{g}_l$$

$$= \overset{4}{\mathbf{I}}{}^* = \mathbf{g}^i \otimes \mathbf{g}^j \otimes \mathbf{g}_i \otimes \mathbf{g}_j \,, \tag{1.7.7}$$

where $\overset{4}{\mathbf{I}}{}^*$ is a tensor depending, similar to $\overset{4}{\mathbf{I}}$ (1.3.9), only on the base vectors $\mathbf{g}_i$.

Chain rule of differentiation. We now assume that the tensor components A_{ij} occuring in Π (1.7.3) depend upon the parameter t. The derivative of Π with respect to t can be carried out by the usual chain rule. By means of (1.7.4) the result can be presented as

$$\dot{\Pi} = \frac{\partial \Pi}{\partial t} = \Pi,_{\mathbf{A}} : \dot{\mathbf{A}} \ , \qquad \dot{\mathbf{A}} = \frac{\partial \, (A_{ij} \, \mathbf{g}^i \otimes \mathbf{g}^j)}{\partial t} = \dot{A}_{ij} \, \mathbf{g}^i \otimes \mathbf{g}^j \tag{1.7.8}$$

in symbolic notation, where the basis $\mathbf{g}_i$ is supposed to be independent of t.

Application. Form the partial derivatives of the following invariant quantities with respect to the tensor $\mathbf{A} = A_{ij} \, \mathbf{g}^i \otimes \mathbf{g}^j$:

$$\mathrm{tr}\,\mathbf{A} = A^{r}_{\cdot r} \ , \qquad (\mathrm{tr}\,\mathbf{A})^2 = A^{r}_{\cdot r} \, A^{s}_{\cdot s} \ ,$$
$$\mathrm{tr}\,\mathbf{A}^2 = A^{r}_{\cdot s} \, A^{s}_{\cdot r} \ , \qquad \mathrm{tr}\,\mathbf{A}^3 = A^{r}_{\cdot s} \, A^{s}_{\cdot t} \, A^{t}_{\cdot r} \ .$$

The first way to solve this problem is to write the above expression in full and to apply then the rule (1.7.4) to the corresponding results. By means of the identity

$$\frac{\partial A^{m}_{\cdot n}}{\partial A^{r}_{\cdot s}} = \delta^{m}_{r} \, \delta^{s}_{n} \ , \tag{1.7.9}$$

which has been already used in (1.7.7), the derivation can be, however, simplified considerably. We obtain

$$(\mathrm{tr}\,\mathbf{A}),_{\mathbf{A}} = \frac{\partial A^{r}_{\cdot r}}{\partial A^{m}_{\cdot s}} \mathbf{g}^m \otimes \mathbf{g}_s = \delta^{r}_{m} \, \delta^{s}_{r} \, \mathbf{g}^m \otimes \mathbf{g}_s = \delta^{s}_{m} \, \mathbf{g}^m \otimes \mathbf{g}_s = \mathbf{I} \ , \tag{1.7.10}$$

$$(\mathrm{tr}\,\mathbf{A})^2,_{\mathbf{A}} = 2 \, \mathrm{tr}\,\mathbf{A} \, (\mathrm{tr}\,\mathbf{A}),_{\mathbf{A}} = 2 \, (\mathrm{tr}\,\mathbf{A}) \, \mathbf{I} \ . \tag{1.7.11}$$

By a similar procedure it can also be shown that

$$(\mathrm{tr}\,\mathbf{A}^2),_{\mathbf{A}} = \frac{\partial \, (A^{r}_{\cdot s} A^{s}_{\cdot r})}{\partial \, A^{m}_{\cdot n}} \mathbf{g}^m \otimes \mathbf{g}_n = 2 \, A^{s}_{\cdot r} \, \mathbf{g}^r \otimes \mathbf{g}_s = 2 \, \mathbf{A}^T \ , \tag{1.7.12}$$

$$(\mathrm{tr}\,\mathbf{A}^3),_{\mathbf{A}} = 3 \, (\mathbf{A}^2)^T \ . \tag{1.7.13}$$

For a symmetrical tensor $\mathbf{C} = \mathbf{C}^T$ the above results become

$$(\mathrm{tr}\,\mathbf{C}),_{\mathbf{C}} = \mathbf{I} \ , \qquad (\mathrm{tr}\,\mathbf{C})^2,_{\mathbf{C}} = 2 \, (\mathrm{tr}\,\mathbf{C}) \, \mathbf{I} \ ,$$
$$(\mathrm{tr}\,\mathbf{C}^2),_{\mathbf{C}} = 2 \, \mathbf{C} \ , \qquad (\mathrm{tr}\,\mathbf{C}^3),_{\mathbf{C}} = 3 \, \mathbf{C}^2 \ . \tag{1.7.14}$$

Application. Under the assumption $\mathbf{S} = \mathbf{S}^T$ construct the partial derivative of the expression $\| \mathbf{S} \|$ with respect to $\mathbf{S}$.

In view of (1.3.58) we have

$$\| \mathbf{S} \| = \sqrt{\mathbf{S} : \mathbf{S}} = \sqrt{\mathrm{tr}\,(\mathbf{S}\,\mathbf{S}^T)} = \sqrt{\mathrm{tr}\,\mathbf{S}^2} \ ,$$

which by virtue of (1.7.4) and (1.7.14) gives:

$$\frac{\partial \| \mathbf{S} \|}{\partial \mathbf{S}} = \frac{1}{2} (\mathrm{tr}\,\mathbf{S}^2)^{-1/2} \, (\mathrm{tr}\,\mathbf{S}^2),_{\mathbf{S}} = \frac{\mathbf{S}}{\| \mathbf{S} \|} \, . \tag{1.7.15}$$

Application. Form the partial derivative of the expression

$$\mathbf{n} \, \mathbf{S} \, \mathbf{n} = S^{ij} \, n_i \, n_j$$

with respect to $\mathbf{n}$ supposing that $\mathbf{S} = \mathbf{S}^T$ and $\mathbf{S}$ is independent of $\mathbf{n}$.

By considering (1.7.9) and the symmetry $S^{ij} = S^{ji}$ we first form

$$\begin{aligned}\frac{\partial (S^{ij}\, n_i\, n_j)}{\partial n_k} &= S^{ij}\frac{\partial n_i}{\partial n_k}\, n_j + S^{ij}\, n_i \frac{\partial n_j}{\partial n_k}\\ &= S^{ij}\,\delta_i^k\, n_j \quad + S^{ij}\, n_i\,\delta_j^k\\ &= S^{kj}\, n_j \quad + S^{ik}\, n_i = 2\, S^{kj}\, n_j\end{aligned} \tag{1.7.16}$$

to obtain according to the rule (1.7.6) the following result presenting a vector:

$$\begin{aligned}(\mathbf{n}\,\mathbf{S}\,\mathbf{n})_{,\mathbf{n}} = \frac{\partial (\mathbf{n}\,\mathbf{S}\,\mathbf{n})}{\partial n_k}\,\mathbf{g}_k &= 2\, S^{kj}\, n_j\, \mathbf{g}_k = 2\,(S^{km}\,\mathbf{g}_k \otimes \mathbf{g}_m)\,(n_j\, \mathbf{g}^j)\\ &= 2\,\mathbf{S}\,\mathbf{n} = 2\,\mathbf{n}\,\mathbf{S}\ .\end{aligned} \tag{1.7.17}$$

1.8 Invariants of a second-order tensor

Definition of invariants. Any second-order tensor $\mathbf{A} = A_{ij}\,\mathbf{g}^i \otimes \mathbf{g}^j$ may be associated with three invariant scalars denoted by $I_\mathbf{A}$, $II_\mathbf{A}$ and $III_\mathbf{A}$. In view of (1.3.50) and (1.3.52) their definitions can be given alternatively in absolute and index notation

$$I_\mathbf{A} = A_{i\cdot}^{\cdot i} = \operatorname{tr}\mathbf{A} = \mathbf{A} : \mathbf{I}\ , \tag{1.8.1}$$

$$II_\mathbf{A} = \frac{1}{2}\,(A_{r\cdot}^{\cdot r}\, A_{s\cdot}^{\cdot s} - A_{\cdot s}^{r\cdot}\, A_{\cdot r}^{s\cdot}) = \frac{1}{2}\left[(\operatorname{tr}\mathbf{A})^2 - \operatorname{tr}\mathbf{A}^2\right]\ , \tag{1.8.2}$$

$$III_\mathbf{A} = e_{rst}\, A_{\cdot 1}^{r\cdot}\, A_{\cdot 2}^{s\cdot}\, A_{\cdot 3}^{t\cdot} = |\, A_{\cdot s}^{r\cdot}\, | = \det\mathbf{A}\ , \tag{1.8.3}$$

where e_{rst} is the permutation tensor associated with orthogonal Cartesian coordinates. We notice that the definition given in (1.8.3) for det $\mathbf{A}$ is in accordance with the previous one (1.3.60) and holds therefore only for the mixed components $A_{\cdot s}^{r\cdot}$. The invariants introduced above are related by the *CAYLEY-HAMILTON theorem*

$$\mathbf{A}^3 - I_\mathbf{A}\,\mathbf{A}^2 + II_\mathbf{A}\,\mathbf{A} - III_\mathbf{A}\,\mathbf{I} = \mathbf{0}\ , \tag{1.8.4}$$

the double contraction of which by the identity tensor $\mathbf{I}$

$$\mathbf{A}^3 : \mathbf{I} - I_\mathbf{A}\,\mathbf{A}^2 : \mathbf{I} + II_\mathbf{A}\,\mathbf{A} : \mathbf{I} - III_\mathbf{A}\,\mathbf{I} : \mathbf{I} = 0 \tag{1.8.5}$$

permits to replace the definition (1.8.3) by a new expression. In view of (1.3.50), (1.8.1) and (1.8.2) the corresponding result is of the form:

$$III_\mathbf{A} = \frac{1}{3}\left[\operatorname{tr}\mathbf{A}^3 - \frac{3}{2}\operatorname{tr}\mathbf{A}^2\operatorname{tr}\mathbf{A} + \frac{1}{2}(\operatorname{tr}\mathbf{A})^3\right]\ . \tag{1.8.6}$$

This result advantageously indicates that $III_\mathbf{A}$ is an invariant scalar, while this fact is obvious for $I_\mathbf{A}$ and $II_\mathbf{A}$ from the definitions (1.8.1) and (1.8.2).

Partial derivatives of the invariants. For later applications the partial derivatives of the invariants I_A, II_A and III_A are needed. They can be calculated by applying the differentiation rule (1.7.4) to equations (1.8.1), (1.8.2) and (1.8.6) leading by virtue of the identities (1.7.10) through (1.7.13) to

$$(I_A),_{\mathbf{A}} = (\mathrm{tr}\,\mathbf{A}),_{\mathbf{A}} = \mathbf{I} \,, \tag{1.8.7}$$

$$(II_A),_{\mathbf{A}} = \frac{1}{2}[(\mathrm{tr}\,\mathbf{A})^2 - \mathrm{tr}\,\mathbf{A}^2],_{\mathbf{A}} = (\mathrm{tr}\,\mathbf{A})\,\mathbf{I} - \mathbf{A}^T \,, \tag{1.8.8}$$

$$\begin{aligned}(III_A),_{\mathbf{A}} &= \left[\frac{1}{3}\mathrm{tr}\,\mathbf{A}^3 - \frac{1}{2}\mathrm{tr}\,\mathbf{A}^2\,\mathrm{tr}\,\mathbf{A} + \frac{1}{6}(\mathrm{tr}\,\mathbf{A})^3\right],_{\mathbf{A}} \\ &= (\mathbf{A}^2)^T - \mathbf{A}^T\,\mathrm{tr}\,\mathbf{A} - \frac{1}{2}(\mathrm{tr}\,\mathbf{A}^2)\,\mathbf{I} + \frac{1}{2}(\mathrm{tr}\,\mathbf{A})^2\,\mathbf{I} \\ &= (\mathbf{A}^2)^T - I_A\,\mathbf{A}^T + II_A\,\mathbf{I} \,. \end{aligned} \tag{1.8.9}$$

By considering (1.8.4), the last result can be further transformed into

$$\begin{aligned}(III_A),_{\mathbf{A}} &= \left[(\mathbf{A}^3 - I_A\,\mathbf{A}^2 + II_A\,\mathbf{A})\,\mathbf{A}^{-1}\right]^T = \left[(III_A\,\mathbf{I})\,\mathbf{A}^{-1}\right]^T \\ &= III_A\,\mathbf{A}^{-T} \,. \end{aligned} \tag{1.8.10}$$

The invariants introduced above are closely related to the eigenvalue problems of second-order tensors which will be discussed in the next section.

Application. Evaluate the invariants of a skew-symmetric tensor $\mathbf{A} = -\mathbf{A}^T$.

Since $\mathbf{A}^3 = -(\mathbf{A}^3)^T$ and the trace of any skew-symmetric tensor vanishes, we have from (1.8.1), (1.8.2) and (1.8.6):

$$\begin{aligned} I_A &= 0 \,, \quad III_A = 0 \,, \\ II_A &= -\frac{1}{2}\mathbf{A}^2 : \mathbf{I} = \frac{1}{2}(\mathbf{A}\,\mathbf{A}^T) : \mathbf{I} = \frac{1}{2}\mathbf{A} : \mathbf{A} = \frac{1}{2}\|\mathbf{A}\|^2 \,, \end{aligned} \tag{1.8.11}$$

in accordance with the rules (1.3.51) and (1.3.58).

1.9 The eigenvalue problem of a second-order tensor

Interpretation of an eigenvalue problem. In this section we shall deal with the question what is to be understood under an *eigenvalue problem* of a second-order tensor $\boldsymbol{\sigma}$ supposing $\boldsymbol{\sigma}$ to be symmetric*. To gain, from the mechanical point of view, a better insight into this problem we identify $\boldsymbol{\sigma} = \sigma^{ij}\,\mathbf{g}_i \otimes \mathbf{g}_j$ with the CAUCHY stress tensor*. This tensor is defined such that the expression

---

* In contrast to the convention in this chapter the second-order tensor $\boldsymbol{\sigma}$ is denoted, in accordance with section 3.1, by a lower case letter.

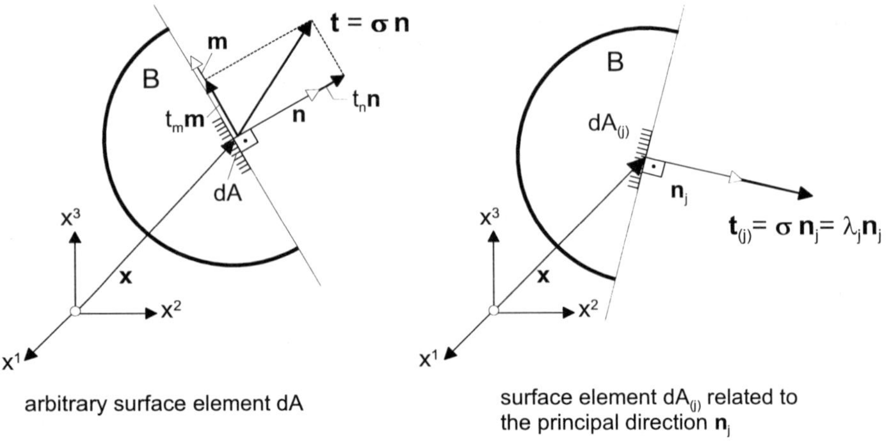

Fig. 1.1. CAUCHY stress vector **t** and $\mathbf{t}_{(j)}$ related to dA and $dA_{(j)}$

$$\mathbf{t} = \boldsymbol{\sigma}\,\mathbf{n} = \sigma^{ij}\, n_j\, \mathbf{g}_i \tag{1.9.1}$$

determines the stress vector **t** acting upon a surface element dA of a body whose unit normal vector is **n**:

$$\mathbf{n} \cdot \mathbf{n} = 1 \ . \tag{1.9.2}$$

The scalar product

$$t_n = \mathbf{n} \cdot \mathbf{t} = \mathbf{n} \cdot (\boldsymbol{\sigma}\,\mathbf{n}) = \mathbf{n}\,\boldsymbol{\sigma}\,\mathbf{n} \quad \rightarrow \quad t_n = \sigma^{ij}\, n_i\, n_j \tag{1.9.3}$$

can be therefore interpreted as the component of **t** in an orthonormal basis (**n**, **m**) in direction **n** as illustrated in Fig. 1.1. Now the question is to determine the positions of the unit vector **n**, for which t_n takes for a given tensor $\boldsymbol{\sigma}$ extreme values. By means of the LAGRANGE-multiplier method this problem is equivalent to determining the extremum of the function

$$\Phi = \mathbf{n}\,\boldsymbol{\sigma}\,\mathbf{n} - \lambda\,(\mathbf{n} \cdot \mathbf{n} - 1) \ , \tag{1.9.4}$$

where λ is an unknown parameter. The requirement that Φ = extremum leads by considering (1.7.17) to

$$\Phi_{,\mathbf{n}} = \frac{\partial \Phi}{\partial \mathbf{n}} = \mathbf{0} \quad \rightarrow \quad (\boldsymbol{\sigma} - \lambda\,\mathbf{I})\,\mathbf{n} = \mathbf{0} \ , \tag{1.9.5}$$

---

* For the definition of $\boldsymbol{\sigma}$ we refer to (3.1.17).

which is called the *eigenvalue problem* of $\boldsymbol{\sigma}$. Equation (1.9.5) has nontrivial solutions $\mathbf{n} \neq \mathbf{0}$, if the *characteristic* equation

$$\det(\boldsymbol{\sigma} - \lambda\,\mathbf{I}) = 0 \quad \rightarrow \quad |\,\sigma^i_j - \lambda\,\delta^i_j\,| = 0 \tag{1.9.6}$$

is satisfied which is of third order in λ:

$$\det(\boldsymbol{\sigma} - \lambda\,\mathbf{I}) = -\lambda^3 + \mathrm{I}_\sigma\,\lambda^2 - \mathrm{II}_\sigma\,\lambda + \mathrm{III}_\sigma = 0 \ . \tag{1.9.7}$$

According to (1.8.1), (1.8.2) and (1.8.3)

$$\mathrm{I}_\sigma = \sigma^i_i = \mathrm{tr}\,\boldsymbol{\sigma} \ ,$$

$$\mathrm{II}_\sigma = \frac{1}{2}\,(\sigma^i_i\,\sigma^j_j - \sigma^i_j\,\sigma^j_i) = \frac{1}{2}\left[(\mathrm{tr}\,\boldsymbol{\sigma})^2 - \mathrm{tr}\,\boldsymbol{\sigma}^2\right] ,$$

$$\mathrm{III}_\sigma = |\,\sigma^i_j\,| = \det\boldsymbol{\sigma} \tag{1.9.8}$$

are the invariants of the symmetric tensor $\boldsymbol{\sigma}$.

Properties of eigenvalues and eigenvectors. Let λ_j ($j = 1, 2, 3$) be the solutions of (1.9.7). We shall now show that the eigenvalues λ_j are equal to the extreme values of t_n. For this purpose we specify (1.9.5) for $\lambda = \lambda_j$ and contract the result from the left side by the eigenvector $\mathbf{n}_j$. This yields by considering (1.9.3)

$$\mathbf{n}_j\,\boldsymbol{\sigma}\,\mathbf{n}_j - \lambda_j\,\mathbf{n}_j\,\mathbf{I}\,\mathbf{n}_j = t_{n(j)} - \lambda_j = 0 \quad \rightarrow \quad \lambda_j = \mathbf{n}_j\,\boldsymbol{\sigma}\,\mathbf{n}_j = t_{n(j)} \tag{1.9.9}$$

confirming the above statement.

Similarly it can be proved that the eigenvectors $\mathbf{n}_j$ ($j = 1, 2, 3$) called the *principal directions* of $\boldsymbol{\sigma}$, are orthogonal. From (1.9.5) we receive similar to (1.9.9)

$$\mathbf{n}_k\,\boldsymbol{\sigma}\,\mathbf{n}_j - \lambda_j\,\mathbf{n}_k\,\mathbf{I}\,\mathbf{n}_j = 0 \ ,$$

$$\mathbf{n}_j\,\boldsymbol{\sigma}\,\mathbf{n}_k - \lambda_k\,\mathbf{n}_j\,\mathbf{I}\,\mathbf{n}_k = 0 \ . \tag{1.9.10}$$

In view of the symmetry $\boldsymbol{\sigma} = \boldsymbol{\sigma}^{\mathrm{T}}$ the difference of the above equations gives

$$(\lambda_k - \lambda_j)\,\mathbf{n}_k \cdot \mathbf{n}_j = 0 \quad \rightarrow \quad \mathbf{n}_k \cdot \mathbf{n}_j = \delta_{kj} \tag{1.9.11}$$

showing that the eigenvectors $\mathbf{n}_j$ ($j = 1, 2, 3$) are mutually orthogonal.

The symmetry property postulated above for the tensor $\boldsymbol{\sigma} = \boldsymbol{\sigma}^{\mathrm{T}}$ implies that the eigenvalues λ_i ($i = 1, 2, 3$) are real and that, as can be deduced from the derivation of (1.9.11), the corresponding eigenvectors $\mathbf{n}_i$ form an *orthonormal* basis so far as they are all distinct*.

---

* Remember that in an orthonormal basis $\mathbf{n}_k$ ($k = 1, 2, 3$) the position of the indices in the vectors $\mathbf{n}_k = \mathbf{n}^k$ is irrevelant.

It remains to show that the tensor $\boldsymbol{\sigma}$ possesses in an orthogonal Cartesian reference frame spanned by the eigenvectors $\mathbf{n}_k = \mathbf{n}^k$ only three components $\bar{\sigma}_{kk}$ (k = 1, 2, 3) identical with the eigenvalues $\bar{\sigma}_{kk} = \lambda_k$. To prove this we assume that the eigenvectors $\mathbf{n}_k$ (of unit length) are all distinct and set

$$\boldsymbol{\sigma} = \sum_{k=1}^{3} \sum_{j=1}^{3} \bar{\sigma}_{kj}\, \mathbf{n}_k \otimes \mathbf{n}_j \ . \tag{1.9.12}$$

The unknown components $\bar{\sigma}_{kj}$ can be calculated by the standard procedure (1.2.13). This leads, by considering the identity

$$\mathbf{n}_k\, \boldsymbol{\sigma}\, \mathbf{n}_j = \lambda_j\, \mathbf{n}_k \cdot \mathbf{n}_j \ , \tag{1.9.13}$$

found from (1.9.10) and by using (1.9.11), to

$$\bar{\sigma}_{kj} = \mathbf{n}_k\, \boldsymbol{\sigma}\, \mathbf{n}_j = \lambda_j\, \mathbf{n}_k \cdot \mathbf{n}_j = \lambda_j\, \delta_{kj} \ , \tag{1.9.14}$$

so that

$$\bar{\sigma}_{kk} = \lambda_k \quad \text{and} \quad \bar{\sigma}_{kj} = 0 \quad \text{for} \quad k \neq j \ . \tag{1.9.15}$$

Thus the relation (1.9.12) takes the form

$$\boldsymbol{\sigma} = \sum_{k=1}^{3} \lambda_k\, \mathbf{n}_k \otimes \mathbf{n}_k \ , \tag{1.9.16}$$

in accordance with the above statement. The formulation (1.9.16) is known as *spectral decomposition* and presents a simple representation of $\boldsymbol{\sigma}$ with only three components λ_k characterizing its invariant property.

According to (1.9.1), $\mathbf{t}_{(j)} = \boldsymbol{\sigma}\, \mathbf{n}_j$ indicates the CAUCHY stress vector acting upon a surface element $dA_{(j)}$ whose unit normal is the eigenvector $\mathbf{n}_j$. From (1.9.13), the following statement can then be deduced.

Remark. The stress vector $\mathbf{t}_{(j)}$ associated with a surface element $dA_{(j)}$, whose unit normal is $\mathbf{n}_j$, is perpendicular to $dA_{(j)}$. If this vector is determined with respect to the basis $\mathbf{n}_j$ (j = 1, 2, 3) it therefore possesses a single component $t_{n(j)} = \lambda_j$ along $\mathbf{n}_j$. Fig. 1.1 and Fig. 1.2 illustrate this important property.

Further important results established above can be summarized as follows:

Remark. Any symmetric tensor $\boldsymbol{\sigma} = \boldsymbol{\sigma}^T$ possesses three real eigenvalues λ_j, the so-called *principle values* connected with three eigenvectors $\mathbf{n}_j$ (j = 1, 2, 3) determining the rectangular *principal axes* of $\boldsymbol{\sigma}$.

Remark. If the tensor $\boldsymbol{\sigma}$ is referred to the orthonormal basis $\mathbf{n}_j \otimes \mathbf{n}_k$, then it has only three components $\bar{\sigma}_{jj} = \lambda_j$ coinciding with the principle values λ_j. This property provides a simple representation (1.9.16) of the tensor $\boldsymbol{\sigma}$ in terms of only three components λ_k called spectral decomposition.

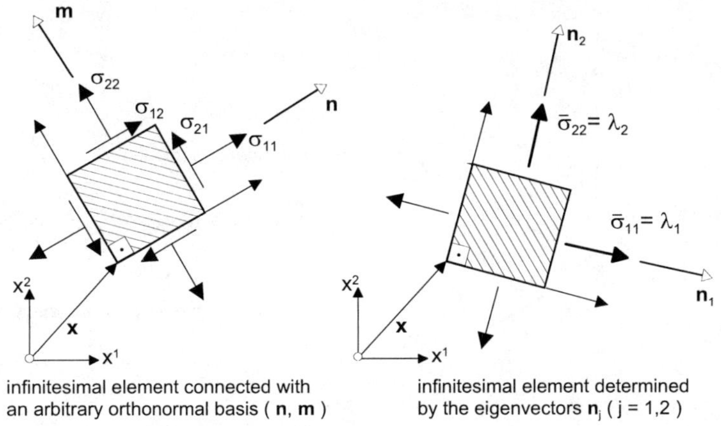

Fig. 1.2. Illustration of the eigenvalues λ_j, the principal stresses, on a two dimensional problem

Remark. According to (1.9.15), the definitions given in (1.9.8) for the invariants I_σ, II_σ and III_σ in terms of tensor components σ^{ij} can now be transformed into

$$I_\sigma = \operatorname{tr}\boldsymbol{\sigma} = \lambda_1 + \lambda_2 + \lambda_3 \ ,$$

$$II_\sigma = \frac{1}{2}\left[(\operatorname{tr}\boldsymbol{\sigma})^2 - \operatorname{tr}\boldsymbol{\sigma}^2\right] = \lambda_1\lambda_2 + \lambda_2\lambda_3 + \lambda_1\lambda_3 \ ,$$

$$III_\sigma = \det\boldsymbol{\sigma} = \lambda_1\lambda_2\lambda_3 \ . \tag{1.9.17}$$

The above equations form a complete set for the determination of the principal values λ_j (j = 1, 2, 3), if the invariants I_σ, II_σ and III_σ are given.

Fig. 1.3 gives a geometrical interpretation of the spectral decomposition (1.9.16) which is, in view of the six constraints (1.9.11) valid for the eigenvectors $\mathbf{n}_i$ (i = 1, 2, 3), determined by six independent variables corresponding to the number of the independent components of the symmetric tensor $\boldsymbol{\sigma}$.

Coaxial tensors. To conclude this section we shall deal briefly with coaxial tensors. Two symmetric second-order tensors **A** and **B** are said to be *coaxial*, if they have the same eigenvectors $\mathbf{n}_i$. Denoting the associated eigenvalues by μ_i and κ_i we may then set

$$\mathbf{A} = \sum_{i=1}^{3} \mu_i\, \mathbf{n}_i \otimes \mathbf{n}_i \ , \quad \mathbf{B} = \sum_{i=1}^{3} \kappa_i\, \mathbf{n}_i \otimes \mathbf{n}_i \ . \tag{1.9.18}$$

By means of this definition and the orthogonality relation (1.9.11) we see that the contraction

$$\mathbf{A}\,\mathbf{B} = \mathbf{B}\,\mathbf{A} = \sum_{i=1}^{3} \mu_i\,\kappa_i\,\mathbf{n}_i \otimes \mathbf{n}_i \tag{1.9.19}$$

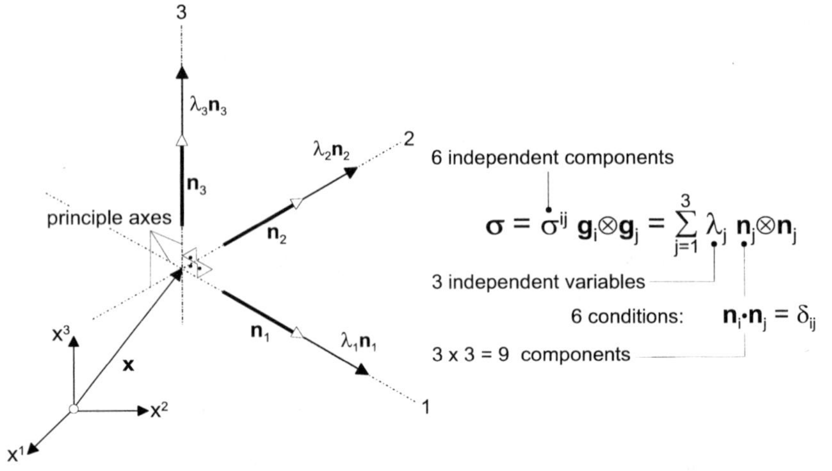

Fig. 1.3. Interpretation of the spectral decomposition of the tensor $\boldsymbol{\sigma} = \boldsymbol{\sigma}^T$

is *commutative*. Conversely, starting from the above relation, it can be shown that the participant tensors **A** and **B** must be coaxial so that we may note:

Remark. A necessary and sufficient condition for the *coaxiality* of two tensors **A** and **B** is the fact that their contraction is *commutative*: **A B** = **B A**.

The definition of coaxial tensors permits to add the following important remarks holding again for symmetric second-order tensors as have been exclusively considered in this section:

Remark. A symmetric tensor **S** and its inverse $\mathbf{S}^{-1}$ are always coaxial and have reciprocal eigenvalues λ_k and $1/\lambda_k$, thus

$$\mathbf{S} = \sum_{k=1}^{3} \lambda_k\, \mathbf{n}_k \otimes \mathbf{n}_k \ , \quad \mathbf{S}^{-1} = \sum_{k=1}^{3} \frac{1}{\lambda_k}\, \mathbf{n}_k \otimes \mathbf{n}_k \ . \tag{1.9.20}$$

Remark. The tensors $\mathbf{C} = \mathbf{U}^2$ and **U** are coaxial and the eigenvalues of **C** are equal to the squares of those of **U**, thus

$$\mathbf{C} = \sum_{k=1}^{3} (\lambda_k)^2\, \mathbf{n}_k \otimes \mathbf{n}_k \ , \quad \mathbf{U} = \sum_{k=1}^{3} \lambda_k\, \mathbf{n}_k \otimes \mathbf{n}_k \ . \tag{1.9.21}$$

It is easy to show that the initial assumptions $\mathbf{S}\,\mathbf{S}^{-1} = \mathbf{I}$ and $\mathbf{C} = \mathbf{U}^2$ are automatically fulfilled by the expressions given in (1.9.20) and (1.9.21) and that, consequently, the statements (1.9.20) and (1.9.21) are true.

Application. Deduce from **A B** = **B A** that the participant tensors must be coaxial:

$$\mathbf{A} = \sum_{i=1}^{3} \mu_i \, \mathbf{n}_i \otimes \mathbf{n}_i \, , \quad \mathbf{B} = \sum_{i=1}^{3} \kappa_i \, \mathbf{n}_i \otimes \mathbf{n}_i \, . \tag{1.9.22}$$

For the proof we use for **B** the above expression and for **A** the usual component relation

$$\mathbf{A} = \sum_{i=1}^{3} \sum_{j=1}^{3} \bar{A}_{ij} \, \mathbf{n}_i \otimes \mathbf{n}_j \tag{1.9.23}$$

with unknown coefficients $\bar{A}_{ij}$. Then we require that

$$\mathbf{A}\,\mathbf{B} = \sum_{i=1}^{3} \sum_{j=1}^{3} \bar{A}_{ij} \, \kappa_j \, \mathbf{n}_i \otimes \mathbf{n}_j = \mathbf{B}\,\mathbf{A} = \sum_{i=1}^{3} \sum_{j=1}^{3} \bar{A}_{ij} \, \kappa_i \, \mathbf{n}_i \otimes \mathbf{n}_j \, .$$

This yields:

$$\sum_{i=1}^{3} \sum_{j=1}^{3} \bar{A}_{ij} \, (\kappa_i - \kappa_j) \, \mathbf{n}_i \otimes \mathbf{n}_j = \mathbf{0} \, . \tag{1.9.24}$$

In view of the assumption $\kappa_i \neq \kappa_j$ for $i \neq j$ the above relation is satisfied only if $\bar{A}_{ij} = 0$ for $i \neq j$. Thus the expression given in (1.9.22) for **A** holds, if **A B** = **B A**.

Application. Verify that the tensors **C** and $\mathbf{E} = \frac{1}{2}(\mathbf{C} - \mathbf{I})$ are coaxial.

In accordance with the requirement (1.9.19) we find:

$$\mathbf{E}\,\mathbf{C} = \frac{1}{2}(\mathbf{C}^2 - \mathbf{C}) = \mathbf{C}\left[\frac{1}{2}(\mathbf{C} - \mathbf{I})\right] = \mathbf{C}\,\mathbf{E} \, . \tag{1.9.25}$$

1.10 Rotation tensor, rotation vector

The rotation tensor **R** plays an important role in the polar decomposition theorem to be introduced in section 2.4. Let $\tilde{\mathbf{g}}_i$ and $\tilde{\mathbf{g}}^i$ be the rotated counterparts of the base vectors $\mathbf{g}_i$ and $\mathbf{g}^i$, respectively. Since a rigid body motion conserves the length of vectors as well as the angle enclosed by each pair of them we have

$$\begin{aligned} \tilde{\mathbf{g}}_i \cdot \tilde{\mathbf{g}}_j &= \tilde{g}_{ij} = \mathbf{g}_i \cdot \mathbf{g}_j = g_{ij} \, , \\ \tilde{\mathbf{g}}^i \cdot \tilde{\mathbf{g}}^j &= \tilde{g}^{ij} = \mathbf{g}^i \cdot \mathbf{g}^j = g^{ij} \end{aligned} \tag{1.10.1}$$

showing the equalities $\tilde{g}_{ij} = g_{ij}$ and $\tilde{g}^{ij} = g^{ij}$ for the metric tensor components related to the bases $\tilde{\mathbf{g}}_i$ and $\mathbf{g}_i$.

A rotation tensor is understood to be a tensor describing the rotation of a complete set of base vectors $\mathbf{g}_i$ (i = 1, 2, 3) into a new one $\tilde{\mathbf{g}}_i$. Thus the rotation of any vector defined with respect to $\mathbf{g}_i$ will be automatically described. As has been shown in section 1.4 in detail such a tensor, denoted in this section by **R**, is orthogonal

$$\mathbf{R}\,\mathbf{R}^T = \mathbf{R}^T\,\mathbf{R} = \mathbf{I} \tag{1.10.2}$$

and describes the rotation of vectors in the form (1.4.3). Accordingly, we may set for the base vectors $\mathbf{g}_i$ and $\mathbf{g}^i$

$$\tilde{\mathbf{g}}_i = \mathbf{R}\,\mathbf{g}_i \ , \quad \mathbf{g}_i = \mathbf{R}^T\,\tilde{\mathbf{g}}_i \ , \tag{1.10.3}$$

$$\tilde{\mathbf{g}}^i = \mathbf{R}\,\mathbf{g}^i \ , \quad \mathbf{g}^i = \mathbf{R}^T\,\tilde{\mathbf{g}}^i \ . \tag{1.10.4}$$

If the rotated counterparts $\tilde{\mathbf{g}}_i$ of $\mathbf{g}_i$ are given, then the rotation tensor $\mathbf{R}$ can be constructed as

$$\mathbf{R} = \tilde{\mathbf{g}}_j \otimes \mathbf{g}^j \ , \quad \mathbf{R}^T = \mathbf{R}^{-1} = \mathbf{g}^j \otimes \tilde{\mathbf{g}}_j \ . \tag{1.10.5}$$

By considering (1.10.1) one can easily verify that the tensor $\mathbf{R}$ defined by (1.10.5) satisfies the relations (1.10.2) to (1.10.4) automatically. To calculate the components of $\mathbf{R}$ defined by

$$\mathbf{R} = \mathrm{R}_{ij}\,\mathbf{g}^i \otimes \mathbf{g}^j = \tilde{\mathrm{R}}_{ij}\,\tilde{\mathbf{g}}^i \otimes \tilde{\mathbf{g}}^j \ , \tag{1.10.6}$$

we use the rule (1.2.13)

$$\mathrm{R}_{ij} = \mathbf{g}_i\,\mathbf{R}\,\mathbf{g}_j \ , \quad \tilde{\mathrm{R}}_{ij} = \tilde{\mathbf{g}}_i\,\mathbf{R}\,\tilde{\mathbf{g}}_j \ . \tag{1.10.7}$$

This gives by considering (1.10.1) and (1.10.5)

$$\mathrm{R}_{ij} = \tilde{\mathrm{R}}_{ij} = \mathbf{g}_i \cdot \tilde{\mathbf{g}}_j \ . \tag{1.10.8}$$

Accordingly, the components R_{ij} and $\tilde{\mathrm{R}}_{ij}$ with respect to the bases $\mathbf{g}^i \otimes \mathbf{g}^j$ and $\tilde{\mathbf{g}}^i \otimes \tilde{\mathbf{g}}^j$ are identical.

To show another important characteristic of $\mathbf{R}$ we refer to the relation (1.10.2), the *variation* of which delivers

$$\delta\mathbf{R}\,\mathbf{R}^T = -\mathbf{R}\,\delta\mathbf{R}^T \ , \tag{1.10.9}$$

or in component form*

$$\delta\,\mathrm{R}_{i\cdot}^{\ m}\,(\mathrm{R}^T)_{mn} = \delta\,\mathrm{R}_{i\cdot}^{\ m}\,\mathrm{R}_{nm} = -\mathrm{R}_{i\cdot}^{\ m}\,\delta\,(\mathrm{R}^T)_{mn} = -\mathrm{R}_{i\cdot}^{\ m}\,\delta\,\mathrm{R}_{nm} \ . \tag{1.10.10}$$

Accordingly, $\delta\,\mathbf{R}\,\mathbf{R}^T$ is a skew-symmetric tensor. This property allows, as has been shown in (1.3.56), to express the contraction of the tensor $\delta\,\mathbf{R}\,\mathbf{R}^T$ with an arbitrary vector $\mathbf{u}$ in the form

$$(\delta\,\mathbf{R}\,\mathbf{R}^T)\,\mathbf{u} = \boldsymbol{\omega} \times \mathbf{u} = \hat{\boldsymbol{\omega}}\,\mathbf{u} \ , \tag{1.10.11}$$

where $\boldsymbol{\omega}$ is the axial vector of $\delta\,\mathbf{R}\,\mathbf{R}^T$ and $\hat{\boldsymbol{\omega}}$ is an abbreviation

$$\hat{\boldsymbol{\omega}} = \boldsymbol{\omega}\times \qquad \rightarrow \hat{\boldsymbol{\omega}}\,\mathbf{u} = \boldsymbol{\omega} \times \mathbf{u} \tag{1.10.12}$$

permitting to omit the symbol $\times$.

---

* It is supposed that the basis $\mathbf{g}_i$ is not affected by the variation δ.

If the set of the rotated vectors $\tilde{\mathbf{g}}_i$ is given, the tensor $\mathbf{R}$ can be determined according to (1.10.5). Now the question is how to construct $\mathbf{R}$ if the rotation is described through a rotation axis $\mathbf{e}$ and a rotation angle ω about $\mathbf{e}$. This task can be accomplished by means of a rotation vector $\boldsymbol{\Omega}$, which is in general form defined by

$$\boldsymbol{\Omega} = \alpha\,\mathbf{e}\ . \tag{1.10.13}$$

Here, the *unit* vector $\mathbf{e}$ indicates the direction of the rotation axis and α is a parameter depending on the rotation angle ω transforming $\mathbf{g}_i$ into $\tilde{\mathbf{g}}_i$ (Fig. 1.4). According to the selection of α the transformation of $\mathbf{g}_i$ into $\tilde{\mathbf{g}}_i$ can be formulated in different manners. In the following, two examples are presented (see e.g. Pietraszkiewicz and Badur 1983; Büchter 1992; Sansour 1992):

$$\boldsymbol{\Omega} = \sin\omega\,\mathbf{e} \;\rightarrow\; \tilde{\mathbf{g}}_i = \mathbf{g}_i + \boldsymbol{\Omega}\times\mathbf{g}_i + \frac{1}{2\cos^2\frac{\omega}{2}}\,\boldsymbol{\Omega}\times(\boldsymbol{\Omega}\times\mathbf{g}_i)\ , \tag{1.10.14}$$

$$\boldsymbol{\Omega} = \omega\,\mathbf{e} \;\rightarrow\; \tilde{\mathbf{g}}_i = \mathbf{g}_i + \frac{\sin\omega}{\omega}\,\boldsymbol{\Omega}\times\mathbf{g}_i + \frac{1-\cos\omega}{\omega^2}\,\boldsymbol{\Omega}\times(\boldsymbol{\Omega}\times\mathbf{g}_i)\ , \tag{1.10.15}$$

where the vector $\boldsymbol{\Omega}$ used in (1.10.15) is known as RODRIGUES *rotation* vector. In the last mentioned case the rotation tensor $\mathbf{R}$ reads as

$$\tilde{\mathbf{g}}_i = \mathbf{R}\,\mathbf{g}_i \;\rightarrow\; \mathbf{R} = \mathbf{I} + \frac{\sin\omega}{\omega}\,\hat{\boldsymbol{\Omega}} + \frac{1-\cos\omega}{\omega^2}\,\hat{\boldsymbol{\Omega}}\hat{\boldsymbol{\Omega}}\ , \tag{1.10.16}$$

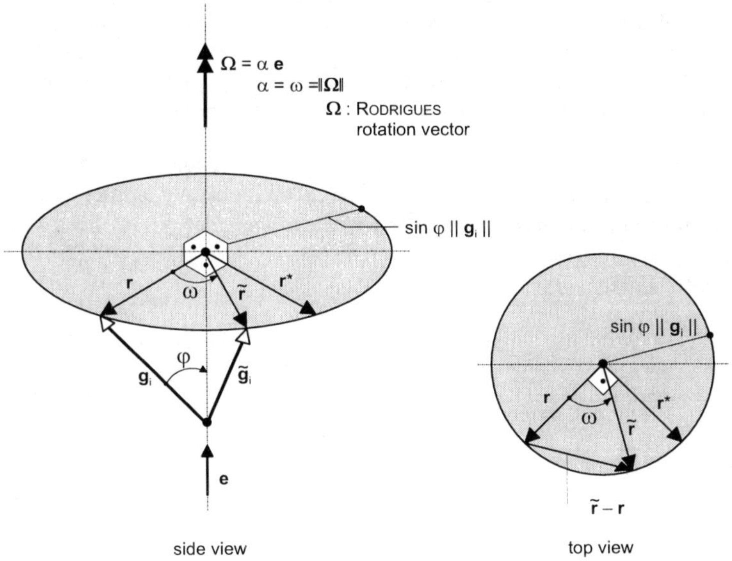

Fig. 1.4. Rotation of any arbitrary base vector $\mathbf{g}_i$ (i = 1, 2, 3) about the axis $\mathbf{e}$ with a rotational angle ω

with the identity tensor **I** and the abbreviation $\hat{\mathbf{\Omega}}$ defined as:

$$\hat{\mathbf{\Omega}} = \mathbf{\Omega}\times \quad \rightarrow \quad \hat{\mathbf{\Omega}}\,\mathbf{g}_i = \mathbf{\Omega}\times\mathbf{g}_i \ . \tag{1.10.17}$$

In view of the identity

$$\mathbf{\Omega}\times(\mathbf{\Omega}\times\mathbf{d}) = \mathbf{\Omega}\,(\mathbf{\Omega}\cdot\mathbf{d}) - \mathbf{d}\,(\mathbf{\Omega}\cdot\mathbf{\Omega}) = (\mathbf{\Omega}\otimes\mathbf{\Omega} - \omega^2\,\mathbf{I})\,\mathbf{d} \tag{1.10.18}$$

holding for the RODRIGUES rotation vector $\mathbf{\Omega}$, where $\mathbf{d}$ is an arbitrary vector, equation (1.10.16) can be expressed equivalently as (Pietraszkiewicz and Badur 1983; Simo et al. 1990; Menzel 1996):

$$\mathbf{R} = \cos\omega\,\mathbf{I} + \frac{\sin\omega}{\omega}\,\hat{\mathbf{\Omega}} + \frac{1-\cos\omega}{\omega^2}\,\mathbf{\Omega}\otimes\mathbf{\Omega}\ , \qquad \omega = \|\mathbf{\Omega}\| \ . \tag{1.10.19}$$

By considering (1.10.18) it can be easily verified that both expressions (1.10.16) and (1.10.19) presented for **R** deliver the same result if they are contracted by an arbitrary vector **d**.

If we contract equation (1.10.16) by the RODRIGUES rotation vector $\mathbf{\Omega} = \omega\,\mathbf{e}$

$$\mathbf{R}\,\mathbf{\Omega} = \mathbf{\Omega} \quad \rightarrow \quad (\mathbf{R} - 1\cdot\mathbf{I})\,\mathbf{\Omega} = \mathbf{0} \tag{1.10.20}$$

and remember that any orthogonal tensor **R** with det **R** = +1 is essentially expressible in the form (1.10.16) or (1.10.19) we discover a basic property of orthogonal tensors:

Remark. The RODRIGUES rotation vector $\mathbf{\Omega} = \omega\,\mathbf{e}$ is the eigenvector of any orthogonal tensor **R** with det **R** = +1 and has the real eigenvalue 1. In this sense any orthogonal tensor describes a rotation about its own eigenvector $\mathbf{\Omega}$ the magnitude of which is the rotation angle $\omega = \|\mathbf{\Omega}\|$. The remaining eigenvalues of **R** are conjugate complex and have the values: $\cos\omega \pm i\sin\omega$ (Pietraszkiewicz and Badur 1983; Menzel 1996).

The rotation tensor (1.10.16) in terms of the RODRIGUES rotation vector $\mathbf{\Omega} = \omega\,\mathbf{e}$ provides a singularity-free description of arbitrarily large rotations and is therefore of major importance for finite-rotation analysis. Particularly in shell theory it has been widely used to describe the rotation of the so-called shell director (Pietraszkiewicz 1979; Simo et al. 1990; Büchter 1992).

Application. The rotation of a base vector $\mathbf{g}_i$ (i = 1, 2, 3) into $\tilde{\mathbf{g}}_i$ is illustrated in Fig. 1.4. The unit vector **e** determines the direction of the rotation axis and ω denotes the rotation angle. Our aim is to prove the transformation (1.10.15) expressed in terms of the RODRIGUES rotation vector $\mathbf{\Omega} = \omega\,\mathbf{e}$.

According to Fig. 1.4 we have

$$\begin{aligned}\tilde{\mathbf{g}}_i &= \mathbf{g}_i + \tilde{\mathbf{r}} - \mathbf{r} = \mathbf{g}_i + \|\tilde{\mathbf{r}}\|\left(\cos\omega\,\frac{\mathbf{r}}{\|\mathbf{r}\|} + \sin\omega\,\frac{\mathbf{r}^*}{\|\mathbf{r}^*\|}\right) - \mathbf{r} \\ &= \mathbf{g}_i + \sin\omega\,\mathbf{r}^* + (\cos\omega - 1)\,\mathbf{r}\end{aligned} \tag{1.10.21}$$

where we have considered that $\|\mathbf{r}\| = \|\tilde{\mathbf{r}}\| = \|\mathbf{r}^*\|$. Now the problem is to express $\mathbf{r}^*$ and $\mathbf{r}$ occuring in (1.10.21) in terms of $\mathbf{g}_i$ and $\mathbf{\Omega}$. By definition the vector $\mathbf{r}^*$ is perpendicular to $\mathbf{g}_i$ and $\mathbf{\Omega}$ and its magnitude is given by $\|\mathbf{r}^*\| = \sin\varphi \, \|\mathbf{g}_i\|$. We therefore may write

$$\mathbf{r}^* = \mathbf{e} \times \mathbf{g}_i = \frac{1}{\omega} \mathbf{\Omega} \times \mathbf{g}_i \; , \tag{1.10.22}$$

and, consequently, by considering again Fig. 1.4

$$\mathbf{r} = -\mathbf{e} \times \mathbf{r}^* = -\frac{1}{\omega^2} \mathbf{\Omega} \times (\mathbf{\Omega} \times \mathbf{g}_i) \; . \tag{1.10.23}$$

Using the above results equation (1.10.21) takes the form

$$\tilde{\mathbf{g}}_i = \mathbf{g}_i + \frac{\sin\omega}{\omega} \mathbf{\Omega} \times \mathbf{g}_i + \frac{1-\cos\omega}{\omega^2} \mathbf{\Omega} \times (\mathbf{\Omega} \times \mathbf{g}_i) \; , \tag{1.10.24}$$

in accordance with (1.10.15).

1.11 Analytical solution of eigenvalue-problems

In section 1.9 we have shown that any symmetric second-order tensor $\mathbf{C}$ has three real eigenvalues Λ_i (i = 1, 2, 3) with eigenvectors $\mathbf{N}_i$ * (of unit length). The eigenvalues Λ_i can be determined according to (1.9.7) from a *characteristic* equation of the form

$$\Lambda^3 - \mathrm{I}_C \Lambda^2 + \mathrm{II}_C \Lambda - \mathrm{III}_C = 0 \; . \tag{1.11.1}$$

Then, as is shown in (1.9.16), the so-called *spectral* representation of $\mathbf{C}$

$$\mathbf{C} = \sum_{i=1}^{3} \Lambda_i \mathbf{N}_i \otimes \mathbf{N}_i = \sum_{i=1}^{3} \Lambda_i \mathbf{C}_i \; , \qquad \mathbf{C}_i = \mathbf{N}_i \otimes \mathbf{N}_i \; , \tag{1.11.2}$$

is possible, where, for convenience, the second order-tensor $\mathbf{C}_i = \mathbf{N}_i \otimes \mathbf{N}_i$ has been used as abbreviation.

Now we present without derivation the analytical solution for the eigenvalues Λ_i and the closed form solution of the tensor $\mathbf{C}_i$ introduced in (1.11.2). The principal values Λ_i satisfying (1.11.1) are given by

$$\Lambda_k = \frac{1}{3}\left[\mathrm{I}_C + 2\left(\mathrm{I}_C^2 - 3\,\mathrm{II}_C\right)^{1/2} \cos\frac{1}{3}(\Theta + 2\pi k)\right] , \quad k = 1, 2, 3 \tag{1.11.3}$$

where the abbreviation

$$\Theta = \arccos\left[\frac{2\,\mathrm{I}_C^3 - 9\,\mathrm{I}_C\,\mathrm{II}_C + 27\,\mathrm{III}_C}{2\,(\mathrm{I}_C^2 - 3\,\mathrm{II}_C)^{3/2}}\right] \tag{1.11.4}$$

* For convenience we denote the eigenvalues by Λ_i and the eigenvectors by $\mathbf{N}_i$ in accordance with the notations to be used in section 2.6.

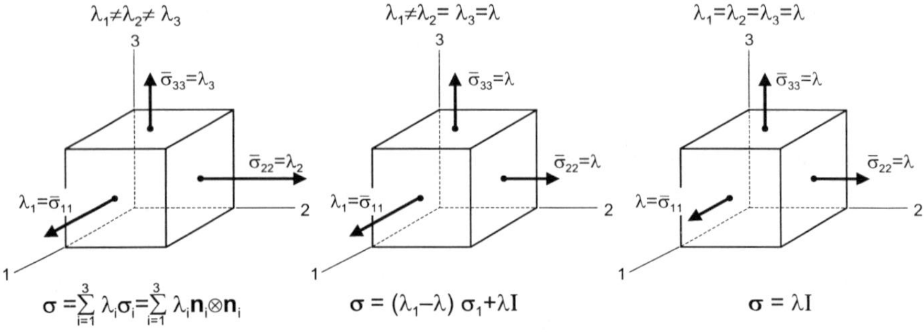

Fig. 1.5. Eigenvalues and spectral decompositions on the example of CAUCHY stress tensor σ

is an invariant function depending upon the invariants I_C, II_C and III_C of $\mathbf{C}$ (Morman 1986). To present the closed form for $\mathbf{C}_i$ one should distinguish between three cases.

- If the eigenvalues Λ_i (i = 1, 2, 3) are all distinct, then

$$\mathbf{C}_i = \Lambda_i \frac{\mathbf{C} - (I_C - \Lambda_i)\,\mathbf{I} + III_C\,\Lambda_i^{-1}\,\mathbf{C}^{-1}}{2\,\Lambda_i^2 - I_C\,\Lambda_i + III_C\,\Lambda_i^{-1}} , \tag{1.11.5}$$

$$\text{or } \mathbf{C}_r = \frac{1}{(\Lambda_r - \Lambda_s)\,(\Lambda_s - \Lambda_t)} (\mathbf{C} - \Lambda_s\,\mathbf{I})\,(\mathbf{C} - \Lambda_t\,\mathbf{I}) , \tag{1.11.6}$$

  where (r, s, t) represents a cyclic permutation of (1, 2, 3).

- In the case of coalescence of two eigenvalues ($\Lambda_1 \neq \Lambda_2 = \Lambda_3 = \Lambda$) we have

$$\mathbf{C} = \sum_{i=1}^{3} \Lambda\,\mathbf{N}_i \otimes \mathbf{N}_i + (\Lambda_1 - \Lambda)\,\mathbf{N}_1 \otimes \mathbf{N}_1 = (\Lambda_1 - \Lambda)\,\mathbf{C}_1 + \Lambda\,\mathbf{I} , \tag{1.11.7}$$

  whereas

$$\mathbf{C}_1 = \frac{1}{(\Lambda_1 - \Lambda)} (\mathbf{C} - \Lambda\,\mathbf{I}) . \tag{1.11.8}$$

- Finally, for the case of coalescence of all eigenvalues ($\Lambda_1 = \Lambda_2 = \Lambda_3 = \Lambda$) the closed form solution becomes

$$\mathbf{C} = \Lambda\,\mathbf{I} . \tag{1.11.9}$$

The solutions presented above for three different cases are illustrated in Fig. 1.5 on the example of the CAUCHY stress tensor $\boldsymbol{\sigma}$.

1.12 Tensor functions on the basis of power series

Up to now we have introduced second-order tensors through tensor products, but more frequently through simple contractions. Tensors may be also introduced by means of power series as has been shown e.g in Ting 1985 or Morman 1986. In this section we consider a symmetric second-order tensor **C** and give some examples for the construction of tensor functions*:

$$\exp \mathbf{C} = e^{\mathbf{C}} = \mathbf{I} + \mathbf{C} + \frac{1}{2!}\mathbf{C}^2 + \frac{1}{3!}\mathbf{C}^3 + \ldots \tag{1.12.1}$$

$$\ln(\mathbf{I} + \mathbf{C}) = \mathbf{C} - \frac{1}{2}\mathbf{C}^2 + \frac{1}{3}\mathbf{C}^3 \ldots \tag{1.12.2}$$

$$\sin \mathbf{C} = \mathbf{C} - \frac{1}{3!}\mathbf{C}^3 + \frac{1}{5!}\mathbf{C}^5 \ldots \,. \tag{1.12.3}$$

The tensors introduced above are said to be isotropic. A tensor-valued tensor function **G**(**C**) in **C** is *isotropic* if the relation

$$\mathbf{Q}\,\mathbf{G}(\mathbf{C})\,\mathbf{Q}^T = \mathbf{G}(\mathbf{Q}\,\mathbf{C}\,\mathbf{Q}^T) \tag{1.12.4}$$

holds for all orthogonal tensors **Q** (Ting 1985). Remembering the well-known property $\mathbf{Q}^T = \mathbf{Q}^{-1}$ of **Q** it can be shown that $\mathbf{C}^m$, where m is an integer, is isotropic. Thus

$$(\mathbf{Q}\,\mathbf{C}\,\mathbf{Q}^T)^m = (\mathbf{Q}\,\mathbf{C}\,\mathbf{Q}^T)(\mathbf{Q}\,\mathbf{C}\,\mathbf{Q}^T)\ldots(\mathbf{Q}\,\mathbf{C}\,\mathbf{Q}^T) = \mathbf{Q}\,\mathbf{C}^m\,\mathbf{Q}^T \tag{1.12.5}$$

in accordance with the requirement (1.12.4). Consequently, exp **C**, ln (**I** + **C**), sin **C** are isotropic tensor functions.

Symmetric tensors play a very important role in continuum mechanics. Their particularity is that they are accessible to *spectral* representations introduced in (1.9.16) in connection with eigenvalue problems. If $\mathbf{C} = \mathbf{C}^T$ is symmetric, then the tensors **I**, $\mathbf{C}^2$, ... occuring in the power series (1.12.1) to (1.12.3) are coaxial with the tensor **C**. Consequently, the infinite power series (1.12.1) and (1.12.2) can be replaced in closed form by the spectral representations

$$\exp \mathbf{C} = \sum_{i=1}^{3} \left(1 + \Lambda_i + \frac{\Lambda_i^2}{2!} + \frac{\Lambda_i^3}{3!} + \ldots\right) \mathbf{n}_i \otimes \mathbf{n}_i = \sum_{i=1}^{3} e^{\Lambda_i}\, \mathbf{n}_i \otimes \mathbf{n}_i \,, \tag{1.12.6}$$

$$\ln(\mathbf{I} + \mathbf{C}) = \sum_{i=1}^{3} \left(\Lambda_i - \frac{1}{2}\Lambda_i^2 + \frac{1}{3}\Lambda_i^3 \ldots\right) \mathbf{n}_i \otimes \mathbf{n}_i = \sum_{i=1}^{3} \ln(1 + \Lambda_i)\, \mathbf{n}_i \otimes \mathbf{n}_i \,, \tag{1.12.7}$$

where Λ_i and $\mathbf{n}_i$ are, respectively, the eigenvalues and the eigenvectors of **C**. The above expressions on the basis of the spectral decomposition suggest to extend the definition

* These definitions are similar to the power series of e^x, ln (1 + x), sin x.

of tensor functions to powers of the form $\mathbf{C}^{1/m}$, where m is an integer. If $\mathbf{C}^{1/m}$ is defined so as to satisfy the relation

$$\mathbf{C} = (\mathbf{C}^{1/m})\,(\mathbf{C}^{1/m}) \ldots (\mathbf{C}^{1/m}) = (\mathbf{C}^{1/m})^m \tag{1.12.8}$$

the corresponding spectral decomposition reads as

$$\mathbf{C}^{1/m} = \sum_{i=1}^{n} (\Lambda_i)^{1/m}\, \mathbf{n}_i \otimes \mathbf{n}_i \tag{1.12.9}$$

permitting an easy evaluation of $\mathbf{C}^{1/m}$, if $\mathbf{n}_i$ and λ_i are given. The tensor functions $\mathbf{C}^{1/m}$ are isotropic (Ting 1985). To show this we use the relation

$$\begin{aligned}(\mathbf{Q}\,\mathbf{C}^{1/m}\,\mathbf{Q}^T)^m &= (\mathbf{Q}\,\mathbf{C}^{1/m}\,\mathbf{Q}^T)\,(\mathbf{Q}\,\mathbf{C}^{1/m}\,\mathbf{Q}^T) \ldots (\mathbf{Q}\,\mathbf{C}^{1/m}\,\mathbf{Q}^T)\\ &= \mathbf{Q}\,(\mathbf{C}^{1/m})^m\,\mathbf{Q}^T = \mathbf{Q}\,\mathbf{C}\,\mathbf{Q}^T\end{aligned} \tag{1.12.10}$$

leading in view of the definition (1.12.8) to

$$\mathbf{Q}\,\mathbf{C}^{1/m}\,\mathbf{Q}^T = (\mathbf{Q}\,\mathbf{C}\,\mathbf{Q}^T)^{1/m} \;. \tag{1.12.11}$$

Hence $\mathbf{C}^{1/m}$ is an *isotropic* tensor function. Notice that if $\mathbf{G}\,(\mathbf{C})$ is isotropic, so is $\mathbf{G}^{-1}(\mathbf{C})$. Therefore, $\mathbf{C}^{-1/m} = (\mathbf{C}^{1/m})^{-1}$ is isotropic. Hence, $\mathbf{C}^{1/2}$ and $\mathbf{C}^{-1/2}$ are examples for isotropic tensor functions of $\mathbf{C}$.

Remark. The representation theorem (Truesdell and Noll 1965) states that $\mathbf{G}\,(\mathbf{C})$ is *isotropic* if and only if it is transformable into the form

$$\mathbf{G}\,(\mathbf{C}) = \Phi_0\,\mathbf{I} + \Phi_1\,\mathbf{C} + \Phi_2\,\mathbf{C}^2 \;, \tag{1.12.12}$$

where Φ_0, Φ_1 and Φ_2 are functions of the invariants or the eigenvalues of $\mathbf{C}$.

The application of the representation theorem is shown in section 2.11 on the example of the tensor function $\mathbf{f}\,(\mathbf{b}) = \mathbf{b}^{m/2}$, where $\mathbf{b}$ is a symmetric second-order tensor.

1.13 Exponential, skew-symmetric tensors

For applications in connection with finite rotation analysis the *exponential* of a skew-symmetric tensor is of major importance. We first recall that a skew-symmetric tensor $\mathbf{W} = -\mathbf{W}^T$ has three independent components and that, due to this property, it is transformable into an axial vector $\boldsymbol{\theta}$

$$\mathbf{W}\,\mathbf{d} = \boldsymbol{\theta} \times \mathbf{d} = \hat{\boldsymbol{\theta}}\,\mathbf{d} \;\;\rightarrow\;\; \mathbf{W} := \hat{\boldsymbol{\theta}} = \boldsymbol{\theta}\times \tag{1.13.1}$$

as has been demonstrated in (1.3.56). Note that $\hat{\boldsymbol{\theta}}$ is the notation for the tensor $\mathbf{W}$, if it is expressed in terms of the axial vector $\boldsymbol{\theta}$. If the tensor $\hat{\boldsymbol{\theta}}$ is referred to the orthonormal basis $\mathbf{i}_i = \mathbf{i}^i$ it is determined by the following components:

$$\hat{\boldsymbol{\theta}} = \hat{\theta}_{ij}\, \mathbf{i}^i \otimes \mathbf{i}^j : \quad \hat{\theta}_{ij} = \mathbf{i}_i\, \hat{\boldsymbol{\theta}}\, \mathbf{i}_j = \mathbf{i}_i \cdot (\boldsymbol{\theta} \times \mathbf{i}_j) = \begin{bmatrix} 0 & -\theta^3 & \theta^2 \\ \theta^3 & 0 & -\theta^1 \\ -\theta^2 & \theta^1 & 0 \end{bmatrix}, \tag{1.13.2}$$

where $\boldsymbol{\theta} = \theta^k\, \mathbf{i}_k$.* The exponential of $\mathbf{W}$, $\exp \mathbf{W} = \mathbf{e}^{\mathbf{W}}$, is described by a power series similar to that introduced in (1.12.1). For convenience, it is suitable to present the result in terms of the axial vector $\boldsymbol{\theta}$ as:

$$\exp \mathbf{W} := \mathbf{e}^{\mathbf{W}} = \mathbf{I} + \hat{\boldsymbol{\theta}} + \frac{1}{2!}\hat{\boldsymbol{\theta}}^2 + \frac{1}{3!}\hat{\boldsymbol{\theta}}^3 + \dots \,. \tag{1.13.3}$$

In section 1.10 we have observed that any orthogonal tensor $\mathbf{R}$ may be constructed by means of a rotation vector $\boldsymbol{\Omega} = \alpha\, \mathbf{e}$ (1.10.13). If the RODRIGUES vector is adopted for this purpose, the result is given by (1.10.16) or (1.10.19). The infinite power series (1.13.3) and the rotation tensor (1.10.16) are, in reality, related to each other. ARGYRIS 1982 has proved that:

Remark. The rotation tensor (1.10.16) in terms of the RODRIGUES vector is the *closed* form of the power series (1.13.3) in terms of the axial vector $\boldsymbol{\theta}$. Evidently, this equality holds, if in (1.10.16) or (1.10.19) the RODRIGUES vector is identified with the axial vector $\boldsymbol{\theta}$.

Using the notation $\boldsymbol{\Omega} \to \boldsymbol{\theta}$ we therefore have:

$$\exp \hat{\boldsymbol{\theta}} = \mathrm{e}^{\hat{\boldsymbol{\theta}}} = \mathbf{R}\,(\boldsymbol{\theta}) \,, \quad \hat{\boldsymbol{\theta}} = \boldsymbol{\theta} \times \,, \tag{1.13.4}$$

where

$$\begin{aligned} \mathbf{R} &= \mathbf{I} + \frac{\sin\theta}{\theta}\hat{\boldsymbol{\theta}} + \frac{1 - \cos\theta}{\theta^2}\hat{\boldsymbol{\theta}}\,\hat{\boldsymbol{\theta}} \\ &= \cos\theta\, \mathbf{I} + \frac{\sin\theta}{\theta}\hat{\boldsymbol{\theta}} + \frac{1 - \cos\theta}{\theta^2}\boldsymbol{\theta} \otimes \boldsymbol{\theta} \end{aligned} \tag{1.13.5}$$

with $\theta = \| \boldsymbol{\theta} \|$.

The equality (1.13.4) in connection with the series expansion (1.13.3) has been widely used in shell theory for the parametrisation of the *inextensible* shell director by the so-called *updated* formulation (Simo et al. 1990; Büchter and Ramm 1992; Menzel 1996). Within an iterative-incremental procedure the finite rotation of the shell director can be described by the tensors $\hat{\boldsymbol{\theta}}$ and $\hat{\boldsymbol{\theta}}\,\hat{\boldsymbol{\theta}}$ occuring in the power series (1.13.3). Once an iteration step is achieved the rotation tensor $\mathbf{R}$ (1.13.5) can be used to determine the final position of the director in an exact form.

---

* In (1.13.2), the notation θ^k does not denote curvilinear coordinates, but the components of $\boldsymbol{\theta}$ with respect to $\mathbf{i}_k$.

1.14 Summary of notations and formulae

Tensor operations

scalar product	$\mathbf{a} \cdot \mathbf{b} = \mathbf{b} \cdot \mathbf{a}$
vector product	$\mathbf{c} = \mathbf{a} \times \mathbf{b} = -\mathbf{b} \times \mathbf{a}$
mixed product	$[\mathbf{a}\ \mathbf{b}\ \mathbf{c}] = \mathbf{a} \cdot (\mathbf{b} \times \mathbf{c})$
tensorial product	$\mathbf{C} = \mathbf{a} \otimes \mathbf{b} \neq \mathbf{b} \otimes \mathbf{a} = \mathbf{C}^T$
	$\mathbf{S} \otimes \mathbf{T} = (S^{ij}\,\mathbf{g}_i \otimes \mathbf{g}_j) \otimes (T^{mn}\,\mathbf{g}_m \otimes \mathbf{g}_n) = S^{ij}\,T^{mn}\,\mathbf{g}_i \otimes \mathbf{g}_j \otimes \mathbf{g}_m \otimes \mathbf{g}_n$
simple contraction	$(\mathbf{a} \otimes \mathbf{b})\,(\mathbf{c} \otimes \mathbf{d}) = (\mathbf{b} \cdot \mathbf{c})\,(\mathbf{a} \otimes \mathbf{d})$
	$\mathbf{S}\,\mathbf{T} = (S^{ij}\,\mathbf{g}_i \otimes \mathbf{g}_j)\,(T^{mn}\,\mathbf{g}_m \otimes \mathbf{g}_n) = S^{ij}\,T_{j\cdot}^{\cdot n}\,\mathbf{g}_i \otimes \mathbf{g}_n$
double contraction	$(\mathbf{a} \otimes \mathbf{b}) : (\mathbf{c} \otimes \mathbf{d} \otimes \mathbf{e}) = (\mathbf{a} \cdot \mathbf{c})\,(\mathbf{b} \cdot \mathbf{d})\,\mathbf{e}$
	$\mathbf{T} : \mathbf{R} = (T^{lmn}\,\mathbf{g}_l \otimes \mathbf{g}_m \otimes \mathbf{g}_n) : (R^{ijk}\,\mathbf{g}_i \otimes \mathbf{g}_j \otimes \mathbf{g}_k) = T^{lmn}\,R_{mn\cdot}^{\cdot\cdot k}\,\mathbf{g}_l \otimes \mathbf{g}_k$
identities	$\mathbf{a} \times (\mathbf{b} \times \mathbf{c}) = \mathbf{b}\,(\mathbf{a} \cdot \mathbf{c}) - \mathbf{c}\,(\mathbf{a} \cdot \mathbf{b})$
	$(\mathbf{a} \otimes \mathbf{b})\,\mathbf{u} = (\mathbf{b} \cdot \mathbf{u})\,\mathbf{a}$
	$\mathbf{S} : (\mathbf{c} \otimes \mathbf{d}) = \mathbf{c}\,\mathbf{S}\,\mathbf{d}$
	$\mathbf{A} : (\mathbf{B}\,\mathbf{C}) = (\mathbf{B}^T\,\mathbf{A}) : \mathbf{C} = (\mathbf{A}\,\mathbf{C}^T) : \mathbf{B}$
	$\mathbf{S} : \mathbf{T} = \mathbf{S}^T : \mathbf{T}^T$

Special tensors

identity tensor	$\mathbf{I} = \mathbf{g}_i \otimes \mathbf{g}^i = \mathbf{g}^i \otimes \mathbf{g}_i$
inverse tensor	$\mathbf{S}^{-1} = (S^{-1})^{ij}\,\mathbf{g}_i \otimes \mathbf{g}_j$
	$\mathbf{S}\,\mathbf{S}^{-1} = \mathbf{S}^{-1}\,\mathbf{S} = \mathbf{I}\ , \quad (\mathbf{S}^{-1})^{-1} = \mathbf{S}\ , \quad (\mathbf{S}\,\mathbf{T})^{-1} = \mathbf{T}^{-1}\,\mathbf{S}^{-1}$
transposed tensor	$\mathbf{S}^T = (S^T)_{ij}\,\mathbf{g}^i \otimes \mathbf{g}^j = S_{ji}\,\mathbf{g}^i \otimes \mathbf{g}^j$
	$\mathbf{u} \cdot (\mathbf{S}\,\mathbf{v}) = \mathbf{v} \cdot (\mathbf{S}^T\,\mathbf{u}) = \mathbf{u}\,\mathbf{S}\,\mathbf{v} = \mathbf{v}\,\mathbf{S}^T\,\mathbf{u}$
	$(\mathbf{T} + \mathbf{S})^T = \mathbf{T}^T + \mathbf{S}^T\ , \quad (\mathbf{T}\,\mathbf{S})^T = \mathbf{S}^T\,\mathbf{T}^T$
symmetric tensor	$\mathbf{S} = S_{ij}\,\mathbf{g}^i \otimes \mathbf{g}^j = S_{ji}\,\mathbf{g}^i \otimes \mathbf{g}^j$
	$\mathbf{S} = \mathbf{S}^T\ , \quad S_{ij} = S_{ji}$
skew-symmetric tensor	$\mathbf{S} = S_{ij}\,\mathbf{g}^i \otimes \mathbf{g}^j = -S_{ji}\,\mathbf{g}^i \otimes \mathbf{g}^j$
	$\mathbf{S} = -\mathbf{S}^T\ , \quad S_{ij} = -S_{ji}$

permutation tensor	$\mathbf{E} = \varepsilon_{ijk}\, \mathbf{g}^i \otimes \mathbf{g}^j \otimes \mathbf{g}^k$
	$\mathbf{g}_i \times \mathbf{g}_j = \mathbf{E} : (\mathbf{g}_i \otimes \mathbf{g}_j) = -\mathbf{g}_i\, \mathbf{E}\, \mathbf{g}_j = \varepsilon_{ijk}\, \mathbf{g}^k$
trace of second-order tensors	$\mathrm{tr}\,(\mathbf{a} \otimes \mathbf{b}) = \mathbf{I} : (\mathbf{a} \otimes \mathbf{b}) = \mathbf{a} \cdot \mathbf{b}$
	$\mathrm{tr}\, \mathbf{S}^T = \mathrm{tr}\, \mathbf{S} = \mathbf{I} : \mathbf{S} = S_{i\cdot}^{\ i} = S_{\cdot i}^{i}$
	$\mathrm{tr}\,(\mathbf{S}\,\mathbf{T}^T) = \mathrm{tr}\,(\mathbf{T}\,\mathbf{S}^T) = \mathbf{T} : \mathbf{S} = \mathrm{tr}\,(\mathbf{S}^T\, \mathbf{T}) = \mathrm{tr}\,(\mathbf{T}^T\, \mathbf{S}) = \mathbf{T}^T : \mathbf{S}^T$
axial vector t	$\mathbf{T}\,\mathbf{u} = \mathbf{t} \times \mathbf{u} = \hat{\mathbf{t}}\,\mathbf{u}\ , \quad \mathbf{T} = -\mathbf{T}^T$
norm of second-order tensors	$\Vert \mathbf{S} \Vert = \sqrt{\mathbf{S} : \mathbf{S}} = \sqrt{\mathrm{tr}\,(\mathbf{S}\,\mathbf{S}^T)}$
orthogonal tensor	$\mathbf{Q}^T = \mathbf{Q}^{-1}$
	$\tilde{\mathbf{a}} = \mathbf{Q}\,\mathbf{a} \qquad \tilde{\mathbf{b}} = \mathbf{Q}\,\mathbf{b} \qquad \tilde{\mathbf{a}} \cdot \tilde{\mathbf{b}} = \mathbf{a} \cdot \mathbf{b}$
spherical tensor	$\mathrm{sph}\, \mathbf{T} = 1/3\, (\mathrm{tr}\, \mathbf{T})\, \mathbf{I}$
deviatoric tensor	$\mathrm{dev}\, \mathbf{T} = \mathbf{T} - \mathrm{sph}\, \mathbf{T}$
	$\mathbf{T} = \mathrm{dev}\, \mathbf{T} + \mathrm{sph}\, \mathbf{T}$

Differentiation rules

partial derivatives with respect to coordinates	$\mathbf{S}_{,i} = \mathbf{S}\vert_i = S^{mn}\vert_i\, \mathbf{g}_m \otimes \mathbf{g}_n$
	$S^{mn}\vert_i = S^{mn}{}_{,i} + \Gamma^m_{ir}\, S^{rn} + \Gamma^n_{ir}\, S^{mr}\ , \quad \Gamma^m_{ir} = \mathbf{g}^m \cdot \mathbf{g}_{i,r}$
	$\mathbf{g}_i\vert_j = \mathbf{g}_{i,j} - \Gamma^r_{ij}\, \mathbf{g}_r = \mathbf{0}$
partial derivatives with respect to a tensor	$\Pi_{,\mathbf{A}} = \dfrac{\partial \Pi}{\partial \mathbf{A}} = \dfrac{\partial \Pi}{\partial A_{ij}}\, \mathbf{g}_i \otimes \mathbf{g}_j$
	$\Pi_{,\mathbf{A}\otimes\mathbf{A}} = \dfrac{\partial^2 \Pi}{\partial \mathbf{A}\, \partial \mathbf{A}} = \dfrac{\partial^2 \Pi}{\partial A_{ij}\, \partial A_{mn}}\, \mathbf{g}_i \otimes \mathbf{g}_j \otimes \mathbf{g}_m \otimes \mathbf{g}_n$
	$\mathbf{B}_{,\mathbf{A}} = \dfrac{\partial \mathbf{B}}{\partial \mathbf{A}} = \dfrac{\partial B_{ij}}{\partial A_{kl}}\, \mathbf{g}^i \otimes \mathbf{g}^j \otimes \mathbf{g}_k \otimes \mathbf{g}_l$
identities for a second-order tensor A	$(\mathrm{tr}\, \mathbf{A})_{,\mathbf{A}} = \mathbf{I}\ , \quad (\mathrm{tr}\, \mathbf{A})^2{}_{,\mathbf{A}} = 2\, (\mathrm{tr}\, \mathbf{A})\, \mathbf{I}\ , \quad (\mathrm{tr}\, \mathbf{A}^2)_{,\mathbf{A}} = 2\, \mathbf{A}^T\ , \quad (\mathrm{tr}\, \mathbf{A}^3)_{,\mathbf{A}} = 3\, (\mathbf{A}^2)^T$
	$\dfrac{\partial \Vert \mathbf{A} \Vert}{\partial \mathbf{A}} = \dfrac{\mathbf{A}}{\Vert \mathbf{A} \Vert}$

Differential operators

gradient of a scalar-valued function	$\mathrm{grad}\, \Phi = \nabla \Phi = \Phi_{,k}\, \mathbf{g}^k$

gradient of a vector $\operatorname{grad}\mathbf{u} = \mathbf{u}_{,k} \otimes \mathbf{g}^k = u_i|_k\, \mathbf{g}^i \otimes \mathbf{g}^k$

divergence of a vector $\operatorname{div}\mathbf{u} = \operatorname{grad}\mathbf{u} : \mathbf{I} = u^k|_k = \mathbf{u}_{,k} \cdot \mathbf{g}^k = \nabla \cdot \mathbf{u}$

divergence of a second-order tensor $\operatorname{div}\mathbf{A} = \operatorname{grad}\mathbf{A} : \mathbf{I} = A^{ij}|_j\, \mathbf{g}_i = \mathbf{A}_{,k}\, \mathbf{g}^k$

identities

$$\operatorname{grad}(\Phi\, \mathbf{u}) = \mathbf{u} \otimes \operatorname{grad}\Phi + \Phi \operatorname{grad}\mathbf{u}$$

$$\operatorname{div}(\mathbf{u}\, \mathbf{A}) = \mathbf{A} : \operatorname{grad}\mathbf{u} + \mathbf{u} \cdot \operatorname{div}\mathbf{A}$$

Invariants of an arbitrary second-order tensor A

definition of the invariants

$$\mathrm{I}_{\mathbf{A}} = \operatorname{tr}\mathbf{A}$$

$$\mathrm{II}_{\mathbf{A}} = \frac{1}{2}\left[(\operatorname{tr}\mathbf{A})^2 - \operatorname{tr}\mathbf{A}^2\right]$$

$$\mathrm{III}_{\mathbf{A}} = \det\mathbf{A} = \frac{1}{3}\left[\operatorname{tr}\mathbf{A}^3 - \frac{3}{2}\operatorname{tr}\mathbf{A}^2 \operatorname{tr}\mathbf{A} + \frac{1}{2}(\operatorname{tr}\mathbf{A})^3\right]$$

partial derivatives of the invariants

$$(\mathrm{I}_{\mathbf{A}})_{,\mathbf{A}} = \mathbf{I}$$

$$(\mathrm{II}_{\mathbf{A}})_{,\mathbf{A}} = (\operatorname{tr}\mathbf{A})\,\mathbf{I} - \mathbf{A}^T$$

$$(\mathrm{III}_{\mathbf{A}})_{,\mathbf{A}} = (\mathbf{A}^2)^T - \mathrm{I}_{\mathbf{A}}\,\mathbf{A}^T + \mathrm{II}_{\mathbf{A}}\,\mathbf{I} = \mathrm{III}_{\mathbf{A}}\,\mathbf{A}^{-T}$$

Eigenvalue problem of a second-order tensor

eigenvalue problem $(\mathbf{C} - \Lambda\,\mathbf{I})\,\mathbf{n} = 0$

characteristic equation $\det(\mathbf{C} - \Lambda\,\mathbf{I}) = -\Lambda^3 + \mathrm{I}_{\mathbf{C}}\,\Lambda^2 - \mathrm{II}_{\mathbf{C}}\,\Lambda + \mathrm{III}_{\mathbf{C}} = 0$

*spectral decomposition for a symmetric tensor **C***

$$\mathbf{C} = \sum_{k=1}^{3} C_k\, \mathbf{n}_k \otimes \mathbf{n}_k = \sum_{k=1}^{3} C_k\, \mathbf{C}_k \,, \quad \mathbf{C}_k = \mathbf{n}_k \otimes \mathbf{n}_k$$

coaxial tensors

$$\mathbf{A} = \sum_{i=1}^{3} \mu_i\, \mathbf{n}_i \otimes \mathbf{n}_i \,, \quad \mathbf{B} = \sum_{i=1}^{3} \kappa_i\, \mathbf{n}_i \otimes \mathbf{n}_i$$

property $\mathbf{A}\,\mathbf{B} = \mathbf{B}\,\mathbf{A}$

Rotation tensor on the basis of the RODRIGUES rotation vector

$$\mathbf{R} = \mathbf{I} + \frac{\sin\omega}{\omega}\hat{\mathbf{\Omega}} + \frac{1-\cos\omega}{\omega^2}\hat{\mathbf{\Omega}}\hat{\mathbf{\Omega}} \,, \quad \hat{\mathbf{\Omega}} = \mathbf{\Omega}\times$$

$$\mathbf{R} = \cos\omega\,\mathbf{I} + \frac{\sin\omega}{\omega}\hat{\mathbf{\Omega}} + \frac{1-\cos\omega}{\omega^2}\mathbf{\Omega}\otimes\mathbf{\Omega} \,, \quad \omega = \|\mathbf{\Omega}\|$$

Exercises

1.1. Which properties have $\mathbf{b}$ and $\mathbf{u}$ if the equality $\mathbf{T}\,\mathbf{u} = \mathbf{0}$ holds for $\mathbf{T} = \mathbf{a} \otimes \mathbf{b}$ where $\mathbf{a}$, $\mathbf{b}$ and $\mathbf{u}$ are arbitrary non-vanishing vectors?

1.2. Let $\mathbf{a}$, $\mathbf{b}$, $\mathbf{u}$ and $\mathbf{v}$ be non-vanishing vectors. Determine the tensor $\mathbf{T}$ which satisfies the equality $\mathbf{T}\,\mathbf{u} = \mathbf{v}$ if $\mathbf{v} = (\mathbf{b}\cdot\mathbf{u})\,\mathbf{a} - (\mathbf{a}\cdot\mathbf{u})\,\mathbf{b}$.

1.3. Let $(\mathbf{a}, \mathbf{b}, \mathbf{c})$ be an orthonormal basis. Construct the simple contraction $\mathbf{T}\,\mathbf{u}$ of the tensor $\mathbf{T} = \mathbf{a} \otimes \mathbf{a} + \mathbf{b} \otimes \mathbf{b} + \mathbf{c} \otimes \mathbf{c}$ with an arbitrary vector $\mathbf{u}$ in order to discover the property of $\mathbf{T}$.

1.4. Find the tensor $\mathbf{A}$ transforming $\mathbf{i}_i$ into $\mathbf{g}_i$ if the equality $\mathbf{g}_i = \mathbf{A}\,\mathbf{i}_i$ holds.

1.5. Verify that $[(\mathbf{a} \otimes \mathbf{b})\,(\mathbf{c} \otimes \mathbf{d})] : \mathbf{I} = (\mathbf{b} \cdot \mathbf{c})\,(\mathbf{a} \cdot \mathbf{d})$, where $\mathbf{I}$ is the identity tensor.

1.6. Show that $\operatorname{tr} \mathbf{T} = 0$ if $\mathbf{T}$ is a skew-symmetric tensor.

1.7. Determine the orthogonal tensor $\mathbf{Q}$ which transforms the basis $\mathbf{i}_j$ into a given set of vectors $\mathbf{i}_j^*$ and determine the components of $\mathbf{Q} = Q^{ij}\,\mathbf{i}_i \otimes \mathbf{i}_j = \overline{Q}^{ij}\,\mathbf{g}_i \otimes \mathbf{g}_j$, where $\mathbf{g}_i$ are base vectors of a cylindrical coordinate system.

1.8. Let $\mathbf{g}_i$ and $\mathbf{G}_i$ be the base vectors relative to two different configurations of a body. Show that, if the transformation $\mathbf{g}_i = \mathbf{F}\,\mathbf{G}_i$ holds for the covariant base vectors, the contravariant ones $\mathbf{g}^i$ and $\mathbf{G}^i$ are related by $\mathbf{g}^i = \mathbf{F}^{-T}\,\mathbf{G}^i$.

1.9. Show that $\mathbf{a}\,\mathbf{C}\,\mathbf{b} = \mathbf{b}\,\mathbf{C}\,\mathbf{a}$ if $\mathbf{C}$ is a symmetric second order tensor and if $\mathbf{a}$, $\mathbf{b}$ are vectors.

1.10. Find a tensor which is orthogonal and symmetric. Confirm your result by using spectral decomposition.

1.11. Show that $\mathbf{S}\,\mathbf{S}^{-1} = \mathbf{I}$ is equivalent to $\mathbf{S}^{-1}\,\mathbf{S} = \mathbf{I}$.

1.12. Find the relation between the eigenvalues of $\mathbf{C}$ and $\mathbf{E}$, if $\mathbf{E} = \frac{1}{2}\,(\mathbf{C} - \mathbf{I})$.

1.13. Let $\mathbf{g}_i$ and $\mathbf{G}_i$ be two sets of base vectors in E3. Evaluate the components of the tensors $\mathbf{F} = \mathbf{g}_i \otimes \mathbf{G}^i$ and $\mathbf{C} = \mathbf{F}^T\,\mathbf{F}$ with respect to the basis $\mathbf{G}_i \otimes \mathbf{G}_j$.

1.14. Starting from $\mathbf{F} = \mathbf{g}_i \otimes \mathbf{G}^i$ express the tensors $\mathbf{F}^T$, $\mathbf{F}^{-1}$ and $\mathbf{F}^{-T}$ in terms of $\mathbf{g}_i$, $\mathbf{g}^i$, $\mathbf{G}_i$ and $\mathbf{G}^i$.

1.15. Show that the partial derivative of a scalar Φ with respect to $\mathbf{A} = \mathbf{A}^T$ is symmetric.

1.16. Verify that the tensors $\mathbf{U}^n$ $(n = 0, 1, 2, \ldots)$ are all coaxial.

1.17. Find a tensor which is coaxial to any symmetric second-order tensor.

1.18. Evaluate $\operatorname{tr} \mathbf{A}^2$, $\operatorname{tr} \mathbf{A}^3$ if $\mathbf{A} = -\mathbf{A}^T$.

1.19. Let $\Psi = \Psi\,(I_C, II_C, III_C)$ be a scalar-valued function depending of the invariants of a second-order symmetric tensor $\mathbf{C}$. Construct the partial derivative of Ψ with respect to $\mathbf{C}$.

1.20. Show that, if $\mathbf{T} = \mathbf{T}^T$, then $\mathbf{T} : \mathbf{S} = \mathbf{T} : \operatorname{sym} \mathbf{S}$.

1.21. Verify that, if $\mathbf{T} = \mathbf{T}^T$, the same property holds also for the inverse tensor $\mathbf{T}^{-1}$.

1.22. Show that $\mathbf{T} = \mathbf{T}^T$ implies that $\mathbf{S} = \mathbf{F}^{-1}\,\mathbf{T}\,\mathbf{F}^{-T}$ is symmetric.

1.23. Evaluate the eigenvalues and eigenvectors of a skew-symmetric tensor $\mathbf{A} = -\mathbf{A}^T$.

1.24. Prove the equality $\boldsymbol{\Omega} \times (\boldsymbol{\Omega} \times \mathbf{d}) = (\boldsymbol{\Omega} \otimes \boldsymbol{\Omega} - \omega^2\,\mathbf{I})\,\mathbf{d}$ where $\boldsymbol{\Omega}$ and $\mathbf{d}$ are vectors and $\omega = \lVert \boldsymbol{\Omega} \rVert$.

1.25. Show that if $\mathbf{C}$ is symmetric so is $\mathbf{C}^{-1}$.

1.26. Show that $\mathbf{C}^{-1}$ is an isotropic function to **C**.

1.27. Starting from the definition of an eigenvalue problem show that **C** and $\mathbf{C}^{-1}$ have the same eigenvectors and reciprocal eigenvalues.

1.28. The rotation of a vector **d** with respect to a given vector $\mathbf{d}_o$ in E3 may be described by $\mathbf{d} = \mathbf{R}\,\mathbf{d}_o$, where **R** is an orthogonal tensor. Construct the variations $\delta^n\,\mathbf{d}$ due to a variation of **R** by using the concept of axial vector. Can the position of **d** at $\mathbf{R} + \delta\,\mathbf{R}$ be determined in closed form?

1.29. Prove that **T**, dev **T** and sph **T** are coaxial.

2 Deformation

This section introduces various deformation, stretch and strain tensors to describe the deformation behaviour of a solid during an arbitrary motion. Their definitions are essentially based on the deformation gradient and the stretch variables introduced by the polar decomposition theorem. Emphasis is given to the eigenvalue problems of stretches presenting a suitable background for the definition of various strain measures within a unified concept. Attention is then dedicated to pull-back and push-forward operations which are of major importance for the construction of the LIE-derivatives. Finally the rate of the deformation tensor and the spin tensor are introduced and their geometrical interpretations are given.

2.1 General backgrounds

Lagrangian coordinates. We consider the configuration B_o of a body at time t_o in a 3D Euclidean space E3. In B_o, the body is supposed to be unloaded, undeformed and unstressed. The position vector of a typical point P_o of B_o relative to the origin 0 of an orthogonal Cartesian coordinate system is denoted by

$$\mathbf{X} = X^i \, \mathbf{i}_i \,, \quad \text{where} \quad X^i : X^1, X^2, X^3 \tag{2.1.1}$$

are *LAGRANGIAN* or *material coordinates* and $\mathbf{i}_i = \mathbf{i}^i$ are unit vectors along the X^i-axes (Fig. 2.1). The position of the index i in $\mathbf{i}_i$ is immaterial and will be selected according to the requirement of the summation rule. The same is valid for the indices of tensor components defined with respect to the basis $\mathbf{i}_i$.

Euler coordinates. We now suppose the body to take at a certain time t a new configuration B in E3, due to the action of external forces. Thus, the point P_o is moved into the position P which will be determined with respect to the same origin 0 by the position vector

$$\mathbf{x} = x^i \, \mathbf{i}_i \,, \quad \text{where} \quad x^i : x^1, x^2, x^3 \tag{2.1.2}$$

are called *EULER* or *spatial coordinates.*

Transformation of coordinates. We now assume the mapping of B_o into B such that the correspondence of the points P_o and P is one to one. Such a deformation is described by a transformation:

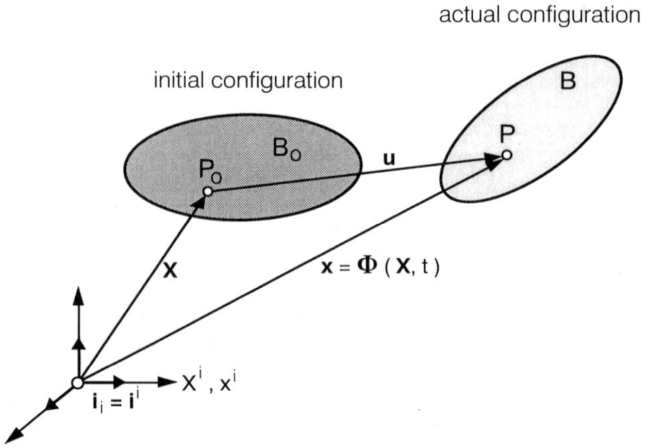

Fig. 2.1. Material and spatial coordinates; displacement vector **u**

$$x^i = x^i(X^1, X^2, X^3, t) \quad \rightarrow \quad \mathbf{x} = \mathbf{x}(\mathbf{X}, t) \ , \tag{2.1.3}$$

which is reversible

$$X^i = X^i(x^1, x^2, x^3, t) \quad \rightarrow \quad \mathbf{X} = \mathbf{X}(\mathbf{x}, t) \ . \tag{2.1.4}$$

This condition is satisfied if the functions x^i and X^i are single-valued and at least once continuously differentiable with respect to their arguments. In addition, the *JACOBIAN* must be positive:

$$J = \left| \frac{\partial x^i}{\partial X^j} \right| > 0 \ . \tag{2.1.5}$$

Note that, in view of the chain rule of differentiation, the relation

$$\frac{\partial x^i}{\partial X^j} \frac{\partial X^j}{\partial x^k} = c^i_{.j} \, \bar{c}^j_{.k} = \delta^i_k \tag{2.1.6}$$

holds permitting to calculate $c^i_{.j}$ when $\bar{c}^j_{.k}$ are given and vice versa.

In the transformation (2.1.3) the symbol **x** appears twice with different meanings. On the right-hand side **x** represents the function whose arguments are **X** and t, while on the left-hand side it represents the value of the function, i.e. the point **x**. A similar interpretation holds for **X** in (2.1.4). This double usage, common in the mechanics literature, may occasionally cause confusion, which can be avoided by using a different symbol **Φ** for the function so that equations (2.1.3) and (2.1.4) may be rewritten as

$$\mathbf{x} = \boldsymbol{\Phi}(\mathbf{X}, t) \ , \quad \mathbf{X} = \boldsymbol{\Phi}^{-1}(\mathbf{x}, t) \ .$$

However, for simplicity we shall often make use of the notations adopted in (2.1.3) and (2.1.4).

Transformation into curvilinear coordinates. In some problems it is preferable to give X^i as function of three independent parameters Θ^i

$$X^i = X^i(\Theta^1, \Theta^2, \Theta^3) \quad \rightarrow \quad \mathbf{X} = \mathbf{X}(\Theta^1, \Theta^2, \Theta^3) \ , \tag{2.1.7}$$

where the functions X^i are supposed to be single-valued and almost everywhere continuously differentiable with respect to their arguments Θ^i as many times as required. We further suppose that the *curvilinear coordinates* Θ^i are *convective*. Such a coordinate system is characterized by the fact that any point of the body is determined in the undeformed as well as the deformed state by the same values of the coordinates Θ^i. From (2.1.3) and (2.1.7) it then follows that

$$x^i = x^i(\Theta^1, \Theta^2, \Theta^3, t) \quad \rightarrow \quad \mathbf{x} = \mathbf{x}(\Theta^1, \Theta^2, \Theta^3, t) \ . \tag{2.1.8}$$

If curvilinear coordinates Θ^i are used, any point of the undeformed body is associated by a set of curvilinear coordinates, the so-called Θ^i-curves (i = 1, 2, 3). Since the coordinates Θ^i are supposed to be convective these curves are subjected to the same deformation as the body and are transformed at time t into a new set of curvilinear coordinates. Note that even if Θ^i are selected as Cartesian coordinates $\Theta^i = X^i$ in B_o the corresponding coordinate lines in B are in general curvilinear. From (2.1.7) we receive

$$d\Theta^i = \frac{\partial \Theta^i}{\partial X^j} dX^j \ , \quad dX^i = \frac{\partial X^i}{\partial \Theta^j} d\Theta^j \tag{2.1.9}$$

and, similarly, from (2.1.8) at a fixed time t

$$d\Theta^i = \frac{\partial \Theta^i}{\partial x^j} dx^j \ , \quad dx^i = \frac{\partial x^i}{\partial \Theta^j} d\Theta^j \tag{2.1.10}$$

indicating that the $d\Theta^i$ remain contravariant tensor components under all transformations including transformations to coordinate systems in B.

Metric properties. We now suppose the mapping B_o into B to be described by curvilinear coordinates Θ^i . From (2.1.7) we find for the base vectors related to P_o

$$\mathbf{G}_i = \mathbf{X}_{,i} = \frac{\partial \mathbf{X}}{\partial \Theta^i} = \frac{\partial X^k}{\partial \Theta^i} \mathbf{i}_k \ , \tag{2.1.11}$$

while the contravariant basis $\mathbf{G}^i$ is given by

$$\mathbf{G}^i = \frac{\partial \Theta^i}{\partial X^k} \mathbf{i}^k \ . \tag{2.1.12}$$

Consequently, the associated metric tensors G_{ij} and G^{ij} can be expressed as

$$G_{ij} = \mathbf{G}_i \cdot \mathbf{G}_j \ , \quad G^{ij} = \mathbf{G}^i \cdot \mathbf{G}^j \ . \tag{2.1.13}$$

For the identity tensor

$$\mathbf{I} = \mathbf{G}_i \otimes \mathbf{G}^i = \mathbf{G} = \mathbf{G}^i \otimes \mathbf{G}_i = \mathbf{I}^{-1} \tag{2.1.14}$$

to be constructed with the basis $\mathbf{G}_i$ of the undeformed configuration we sometimes use the notation $\mathbf{G}$ in order to distinguish it from its deformed counterpart $\mathbf{g}$ to be defined similarly in terms of the base vectors $\mathbf{g}_i$ of the deformed configuration. Note that the distinction between $\mathbf{G}$ and $\mathbf{g}$ is purely formal since $\mathbf{G}=\mathbf{g}$.

Starting from (2.1.8), similar relations can be derived for the deformed state B. For later use we note

$$\mathbf{g}_i = \frac{\partial \mathbf{x}}{\partial \Theta^i} = \mathbf{x},_i = \frac{\partial x^k}{\partial \Theta^i}\,\mathbf{i}_k , \tag{2.1.15}$$

$$\mathbf{g}^i = \frac{\partial \Theta^i}{\partial x^k}\,\mathbf{i}^k , \tag{2.1.16}$$

$$g_{ij} = \mathbf{g}_i \cdot \mathbf{g}_j , \quad g^{ij} = \mathbf{g}^i \cdot \mathbf{g}^j , \tag{2.1.17}$$

$$\mathbf{I} = \mathbf{g}_i \otimes \mathbf{g}^i = \mathbf{g} = \mathbf{g}^i \otimes \mathbf{g}_i = \mathbf{I}^{-1} . \tag{2.1.18}$$

In the sequel upper case letters will be used to denote geometrical variables associated with the undeformed state B_0 while lower case letters indicate geometrical variables related to the deformed state B.

Displacement vector. The position vector of the point P relative to P_0 is called the *displacement vector* $\mathbf{u}$ (Fig. 2.1). For later use we introduce for $\mathbf{u}$ two sets of components U_i and u_i

$$\mathbf{u} = \mathbf{u}\,(\Theta^i, t) = \mathbf{x} - \mathbf{X} = U_i\,\mathbf{G}^i = u_i\,\mathbf{g}^i , \tag{2.1.19}$$

defined with respect to the undeformed basis $\mathbf{G}^i$ and the deformed basis $\mathbf{g}^i$, respectively. By means of the well-known relations for the partial derivative $\mathbf{u},_i$

$$\mathbf{u},_i = \frac{\partial \mathbf{u}}{\partial \Theta^i} = \mathbf{g}_i - \mathbf{G}_i = U_m|_i\,\mathbf{G}^m = u_m||_i\,\mathbf{g}^m \tag{2.1.20}$$

the base vectors $\mathbf{g}_i$ and $\mathbf{G}_i$ can be expressed by

$$\mathbf{g}_i = \mathbf{G}_i + \mathbf{u},_i = (\delta_i^m + U^m|_i)\,\mathbf{G}_m = F^m_{\cdot\,i}\,\mathbf{G}_m , \tag{2.1.21}$$

$$\mathbf{G}_i = \mathbf{g}_i - \mathbf{u},_i = (\delta_i^m - u^m||_i)\,\mathbf{g}_m = (f^{-1})^m_{\cdot\,i}\,\mathbf{g}_m , \tag{2.1.22}$$

where, as abbreviations, $F^m_{\cdot\,i}$ and $(f^{-1})^m_{\cdot\,i}$ have been introduced transforming the base vectors $\mathbf{g}_i$ and $\mathbf{G}_i$ into each other. The notations $(\ldots)|_i$ and $(\ldots)||_i$ indicate the covariant derivatives with respect to the undeformed basis $\mathbf{G}_i$ and the deformed basis $\mathbf{g}_i$, respectively.

Coordinates. The above presentation shows that there are three possibilities for the selection of independent coordinates to describe a deformation process, namely

$$\begin{aligned}
&\textit{material coordinates:} && X^i && \rightarrow && \mathbf{x} = \mathbf{x}\,(X^1, X^2, X^3, t)\ ,\\
&\textit{spatial coordinates:} && x^i && \rightarrow && \mathbf{X} = \mathbf{X}\,(x^1, x^2, x^3, t)\ ,\\
&\textit{curvilinear coordinates:} && \Theta^i && \rightarrow && \mathbf{X} = \mathbf{X}\,(\Theta^1, \Theta^2, \Theta^3)\ ,\\
& && && && \mathbf{x} = \mathbf{x}\,(\Theta^1, \Theta^2, \Theta^3, t)\ .
\end{aligned} \tag{2.1.23}$$

The partial derivative of a function F with respect to X^i, x^i and Θ^i will be denoted respectively by

$$\frac{\partial F}{\partial X^i}\,,\ \frac{\partial F}{\partial x^i}\quad \text{and}\quad F_{,i} = \frac{\partial F}{\partial \Theta^i}\ . \tag{2.1.24}$$

As is indicated in (2.1.23) the problem is, if material coordinates X^i are used as independent coordinates, to determine the spatial coordinates $\mathbf{x}$ in terms of X^i. On the contrary, if spatial coordinates x^i are selected for the description the purpose is the determination of the material coordinates $\mathbf{X}$ in terms of x^i. Curvilinear coordinates Θ^i can be essentially introduced in the undeformed B_o or the deformed state B. Mostly they are selected in the initial configuration B_o. In this case the problem is to evaluate the spatial coordinates $\mathbf{x}$ in terms of Θ^i at an arbitrary time t.

2.2 Deformation gradient

Vectorial line elements. The *vectorial line elements* $d\mathbf{X}$ and $d\mathbf{x}$ related respectively to the *material* coordinates X^i and the *spatial* coordinates x^i play an important role in this section. If we use convective curvilinear coordinates Θ^i, they are given according to (2.1.7), (2.1.11) and (2.1.8), (2.1.15) by:

$$d\mathbf{X} = \frac{\partial \mathbf{X}}{\partial \Theta^i}\, d\Theta^i = \mathbf{G}_i\, d\Theta^i\ , \tag{2.2.1}$$

$$d\mathbf{x} = \frac{\partial \mathbf{x}}{\partial \Theta^i}\, d\Theta^i = \mathbf{g}_i\, d\Theta^i\ . \tag{2.2.2}$$

This clearly indicates that a main characteristic of the deformation is the mapping of the basis $\mathbf{G}_i$ of the initial configuration into the basis of the deformed (actual) configuration $\mathbf{g}_i$. Our aim in this section is to establish the relations between $(\mathbf{G}_i)\ d\mathbf{X}$ and $(\mathbf{g}_i)\ d\mathbf{x}$ by means of a second-order tensor $\mathbf{F}$ called *deformation gradient.*

Deformation gradient. The *deformation gradient* $\mathbf{F}$ is defined through $\mathbf{g}_i$ and $\mathbf{G}_i$ by:

$$\mathbf{F} = \mathbf{g}_i \otimes \mathbf{G}^i\ ,\quad \mathbf{F}^T = \mathbf{G}^i \otimes \mathbf{g}_i\ . \tag{2.2.3}$$

Then, the inverse tensors $\mathbf{F}^{-1}$ and $(\mathbf{F}^T)^{-1} = \mathbf{F}^{-T}$ are given by

$$\mathbf{F}^{-1} = \mathbf{G}_i \otimes \mathbf{g}^i \ , \quad \mathbf{F}^{-T} = \mathbf{g}^i \otimes \mathbf{G}_i \tag{2.2.4}$$

as can be easily confirmed by considering (2.2.3) and (2.2.4)

$$\mathbf{F}\,\mathbf{F}^{-1} = \mathbf{g}_i \otimes \mathbf{g}^i = \mathbf{g} \ , \quad \mathbf{F}^{T}\,\mathbf{F}^{-T} = \mathbf{G}^i \otimes \mathbf{G}_i = \mathbf{G} \ . \tag{2.2.5}$$

The deformation gradient $\mathbf{F}$ is therefore a tensor which maps the undeformed basis $\mathbf{G}_i$ into the deformed one $\mathbf{g}_i$ by

$$\mathbf{g}_i = \mathbf{F}\,\mathbf{G}_i \ , \quad \mathbf{G}_i = \mathbf{F}^{-1}\,\mathbf{g}_i \ , \tag{2.2.6}$$

the corresponding relations between the contravariant base vectors $\mathbf{g}^i$ and $\mathbf{G}^i$ being of the form

$$\mathbf{g}^i = \mathbf{F}^{-T}\,\mathbf{G}^i \ , \quad \mathbf{G}^i = \mathbf{F}^{T}\,\mathbf{g}^i \ . \tag{2.2.7}$$

The validity of the relations (2.2.6) and (2.2.7) can be easily proved. By means of (2.2.3) and (2.2.4) we receive e.g.:

$$\mathbf{F}\,\mathbf{G}_i = (\mathbf{g}_j \otimes \mathbf{G}^j)\,\mathbf{G}_i = \delta^j_i\,\mathbf{g}_j = \mathbf{g}_i \ , \tag{2.2.8}$$

$$\mathbf{F}^{-T}\,\mathbf{G}^i = (\mathbf{g}^j \otimes \mathbf{G}_j)\,\mathbf{G}^i = \delta^i_j\,\mathbf{g}^j = \mathbf{g}^i \ , \tag{2.2.9}$$

in accordance with (2.2.6) and (2.2.7). The transformations (2.2.6) and (2.2.7) are illustrated in Fig. 2.2. We observe that the connections between the covariant base vectors $\mathbf{g}_i$ and $\mathbf{G}_i$ are described by $\mathbf{F}$ and $\mathbf{F}^{-1}$, while the transposed tensors $\mathbf{F}^T$ and $\mathbf{F}^{-T}$ transform the contravariant bases $\mathbf{g}^i$ and $\mathbf{G}^i$ into each other.

By contracting equations (2.2.6) by $d\Theta^i$ we obtain the expressions

$$d\mathbf{x} = \mathbf{F}\,d\mathbf{X} \ , \quad d\mathbf{X} = \mathbf{F}^{-1}\,d\mathbf{x} \tag{2.2.10}$$

for the line elements $d\mathbf{X}$ and $d\mathbf{x}$ introduced respectively in (2.2.1) and (2.2.2). Thus, the deformation gradient $\mathbf{F}$ can also be seen as a tensor which transforms an arbitrary line element $d\mathbf{X}$ at $\mathbf{X}$ into its deformed counterpart $d\mathbf{x}$ at $\mathbf{x}$ (Fig. 2.3). The deformation gradient as link between different configurations is a key notion for the mathematical description of the geometrical aspects of deformations of solids. Note that the transformations (2.2.10) are similar to those given in (2.2.6) for covariant base vectors $\mathbf{g}_i$ and $\mathbf{G}_i$.

Gradients of the coordinates. For later use we also introduce the *material gradient* of $\mathbf{x}$

$$\mathrm{GRAD}\,\mathbf{x} := \frac{\partial \mathbf{x}}{\partial X^j} \otimes \mathbf{i}^j = \frac{\partial \mathbf{x}}{\partial \Theta^i} \otimes \frac{\partial \Theta^i}{\partial X^j}\,\mathbf{i}^j = \mathbf{g}_i \otimes \mathbf{G}^i = \mathbf{F} \tag{2.2.11}$$

and the *spatial gradient* of $\mathbf{X}$

$$\mathrm{grad}\,\mathbf{X} := \frac{\partial \mathbf{X}}{\partial x^j} \otimes \mathbf{i}^j = \frac{\partial \mathbf{X}}{\partial \Theta^i} \otimes \frac{\partial \Theta^i}{\partial x^j}\,\mathbf{i}^j = \mathbf{G}_i \otimes \mathbf{g}^i = \mathbf{F}^{-1} \ , \tag{2.2.12}$$

which, by virtue of (2.1.12) and (2.1.16), turn out to be identical with $\mathbf{F}$ and $\mathbf{F}^{-1}$, respectively. In this sense, GRAD $\mathbf{x}$ can be seen as a new notation for $\mathbf{F}$ to be used if material coordinates X^i are selected for the description of the deformation process. A similar interpretation holds for the spatial gradient grad $\mathbf{X}$ which characterizes the use of the spatial coordinates x^i. If we now set according to (2.2.11) and (2.2.12)

$$\mathrm{GRAD}\,\mathbf{x} := \frac{\partial \mathbf{x}}{\partial \mathbf{X}} = \mathbf{F}\ , \quad \mathrm{grad}\,\mathbf{X} := \frac{\partial \mathbf{X}}{\partial \mathbf{x}} = \mathbf{F}^{-1} \tag{2.2.13}$$

equations (2.2.10) can be given alternatively as

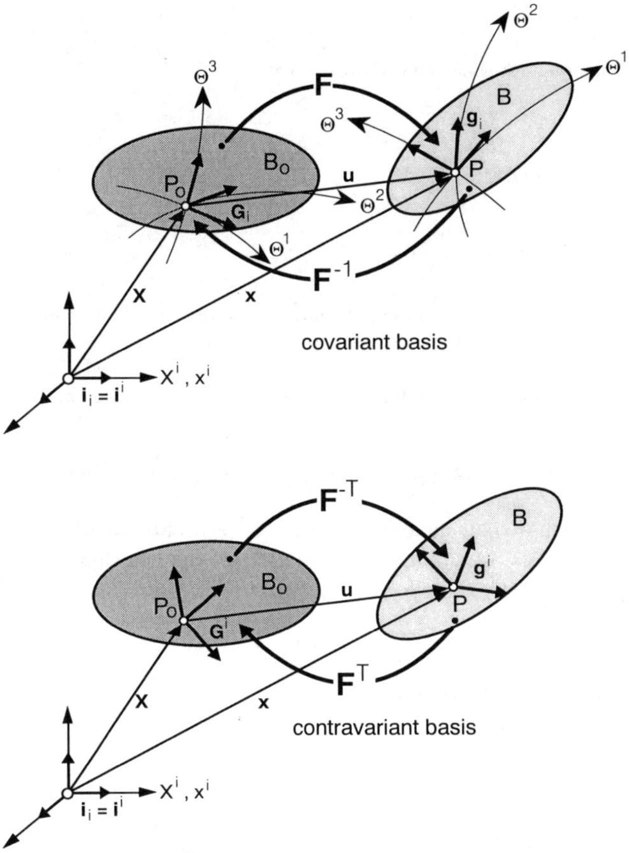

Fig. 2.2. Transformation of covariant and contravariant bases by $\mathbf{F}$, $\mathbf{F}^{-1}$and $\mathbf{F}^{T}$, $\mathbf{F}^{-T}$

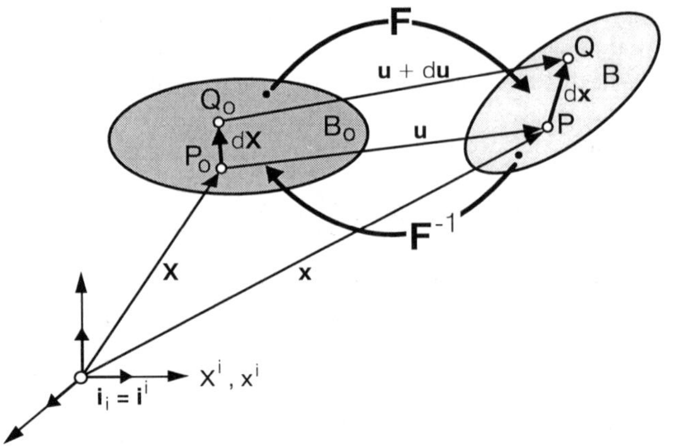

Fig. 2.3. Deformation of the line element $\mathbf{dX}$ into $\mathbf{dx}$

$$d\mathbf{x} = (\mathrm{GRAD}\,\mathbf{x})\,d\mathbf{X} = \frac{\partial \mathbf{x}}{\partial \mathbf{X}}\,d\mathbf{X} \ , \qquad (2.2.14)$$

$$d\mathbf{X} = (\mathrm{grad}\,\mathbf{X})\,d\mathbf{x} = \frac{\partial \mathbf{X}}{\partial \mathbf{x}}\,d\mathbf{x} \ . \qquad (2.2.15)$$

The above relations are of convenience if material coordinates X^i and spatial coordinates x^i are selected as independent coordinates, respectively.

Evaluation of components. For later use we define the components of $\mathbf{F}$ and $\mathbf{F}^{-1}$ as follows:

$$\mathbf{F} = F_{ij}\,\mathbf{G}^i \otimes \mathbf{G}^j = f_{ij}\,\mathbf{g}^i \otimes \mathbf{g}^j \ , \quad \mathbf{F}^{-1} = (F^{-1})_{ij}\,\mathbf{G}^i \otimes \mathbf{G}^j = (f^{-1})_{ij}\,\mathbf{g}^i \otimes \mathbf{g}^j \ , \qquad (2.2.16)$$

where lower case letters indicate components with respect to the basis of the deformed configuration. The components introduced in (2.2.16) can be evaluated by the standard procedure (1.2.13). As an example we present the results for F_{ij} and $(f^{-1})_{ij}$

$$F_{ij} = \mathbf{G}_i\,\mathbf{F}\,\mathbf{G}_j = \mathbf{G}_i \cdot \mathbf{g}_j \ , \quad (f^{-1})_{ij} = \mathbf{g}_i\,\mathbf{F}^{-1}\,\mathbf{g}_j = \mathbf{g}_i \cdot \mathbf{G}_j \ , \qquad (2.2.17)$$

obtained by considering (2.2.3) and (2.2.4). It is possible to collect the results given in (2.2.17) into a single formula

$$F_{ij} = (f^{-1})_{ji} = \mathbf{G}_i \cdot \mathbf{g}_j \ . \qquad (2.2.18)$$

But note that F_{ij} and $(f^{-1})_{ji}$ are defined with respect to different bases and, consequently, equalities similar to (2.2.18) do not hold for mixed and contravariant components:

$$F_{i\cdot}^{\ j} \neq (f^{-1})^{j}_{\cdot i} \ , \quad F^{ij} \neq (f^{-1})^{ij} \ . \qquad (2.2.19)$$

In view of (2.1.21), (2.1.22) the results (2.2.17) can be given in terms of the displacement components U_i and u_i introduced in (2.1.19), thus

$$F_{ij} = \mathbf{G}_i \cdot \mathbf{g}_j = G_{ij} + U_i|_j \ , \tag{2.2.20}$$

$$(f^{-1})_{ij} = \mathbf{g}_i \cdot \mathbf{G}_j = g_{ij} - u_i||_j \ . \tag{2.2.21}$$

We now use (2.2.16) to transform the relations (2.2.6) into component form. By considering (2.2.17) we find

$$\mathbf{g}_i = F_{mj}\,(\mathbf{G}^m \otimes \mathbf{G}^j)\,\mathbf{G}_i = F^m_{\cdot i}\,\mathbf{G}_m = (\mathbf{G}^m \cdot \mathbf{g}_i)\,\mathbf{G}_m \ , \tag{2.2.22}$$

$$\mathbf{G}_i = (f^{-1})_{mj}\,(\mathbf{g}^m \otimes \mathbf{g}^j)\,\mathbf{g}_i = (f^{-1})^m_{\cdot i}\,\mathbf{g}_m = (\mathbf{g}^m \cdot \mathbf{G}_i)\,\mathbf{g}_m \ . \tag{2.2.23}$$

Similarily we receive from (2.2.7) and (2.2.16), (2.2.17)

$$\mathbf{g}^i = (F^{-T})_{m\cdot}^{\ \ i}\,\mathbf{G}^m = (\mathbf{G}_m \cdot \mathbf{g}^i)\,\mathbf{G}^m \ , \tag{2.2.24}$$

$$\mathbf{G}^i = (f^T)_{m\cdot}^{\ \ i}\,\mathbf{g}^m = (\mathbf{g}_m \cdot \mathbf{G}^i)\,\mathbf{g}^m \ . \tag{2.2.25}$$

Differential volume element. Since **F** determines the deformed base vectors $\mathbf{g}_i$ in terms of the undeformed ones $\mathbf{G}_i$ it is possible to transform the geometrical elements associated with the deformed state B into those of the undeformed state B_o through the deformation gradient **F**. We consider an infinitesimal parallelepiped of the deformed body (Fig. 2.4) whose edges are determined by the vectorial line elements $\mathbf{g}_i d\Theta^i$ (no summation over i). Its volume dV is given by

$$dV = \sqrt{g}\, d\Theta^1\, d\Theta^2\, d\Theta^3 \ . \tag{2.2.26}$$

In the undeformed state the same parallelepiped is determined by the vectors $\mathbf{G}_i d\Theta^i$ (no summation over i) and its volume dV_o is

$$dV_o = \sqrt{G}\, d\Theta^1\, d\Theta^2\, d\Theta^3 \ . \tag{2.2.27}$$

Our aim is to express the volume ratio dV/dV_o in terms of the deformation gradient **F**. For the derivation we start from the identity

$$\sqrt{g} = [\mathbf{g}_1\ \mathbf{g}_2\ \mathbf{g}_3] = \mathbf{g}_1 \cdot (\mathbf{g}_2 \times \mathbf{g}_3) = \bar{\varepsilon}_{123} \ , \tag{2.2.28}$$

which, by considering (2.2.22), yields

$$\sqrt{g} = F^i_{\cdot 1}\, F^j_{\cdot 2}\, F^k_{\cdot 3}\ [\mathbf{G}_i\ \mathbf{G}_j\ \mathbf{G}_k] = \sqrt{G}\, e_{ijk}\, F^i_{\cdot 1}\, F^j_{\cdot 2}\, F^k_{\cdot 3} = \sqrt{G}\ |F^{\ i}_{\cdot j}| \ . \tag{2.2.29}$$

Herein e_{ijk} denotes the components of the permutation tensor associated with the basis $\mathbf{i}_i$, while the components $\bar{\varepsilon}_{ijk}$ are referred to the basis $\mathbf{g}_i$. We now introduce by using (1.3.60) and (2.2.20) the determinant

$$\det \mathbf{F} = |F^{\ i}_{\cdot j}| = |\delta^i_j + U^i|_j| \ . \tag{2.2.30}$$

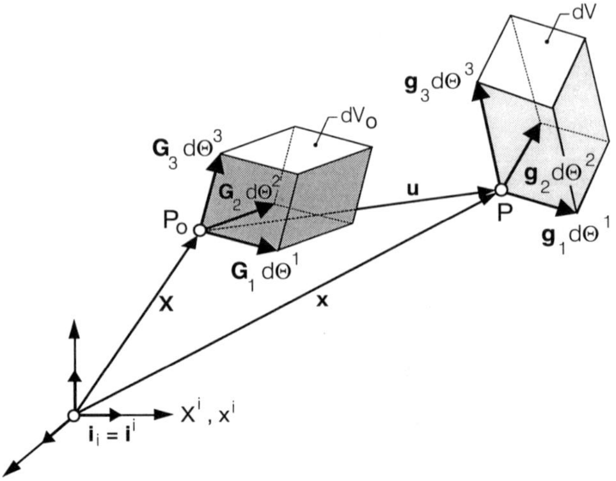

Fig. 2.4. The undeformed and deformed volume elements dV_0 and dV

If, in addition, we consider (2.2.26) and (2.2.27), equation (2.2.29) delivers*

$$J = \frac{dV}{dV_o} = \sqrt{\frac{g}{G}} = \det \mathbf{F} = | F^{i}_{\cdot j} | , \tag{2.2.31}$$

as final result. Since, in view of (1.8.3), det **F** corresponds to the third invariant of **F**, the abbreviation J introduced in (2.2.31), the JACOBIAN, is an invariant scalar.

Differential area element. To describe the change of area during the deformation we consider the area dA_o of a parallelogram with the edges $\delta\mathbf{X}$ and $d\mathbf{X}$ in the undeformed configuration B_o (Fig. 2.5). If **N** is the unit normal vector to dA_o, then according to (2.2.1)

$$\mathbf{N}\, dA_o = d\mathbf{X} \times \delta\mathbf{X} = \mathbf{G}_i \times \mathbf{G}_j\, d\Theta^i\, \delta\Theta^j = \varepsilon_{ijk}\, \mathbf{G}^k\, d\Theta^i\, \delta\Theta^j , \tag{2.2.32}$$

where ε_{ijk} is the permutation tensor referring to the basis $\mathbf{G}_i$. In the deformed state B the considered parallelogram is determined by $d\mathbf{x}$ and $\delta\mathbf{x}$, and has the area dA. With the unit vector **n** normal to dA, equation (2.2.32) is then replaced by

$$\mathbf{n}\, dA = d\mathbf{x} \times \delta\mathbf{x} = \bar{\varepsilon}_{ijk}\, \mathbf{g}^k\, d\Theta^i\, \delta\Theta^j , \tag{2.2.33}$$

where the permutation tensor $\bar{\varepsilon}_{ijk}$

* The definition (2.3.31) for J is identical with (2.1.5). If $\Theta^i = X^i$, then we see from (2.2.11) that $F^{i}_{\cdot j} = \frac{\partial x^i}{\partial X^j}$.

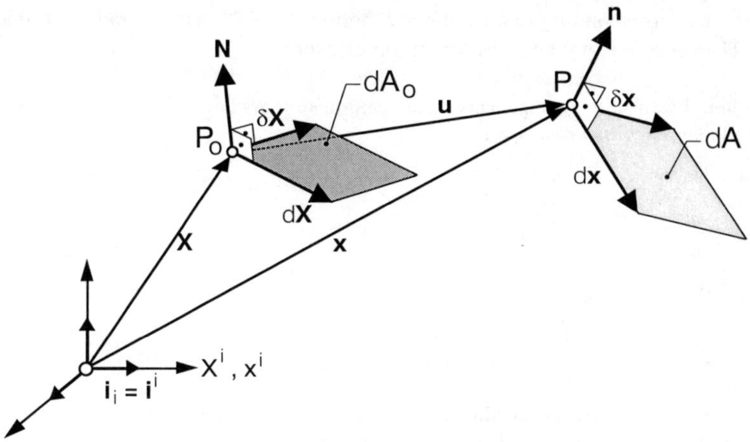

Fig. 2.5. The undeformed and deformed surface elements dA_0 and dA

$$\bar{\varepsilon}_{ijk} = \sqrt{\frac{g}{G}}\, \varepsilon_{ijk} = \det \mathbf{F}\, \varepsilon_{ijk} \tag{2.2.34}$$

is now associated with the deformed basis $\mathbf{g}_i$. We substitute (2.2.34) into (2.2.33) and replace $\mathbf{g}^k$ by the transformation (2.2.7). The comparison of the corresponding result with (2.2.32) yields finally:

$$\mathbf{n}\, dA = \sqrt{\frac{g}{G}}\, \mathbf{N}\, \mathbf{F}^{-1}\, dA_o = \det \mathbf{F}\, \mathbf{F}^{-T}\, \mathbf{N}\, dA_o = \det \mathbf{F}\, \mathbf{N}\, \mathbf{F}^{-1}\, dA_o \ . \tag{2.2.35}$$

This relation establishes the connection between the areas dA_o and dA in terms of the unit normal vectors **N** and **n** as well as det **F** introduced in (2.2.30) and will be used later for the definition of various stress tensors.

Objective strain measure. Finally we deal with the question if the deformation gradient **F** is an objective strain measure. A deformation variable is said to be *objective* if it vanishes identically for rigid body movements of a body which may consist of a rotation and a translation. We assume the body in its initial position B_o to be subjected to a pure rotation by means of an arbitrary rotation tensor **R**. Then we have

$$\mathbf{g}_i = \mathbf{R}\, \mathbf{G}_i \quad \rightarrow \quad \mathbf{F} = \mathbf{g}_i \otimes \mathbf{G}^i = \mathbf{R}\, (\mathbf{G}_i \otimes \mathbf{G}^i) = \mathbf{R} \ . \tag{2.2.36}$$

If the body in its undeformed configuration is subjected to a translation described by a constant displacement vector **u**, then from (2.1.21)

$$\mathbf{g}_i = \mathbf{G}_i + \mathbf{u},_i = \mathbf{G}_i \quad \rightarrow \quad \mathbf{F} = \mathbf{g}_i \otimes \mathbf{G}^i = \mathbf{G}_i \otimes \mathbf{G}^i = \mathbf{I} \ . \tag{2.2.37}$$

In both cases corresponding to a rigid body motion the deformation gradient **F** does not vanish. Therefore it is not an *objective* strain measure.

Application. Evaluate the components of the deformation gradient **F** and its inverse $\mathbf{F}^{-1}$ with respect to the orthonormal basis $\mathbf{i}_i = \mathbf{i}^i$.

Using the relations (2.1.12) and (2.1.15)

$$\mathbf{G}^i = \frac{\partial \Theta^i}{\partial X^j} \mathbf{i}^j , \quad \mathbf{g}_i = \frac{\partial x^k}{\partial \Theta^i} \mathbf{i}_k \tag{2.2.38}$$

we find for **F**:

$$\mathbf{F} = \mathbf{g}_i \otimes \mathbf{G}^i = \frac{\partial x^k}{\partial \Theta^i} \frac{\partial \Theta^i}{\partial X^j} \mathbf{i}_k \otimes \mathbf{i}^j = \frac{\partial x^k}{\partial X^j} \mathbf{i}_k \otimes \mathbf{i}^j \tag{2.2.39}$$

the last expression being identical with GRAD **x** (2.2.11) which characterizes the use of material coordinates X^j. Similarly, by means of the transformations

$$\mathbf{G}_i = \frac{\partial X^k}{\partial \Theta^i} \mathbf{i}_k , \quad \mathbf{g}^i = \frac{\partial \Theta^i}{\partial x^j} \mathbf{i}^j \tag{2.2.40}$$

given in (2.1.11) and (2.1.16) we obtain for $\mathbf{F}^{-1}$

$$\mathbf{F}^{-1} = \mathbf{G}_i \otimes \mathbf{g}^i = \frac{\partial X^k}{\partial \Theta^i} \frac{\partial \Theta^i}{\partial x^j} \mathbf{i}_k \otimes \mathbf{i}^j = \frac{\partial X^k}{\partial x^j} \mathbf{i}_k \otimes \mathbf{i}^j \tag{2.2.41}$$

the last expression corresponding to grad **X** given in (2.2.12).

2.3 Deformation gradient in material and spatial coordinates

For generality, relations in connection with the deformation gradient **F** have been established in section 2.2 for that case where the deformation $B_o \rightarrow B$ is described by convective *curvilinear* coordinates Θ^i. As already emphasized in section 2.2 this can be also performed by using *material* coordinates X^i or *spatial* coordinates x^i as independent variables presenting both orthogonal Cartesian coordinates. We recall that X^i determine the position of a point P_o in the initial state B_o while x^i are the coordinates of its actual position P in the deformed state B. We also note that the consideration of curvilinear coordinates ξ^i as *intermediate* variables may be useful or necessary even if X^i or x^i serve as independent variables.

It may be of course useful to compare the mentioned three possibilities concerning the selection of independent coordinates on the example of the deformation gradient **F** in order to discover the characteristic properties of the corresponding representations. Thus we shall derive in this section the basic relations of each individual case starting for this purpose from the general relations of section 2.2.

Curvilinear coordinates. Notations already used for the variables remain unchanged in this case. Thus we have immediately from (2.1.11), (2.1.12) and (2.1.15), (2.2.16):

$$B_o: \ \mathbf{G}_i = \frac{\partial X^k}{\partial \Theta^i}\mathbf{i}_k \,, \qquad \mathbf{G}^i = \frac{\partial \Theta^i}{\partial X^k}\mathbf{i}^k \,, \tag{2.3.1}$$

$$B\ : \ \mathbf{g}_i = \frac{\partial x^k}{\partial \Theta^i}\mathbf{i}_k \,, \qquad \mathbf{g}^i = \frac{\partial \Theta^i}{\partial x^k}\mathbf{i}^k \,, \tag{2.3.2}$$

and from (2.2.3) and (2.2. 4)

$$B_o \leftrightarrow B: \quad \begin{aligned} \mathbf{F} &= \mathbf{g}_i \otimes \mathbf{G}^i \,, \\ \mathbf{F}^{-1} &= \mathbf{G}_i \otimes \mathbf{g}^i \,. \end{aligned} \tag{2.3.3}$$

Material coordinates. The relations for this case can be obtained from the above generally applicable equations simply by setting $\Theta^i = X^i$. To distinguish the variables of the present case from the above ones we use the notation $(\overset{\bullet}{\ldots})$. Thus we have

$$B_o \begin{cases} \overset{\bullet}{\mathbf{G}}_i = \dfrac{\partial X^k}{\partial X^i}\mathbf{i}_k = \delta_i^k\,\mathbf{i}_k = \mathbf{i}_i \,, \\ \overset{\bullet}{\mathbf{G}}{}^i = \dfrac{\partial X^i}{\partial X^k}\mathbf{i}^k = \delta_k^i\,\mathbf{i}^k = \mathbf{i}^i \,, \end{cases} \tag{2.3.4}$$

$$B \begin{cases} \overset{\bullet}{\mathbf{g}}_i = \dfrac{\partial x^k}{\partial X^i}\mathbf{i}_k = \dfrac{\partial x^k}{\partial \xi^m}\dfrac{\partial \xi^m}{\partial X^i}\mathbf{i}_k = \dfrac{\partial \xi^m}{\partial X^i}\mathbf{g}_m \,, \\ \overset{\bullet}{\mathbf{g}}{}^i = \dfrac{\partial X^i}{\partial x^k}\mathbf{i}^k = \dfrac{\partial X^i}{\partial \xi^m}\dfrac{\partial \xi^m}{\partial x^k}\mathbf{i}^k = \dfrac{\partial X^i}{\partial \xi^m}\mathbf{g}^m \,, \end{cases} \tag{2.3.5}$$

and by virtue of (2.2.11), (2.2.12)

$$B_o \leftrightarrow B: \quad \begin{aligned} \mathbf{F} &= \overset{\bullet}{\mathbf{g}}_k \otimes \mathbf{i}^k = \frac{\partial x^i}{\partial X^k}\mathbf{i}_i \otimes \mathbf{i}^k = \mathrm{GRAD}\,\mathbf{x} \,, \\ \mathbf{F}^{-1} &= \mathbf{i}_k \otimes \overset{\bullet}{\mathbf{g}}{}^k = \frac{\partial X^i}{\partial x^k}\mathbf{i}_i \otimes \mathbf{i}^k = \mathrm{grad}\,\mathbf{X} \,. \end{aligned} \tag{2.3.6}$$

In some cases e.g. in the isoparametric finite element formulation it may be suitable to use curvilinear coordinates ξ^i as intermediate variables. The expressions to be used in this case to calculate the unknown coefficients $\partial x^i/\partial X^k$ and $\partial X^i/\partial x^k$ are given in (2.3.5).

Spatial coordinates. Equations (2.3.1) to (2.3.3) can be specified to the present case by setting $\Theta^i = x^i$. Using the notation $(\overset{*}{\ldots})$ for the corresponding variables we obtain

$$B_o \begin{cases} \overset{*}{\mathbf{G}}_i = \dfrac{\partial X^k}{\partial x^i}\mathbf{i}_k = \dfrac{\partial X^k}{\partial \xi^m}\dfrac{\partial \xi^m}{\partial x^i}\mathbf{i}_k = \dfrac{\partial \xi^m}{\partial x^i}\mathbf{G}_m \,, \\ \overset{*}{\mathbf{G}}{}^i = \dfrac{\partial x^i}{\partial X^k}\mathbf{i}^k = \dfrac{\partial x^i}{\partial \xi^m}\dfrac{\partial \xi^m}{\partial X^k}\mathbf{i}^k = \dfrac{\partial x^i}{\partial \xi^m}\mathbf{G}^m \,, \end{cases} \tag{2.3.7}$$

$$B \left\{ \begin{aligned} \overset{*}{\mathbf{g}}_i &= \frac{\partial x^k}{\partial x^i}\,\mathbf{i}_k = \delta_i^k\,\mathbf{i}_k = \mathbf{i}_i \ , \\ \overset{*}{\mathbf{g}}{}^i &= \frac{\partial x^i}{\partial x^k}\,\mathbf{i}^k = \delta_k^i\,\mathbf{i}^k = \mathbf{i}^i \ , \end{aligned} \right. \tag{2.3.8}$$

and by considering (2.2.11), (2.2.12)

$$B_o \leftrightarrow B: \quad \begin{aligned} \mathbf{F} &= \mathbf{i}_k \otimes \overset{*}{\mathbf{G}}{}^k = \frac{\partial x^k}{\partial X^j}\,\mathbf{i}_k \otimes \mathbf{i}^j = \mathrm{GRAD}\,\mathbf{x} \ , \\ \mathbf{F}^{-1} &= \overset{*}{\mathbf{G}}_k \otimes \mathbf{i}^k = \frac{\partial X^i}{\partial x^k}\,\mathbf{i}_i \otimes \mathbf{i}^k = \mathrm{grad}\,\mathbf{X} \ . \end{aligned} \tag{2.3.9}$$

A geometrical interpretation of the tensors derived above is given in Fig. 2.6 for each considered case. If curvilinear coordinates Θ^i are used, a volume element defined by curvilinear coordinate lines is compared with an element defined similarly in the deformed state. In the case of material coordinates the edges of the volume element are spanned in the undeformed state by the orthonormal basis $\mathbf{i}_i$ and in the deformed state by $\overset{\bullet}{\mathbf{g}}_i$. Finally, if spatial coordinates x^i are used attention is given to an element defined in the deformed state by the orthonormal basis $\mathbf{i}_j$. The problem in this case is to determine its configuration before deformation through the evaluation of the base vectors $\overset{*}{\mathbf{G}}_i$.

Interpretation of the results. Table 2.1 shows the coefficients to be calculated for forming the base vectors entering the definition of $\mathbf{F}$ and $\mathbf{F}^{-1}$ for five cases:

- if X^i or x^i or Θ^i are used as independent coordinates
- if X^i or x^i are used as independent coordinates, combined with curvilinear ones ξ^i as intermediate variables.

The last two cases may be relevant for an isoparametric finite element formulation, where for certain reasons it may be suitable to obtain results referring to material or spatial coordinates. By comparing the second and third line in Table 2.1 we see that the computation of $\overset{\bullet}{\mathbf{g}}_i$ (combined usage of X^i and ξ^i) requires decisively more efforts than that of $\mathbf{g}_i$ (usage of Θ^i). In the first case, the coefficients $\partial\xi^j/\partial X^i$ are, in addition to $\partial x^k/\partial\xi^i$, to be evaluated by solving the following equation system:

$$\frac{\partial X^i}{\partial\xi^k}\frac{\partial\xi^k}{\partial X^j} = \delta_j^i \ ,$$

where the coefficients $\partial X^i/\partial\xi^j$ are known from the isoparametric finite element procedure. A similar conclusion can be drawn by comparing $\overset{*}{\mathbf{G}}_i$ (combined usage of x^i and ξ^i) with $\mathbf{G}_i$ (usage of Θ^i). It is apparent that the computation of $\overset{*}{\mathbf{G}}_i$ requires more efforts than that of $\mathbf{G}_i$.

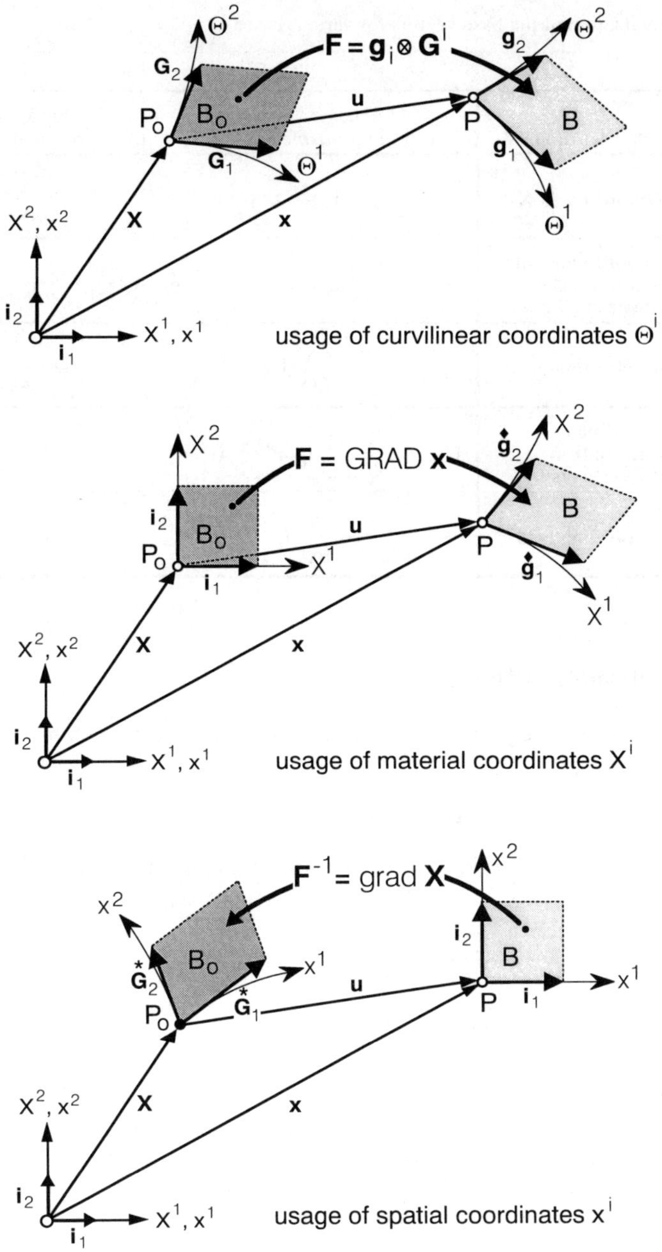

Fig. 2.6. Usage of curvilinear, material and spatial coordinates with the corresponding undeformed and deformed base vectors as 2D illustration

Table 2.1. Evaluation of the base vectors for various cases

independent coordinates	basis in the *undeformed* state	basis in the *deformed* state
material coordinates X^i	$\mathbf{i}_k$	$\dot{\mathbf{g}}_i = \dfrac{\partial x^k}{\partial X^i}\,\mathbf{i}_k$
material coordinates in combination with curvilinear ones ξ^i	$\mathbf{i}_k$	$\dot{\mathbf{g}}_i = \dfrac{\partial x^k}{\partial \xi^j}\dfrac{\partial \xi^j}{\partial X^i}\,\mathbf{i}_k = \dfrac{\partial \xi^j}{\partial X^i}\,\mathbf{g}_j$
curvilinear coordinates Θ^i	$\mathbf{G}_i = \dfrac{\partial X^k}{\partial \Theta^i}\,\mathbf{i}_k$	$\mathbf{g}_i = \dfrac{\partial x^k}{\partial \Theta^i}\,\mathbf{i}_k$
spatial coordinates x^i in combination with curvilinear ones ξ^i	$\overset{*}{\mathbf{G}}_i = \dfrac{\partial X^k}{\partial \xi^j}\dfrac{\partial \xi^j}{\partial x^i}\,\mathbf{i}_k = \dfrac{\partial \xi^j}{\partial x^i}\,\mathbf{G}_j$	$\mathbf{i}_k$
spatial coordinates	$\overset{*}{\mathbf{G}}_i = \dfrac{\partial X^k}{\partial x^i}\,\mathbf{i}_k$	$\mathbf{i}_k$

2.4 Polar decomposition

In section 2.2 we have established relations between the undeformed basis $\mathbf{G}_i$ and the deformed basis $\mathbf{g}_i$

$$\mathbf{g}_i = \mathbf{F}\,\mathbf{G}_i\,, \quad \mathbf{G}_i = \mathbf{F}^{-1}\,\mathbf{g}_i \tag{2.4.1}$$

in terms of the *deformation gradient* $\mathbf{F}$ and its inverse tensor $\mathbf{F}^{-1}$ defined by (2.2.3) and (2.2.4):

$$\mathbf{F} = \mathbf{g}_i \otimes \mathbf{G}^i = \mathrm{GRAD}\,\mathbf{x} = \frac{\partial \mathbf{x}}{\partial \mathbf{X}}\,, \tag{2.4.2}$$

$$\mathbf{F}^{-1} = \mathbf{G}_i \otimes \mathbf{g}^i = \mathrm{grad}\,\mathbf{X} = \frac{\partial \mathbf{X}}{\partial \mathbf{x}}\,. \tag{2.4.3}$$

We have also observed that $\mathbf{F}$ is not an *objective* strain measure since it does not vanish if the body is subjected to a rigid body motion. Our aim is now to introduce stretch variables $\mathbf{U}$ and $\mathbf{v}$ which will be used for the definition of objective strain measures. Furthermore we will show that any arbitrary deformation can be presented as a sum of a translation, a rotation and a deformation (length and angle changes).

To start with, we recall that any second-order tensor can be, according to (1.4.8), decomposed multiplicatively, but that such a decomposition is not unique as long as no suitable conditions are imposed on the rotation tensor $\mathbf{R}$.

Stretch tensors. The *polar decomposition theorem* states that the deformation gradient **F** can be multiplicatively decomposed in the form

$$\mathbf{F} = \mathbf{R}\,\mathbf{U} = \mathbf{v}\,\mathbf{R} \tag{2.4.4}$$

into a rotation tensor **R** and a stretch tensor **U** or **v**. From (2.4.4) we receive for the inverse tensor $\mathbf{F}^{-1}$

$$\mathbf{F}^{-1} = \mathbf{U}^{-1}\,\mathbf{R}^{\mathrm{T}} = \mathbf{R}^{\mathrm{T}}\,\mathbf{v}^{-1} \;, \tag{2.4.5}$$

since $\mathbf{R}^{\mathrm{T}} = \mathbf{R}^{-1}$.The second-order tensors **U** and **v** are called the *right stretch tensor* and the *left stretch tensor*, respectively. Both of them are *positive definite* and supposed to be *symmetric* such that

$$\mathbf{U} = \mathbf{U}^{\mathrm{T}} \;, \qquad \mathbf{v} = \mathbf{v}^{\mathrm{T}} \;. \tag{2.4.6}$$

The above requirements provide that the decompositions in (2.4.4) are defined in a unique manner as is proved e.g. in de Boer 1982. Note that a second-order tensor **U** is said to be positive definite if the inequality

$$\mathbf{y} \cdot (\mathbf{U}\,\mathbf{y}) > 0 \tag{2.4.7}$$

holds for arbitrary non-vanishing vectors **y**.

Attention is now focused on the rotation tensor **R** used in (2.4.4). By definition this tensor is *orthogonal*

$$\mathbf{R}\,\mathbf{R}^{\mathrm{T}} = \mathbf{R}^{\mathrm{T}}\,\mathbf{R} = \mathbf{I} \;. \tag{2.4.8}$$

and describes in the form

$$\tilde{\mathbf{G}}_i = \mathbf{R}\,\mathbf{G}_i \;, \qquad \tilde{\mathbf{G}}^i = \mathbf{R}\,\mathbf{G}^i \tag{2.4.9}$$

the rotation of the basis $\mathbf{G}_i$ into $\tilde{\mathbf{G}}_i$. Accordingly, **R** may be expressed as

$$\mathbf{R} = \tilde{\mathbf{G}}_i \otimes \mathbf{G}^i = \tilde{\mathbf{G}}^i \otimes \mathbf{G}_i \;. \tag{2.4.10}$$

To replace this definition by an alternative one we introduce with the help of the stretch tensor **U** the base vectors

$$\hat{\mathbf{g}}_i = \mathbf{U}\,\mathbf{G}_i \;, \qquad \hat{\mathbf{g}}^i = \mathbf{U}^{-1}\,\mathbf{G}^i \tag{2.4.11}$$

satisfying in view of the symmetry $\mathbf{U} = \mathbf{U}^{\mathrm{T}}$ the usual connection:

$$\hat{\mathbf{g}}_i \cdot \hat{\mathbf{g}}^j = (\mathbf{U}\,\mathbf{G}_i) \cdot (\mathbf{U}^{-1}\,\mathbf{G}^j) = \mathbf{G}_i \cdot (\mathbf{U}\,\mathbf{U}^{-1}\,\mathbf{G}^j) = \mathbf{G}_i \cdot \mathbf{G}^j = \delta_i^j \;. \tag{2.4.12}$$

If we consider (2.4.11) together with (2.4.4) and (2.4.1) we obtain

$$\mathbf{g}_i = \mathbf{F}\,\mathbf{G}_i = \mathbf{R}\,\mathbf{U}\,\mathbf{G}_i = \mathbf{R}\,\hat{\mathbf{g}}_i \;, \tag{2.4.13}$$

which, by virtue of (2.4.4) and (2.4.9), can be also expressed as

$$\mathbf{g}_i = \mathbf{F}\,\mathbf{G}_i = \mathbf{v}\,\mathbf{R}\,\mathbf{G}_i = \mathbf{v}\,\tilde{\mathbf{G}}_i \ . \tag{2.4.14}$$

From (2.4.13) we see that $\mathbf{g}_i$ is the rotated counterpart of $\hat{\mathbf{g}}_i$. This permits to write

$$\mathbf{R} = \mathbf{g}_i \otimes \hat{\mathbf{g}}^i = \mathbf{g}^i \otimes \hat{\mathbf{g}}_i \ . \tag{2.4.15}$$

as an alternative definition for **R**. We finally note that any orthogonal tensor may be expressed in terms of a rotation vector. Thus according to (1.10.16):

$$\mathbf{R} = \mathbf{I} + \frac{\sin\omega}{\omega}\hat{\mathbf{\Omega}} + \frac{1-\cos\omega}{\omega^2}\hat{\mathbf{\Omega}}\hat{\mathbf{\Omega}} \ , \quad \hat{\mathbf{\Omega}} = \mathbf{\Omega}\,\mathrm{x} \ , \tag{2.4.16}$$

where $\mathbf{\Omega}$ is known as RODRIGUES rotation vector.

If (2.4.10) is used to construct the rotation tensor $\mathbf{R} = \mathbf{R}(\tilde{\mathbf{G}}_i)$ the unknown base vectors $\tilde{\mathbf{G}}_i$ involved in **R** can be determined from the symmetry condition

$$\mathbf{U} = \mathbf{R}^T\mathbf{F} = \mathbf{U}^T = \mathbf{F}^T\mathbf{R} \quad \rightarrow \quad U_{ij} = U_{ji} \quad \text{for} \quad i \neq j \tag{2.4.17}$$

and the relation

$$\tilde{G}_{ij} = \tilde{\mathbf{G}}_i \cdot \tilde{\mathbf{G}}_j = \mathbf{G}_i \cdot \mathbf{G}_j = G_{ij} \tag{2.4.18}$$

due to the orthogonality of **R**. The above relations involve nine equations in accordance with the number of components determining the basis $\tilde{\mathbf{G}}_i$. Relations similar to (2.4.17) and (2.4.18) are to be used for determining $\hat{\mathbf{g}}_i$ if (2.4.15) is selected to construct **R**. On the contrary, the last formulation (2.4.16) requires only the consideration of the symmetry conditions (2.4.17) from which the three unknown components Ω_i of $\mathbf{\Omega}$ can be determined.

Interpretation of the polar decomposition theorem. We repeat the transformations (2.4.13) and (2.4.14) in the form

$$\mathbf{R}\,\mathbf{U}\,\mathbf{G}_i = \mathbf{R}\,\hat{\mathbf{g}}_i = \mathbf{g}_i \tag{2.4.19}$$

$$\mathbf{v}\,\mathbf{R}\,\mathbf{G}_i = \mathbf{v}\,\tilde{\mathbf{G}}_i = \mathbf{g}_i \tag{2.4.20}$$

to give a geometrical interpretation of the polar decomposition theorem (Fig. 2.7). From (2.4.19) it follows that the deformation of an infinitesimal volume element at **X** can be considered as the successive application of:

a stretch by the tensor **U** : $\mathbf{G}_i \rightarrow \hat{\mathbf{g}}_i$,

a rigid body rotation by **R** : $\hat{\mathbf{g}}_i \rightarrow \mathbf{g}_i$,

a translation by **u** : $\mathbf{X} \rightarrow \mathbf{x}$.

The above first two steps determine the base vectors $\mathbf{g}_i$ of the deformed state and the last step determines the origin of the base vectors $\mathbf{g}_i$.

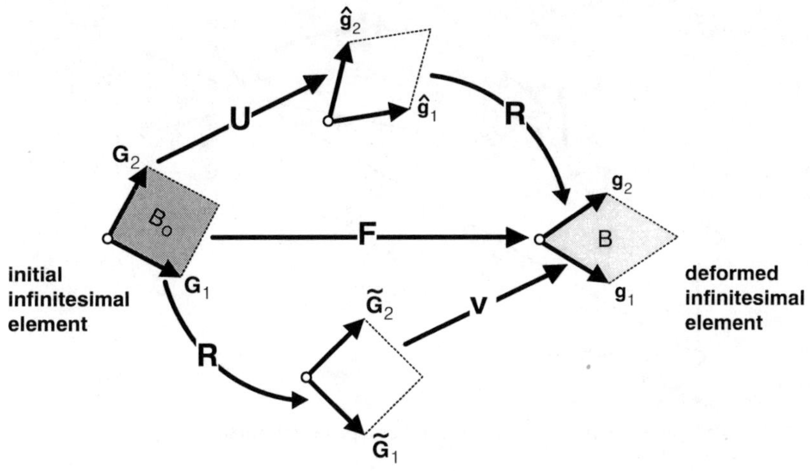

Fig. 2.7. 2D illustration of the polar decompositions $\mathbf{F} = \mathbf{RU} = \mathbf{vR}$

Alternatively, relation (2.4.20) indicates that the same deformation can be considered as the result of the successive application of

a translation by $\mathbf{u}$:	$\mathbf{X} \rightarrow \mathbf{x}$,
a rigid body rotation by $\mathbf{R}$:	$\mathbf{G}_i \rightarrow \tilde{\mathbf{G}}_i$,
a stretch by $\mathbf{v}$:	$\tilde{\mathbf{G}}_i \rightarrow \mathbf{g}_i$.

These mappings are illustrated in Fig. 2.7.

Fig. 2.8 illustrates the construction of the stretch tensor $\mathbf{U}$ according to (2.4.17). As example, a 2D deformation problem is considered, where $\mathbf{G}_3 = \mathbf{i}_3$ is supposed to be orthogonal to $\mathbf{G}_\alpha$ ($\alpha = 1, 2$) and to remain unchanged during the deformation process so that $\mathbf{g}_3 = \mathbf{G}_3 = \mathbf{i}_3$. In this case it is possible to express $\mathbf{R}$ by a rotation vector $\boldsymbol{\Omega} = \omega\, \mathbf{i}_3$ which is perpendicular to the plane (X^1–X^2) and which is determined by a single independent parameter, the rotation angle ω. Then, according to (2.4.9) and (2.4.16)

$$\tilde{\mathbf{G}}_\alpha = \mathbf{R}\,\mathbf{G}_\alpha = \mathbf{G}_\alpha + \sin\omega\,(\mathbf{i}_3 \times \mathbf{G}_\alpha) + (1 - \cos\omega)\,\mathbf{i}_3 \times (\mathbf{i}_3 \times \mathbf{G}_\alpha)\,, \quad \alpha = 1, 2\,. \quad (2.4.21)$$

By means of the deformation gradient $\mathbf{F}$ the initial basis $\mathbf{G}_\alpha$ is transformed into the deformed one, $\mathbf{g}_\alpha$. Now the question is to determine the value ω in (2.4.21) such that the following symmetry condition is fulfilled:

$$U_{12} = \tilde{\mathbf{G}}_1 \cdot \mathbf{g}_2 = U_{21} = \tilde{\mathbf{G}}_2 \cdot \mathbf{g}_1\,. \quad (2.4.22)$$

The fact that this procedure leads to a unique solution for ω can be deduced in the present simple case from geometrical considerations. Note that the expressions given in (2.4.22) for U_{12} and U_{21} will be derived in (2.4.26).

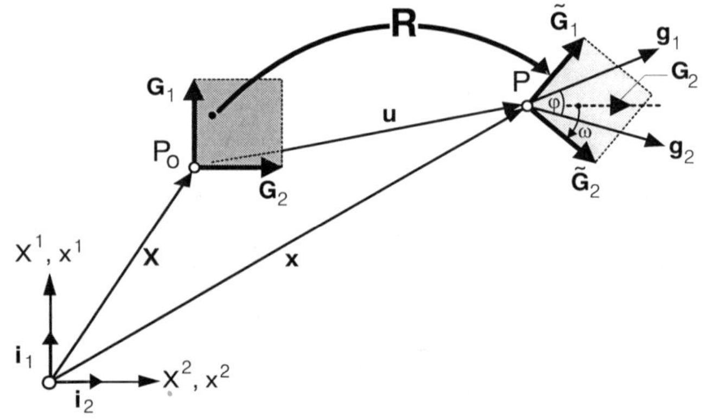

Fig. 2.8. Illustration of the polar decomposition for a plane deformation state

Application. It is to be shown that both symmetry conditions $\mathbf{U} = \mathbf{U}^T$ and $\mathbf{v} = \mathbf{v}^T$ assumed in (2.4.4) can be fulfilled for the same rotation tensor $\mathbf{R}$.

To prove this we suppose that $\mathbf{U} = \mathbf{U}^T$ with a given tensor $\mathbf{R}$ and have therefore to show that $\mathbf{v} = \mathbf{v}^T$ is valid for the same tensor $\mathbf{R}$. From (2.4.4) we have

$$\mathbf{v} = \mathbf{F}\,\mathbf{R}^T = \mathbf{R}\,\mathbf{U}\,\mathbf{R}^T$$

indicating $\mathbf{v}$ to be symmetric if $\mathbf{U} = \mathbf{U}^T$.

Component relations for the stretch tensors. The next goal is to evaluate the components of the stretch tensors

$$\mathbf{U} = \mathbf{R}^T\,\mathbf{F}\ , \tag{2.4.23}$$

$$\mathbf{v} = \mathbf{F}\,\mathbf{R}^T \tag{2.4.24}$$

introduced in (2.4.4). Inserting (2.4.2) and (2.4.10) into (2.4.23)

$$\mathbf{U} = (\mathbf{G}^i \otimes \tilde{\mathbf{G}}_i)\,(\mathbf{g}_j \otimes \mathbf{G}^j) = \tilde{\mathbf{G}}_i \cdot \mathbf{g}_j\,(\mathbf{G}^i \otimes \mathbf{G}^j) \tag{2.4.25}$$

yields for $\mathbf{U}$

$$\mathbf{U} = \mathrm{U}_{ij}\,\mathbf{G}^i \otimes \mathbf{G}^j = \tilde{\mathbf{G}}_i \cdot \mathbf{g}_j\,(\mathbf{G}^i \otimes \mathbf{G}^j)\ , \tag{2.4.26}$$

the components referring to the undeformed basis. Similarly we obtain by using (2.4.2), (2.4.3) and (2.4.15) from (2.4.24)

$$\mathbf{v} = \mathrm{v}^{ij}\,\mathbf{g}_i \otimes \mathbf{g}_j = \mathbf{G}^i \cdot \hat{\mathbf{g}}^j\,(\mathbf{g}_i \otimes \mathbf{g}_j)\ , \tag{2.4.27}$$

$$\mathbf{v}^{-1} = (\mathrm{v}^{-1})_{ij}\,\mathbf{g}^i \otimes \mathbf{g}^j = \hat{\mathbf{g}}_i \cdot \mathbf{G}_j\,(\mathbf{g}^i \otimes \mathbf{g}^j)\ . \tag{2.4.28}$$

Note that, in view of (2.4.9), (2.4.13), (2.4.26) and the symmetry $\mathrm{U}_{ij} = \mathrm{U}_{ji}$ the equality

$$(v^{-1})_{ij} = \hat{g}_i \cdot G_j = (R^T g_i) \cdot G_j = g_i \cdot (R\, G_j) = g_i \cdot \tilde{G}_j = U_{ji} = U_{ij} \qquad (2.4.29)$$

holds for the pure covariant components U_{ij} and $(v^{-1})_{ij}$. In contrast to U_{ij} the components introduced in (2.4.27) and (2.4.28) are defined with respect to the deformed basis. It is in principle possible to evaluate also for **U** components referring to the deformed basis. But the tensor **U** will be only used with components U_{ij} defined in the undeformed basis and it is therefore called *material* tensor. In this sense, **v** forms a *spatial* variable and will be used with components v^{ij} referring to the deformed basis. The basis used for the definition of components is an essential characteristics to distinguish between *material* and *spatial* variables.

Strain tensors related to the stretch tensors. The right stretch tensor **U** is the basis for the definition of the *BIOT strain tensor* **H**:

$$\mathbf{H} := H_{ij}\, \mathbf{G}^i \otimes \mathbf{G}^j = \mathbf{U} - \mathbf{G} \ , \qquad (2.4.30)$$

whose components are according to (2.4.26) given by

$$H_{ij} = U_{ij} - G_{ij} = \tilde{\mathbf{G}}_i \cdot \mathbf{g}_j - \mathbf{G}_i \cdot \mathbf{G}_j \ . \qquad (2.4.31)$$

Similarly, the left stretch tensor $\mathbf{v}^{-1}$ may be used to define the spatial counterpart of **H**

$$\mathbf{h} := h_{ij}\, \mathbf{g}^i \otimes \mathbf{g}^j = \mathbf{g} - \mathbf{v}^{-1} \qquad (2.4.32)$$

the components of which, in view of (2.4.28), are

$$h_{ij} = g_{ij} - (v^{-1})_{ij} = \mathbf{g}_i \cdot \mathbf{g}_j - \hat{\mathbf{g}}_i \cdot \mathbf{G}_j \ . \qquad (2.4.33)$$

Both tensors **H** and **h** are *symmetric* and *objective* strain measures.

Right and left CAUCHY-GREEN tensors. We again refer to the decompositions (2.4.4) in terms of the symmetric stretches **U** and **v** in order to introduce by considering the orthogonality condition $\mathbf{R}^T = \mathbf{R}^{-1}$ the following symmetric second-order tensors

$$\mathbf{C} := \mathbf{F}^T \mathbf{F} = (\mathbf{R}\,\mathbf{U})^T \mathbf{R}\,\mathbf{U} = \mathbf{U}^T \mathbf{R}^T \mathbf{R}\,\mathbf{U} = \mathbf{U}^T \mathbf{U} = \mathbf{U}^2 \ , \qquad (2.4.34)$$

$$\mathbf{b} := \mathbf{F}\,\mathbf{F}^T = \mathbf{v}\,\mathbf{R}\,(\mathbf{v}\,\mathbf{R})^T = \mathbf{v}\,\mathbf{R}\,\mathbf{R}^T \mathbf{v}^T = \mathbf{v}\,\mathbf{v}^T = \mathbf{v}^2 \ , \qquad (2.4.35)$$

which are called the *right CAUCHY-GREEN tensor* **C** and the *left CAUCHY-GREEN tensor* **b**. The last one is also referred to as *FINGER* tensor. From the above definitions it is obvious that both tensors **b** and **C** describe deformations without being influenced by a pure rotation. Therefore they are suitable variables for the definition of objective strain measures which will be introduced in the next section.

In view of (2.4.34), (2.4.35) the tensors **C** and **b** are related by

$$\mathbf{C} = \mathbf{F}^{-1}\,\mathbf{b}\,\mathbf{F} \ , \quad \mathbf{C}^{-1} = \mathbf{F}^{-1}\,\mathbf{b}^{-1}\,\mathbf{F} \ , \qquad (2.4.36)$$

$$\mathbf{b} = \mathbf{F}\,\mathbf{C}\,\mathbf{F}^{-1} \ , \quad \mathbf{b}^{-1} = \mathbf{F}\,\mathbf{C}^{-1}\,\mathbf{F}^{-1} \ . \qquad (2.4.37)$$

By considering in addition the identity (1.3.51)

$$\mathrm{tr}\,(\mathbf{S}\,\mathbf{T}) = \mathrm{tr}\,(\mathbf{T}\,\mathbf{S}) \tag{2.4.38}$$

it can be also verified that

$$\mathrm{tr}\,\mathbf{C} = \mathrm{tr}\,(\mathbf{F}^{T}\,\mathbf{F}) = \mathrm{tr}\,(\mathbf{F}\,\mathbf{F}^{T}) = \mathrm{tr}\,\mathbf{b}\ , \tag{2.4.39}$$

which holds in a similar form for arbitrary powers $\mathbf{C}^n$ and $\mathbf{b}^n$

$$\mathrm{tr}\,\mathbf{C}^{n} = \mathrm{tr}\,\mathbf{b}^{n}\ . \tag{2.4.40}$$

By considering this result it can be deduced from (1.8.1) to (1.8.3) and (1.8.6) that the invariants of **C** and **b** are equal:

$$\begin{aligned} I_{\mathbf{b}} &= I_{\mathbf{C}} = \mathrm{tr}\,\mathbf{C}\ , \\ II_{\mathbf{b}} &= II_{\mathbf{C}} = \frac{1}{2}\left[(\mathrm{tr}\,\mathbf{C})^{2} - \mathrm{tr}\,\mathbf{C}^{2}\right]\ , \\ III_{\mathbf{b}} &= III_{\mathbf{C}} = \det\mathbf{C} = \frac{1}{3}\left[\mathrm{tr}\,\mathbf{C}^{3} - \frac{3}{2}\,\mathrm{tr}\,\mathbf{C}^{2}\,\mathrm{tr}\,\mathbf{C} + \frac{1}{2}\,(\mathrm{tr}\,\mathbf{C})^{3}\right]\ . \end{aligned} \tag{2.4.41}$$

Replacing **C** by **b** the above expressions hold for $I_{\mathbf{b}}$, $II_{\mathbf{b}}$ and $III_{\mathbf{b}}$. By considering the identity $\det\mathbf{A} = \det\mathbf{A}^{T}$ valid for any second-order tensor **A** and (2.4.34), $III_{\mathbf{C}}$ may be given alternatively by

$$III_{\mathbf{C}} = \det\mathbf{C} = \det(\mathbf{F}^{T}\,\mathbf{F}) = \det\mathbf{F}^{T}\det\mathbf{F} = (\det\mathbf{F})^{2} = III_{\mathbf{b}}\ . \tag{2.4.42}$$

This permits to express the Jacobian (2.2.31) in the following forms

$$J = \frac{dV}{dV_{o}} = \sqrt{\frac{g}{G}} = \det\mathbf{F} = \sqrt{III_{\mathbf{C}}} = \sqrt{III_{\mathbf{b}}}\ . \tag{2.4.43}$$

From the equalities (2.4.41) we deduce that **b** and **C** have the same eigenvalues to be denoted by Λ_i (i = 1, 2, 3).

A further important relation is according to (2.4.34):

$$\mathbf{C}\,\mathbf{U} = \mathbf{U}^{3} = \mathbf{U}\,\mathbf{C}\ , \tag{2.4.44}$$

which, in view of (1.9.19), indicates that the tensors **C** and **U** are *coaxial*. Thus we may state the following:

Remark. The principal directions of **U** coincide with those of **C**; each principal value λ_i of **U** is, in view of $\mathbf{C} = \mathbf{U}^2$, the positive square root of the corresponding (positive) principal value Λ_i of **C**: $\lambda_i = \sqrt{\Lambda_i}$. Similarly the principal values λ_i of **v** are, in view of $\mathbf{b} = \mathbf{v}^2$, given by $\lambda_i = \sqrt{\Lambda_i}$ and the principal axes of **v** coincide with those of **b** (and $\mathbf{b}^{-1}$).

Components of C and b. The components of the tensors $\mathbf{C}$, $\mathbf{b}$ and $\mathbf{C}^{-1}$, $\mathbf{b}^{-1}$ can be evaluated from (2.4.34) and (2.4.35) using the expressions given in (2.4.2), (2.4.3) for $\mathbf{F}$ and $\mathbf{F}^{-1}$. The results are

$$\mathbf{C} := C_{ij}\,\mathbf{G}^i \otimes \mathbf{G}^j = \mathbf{g}_i \cdot \mathbf{g}_j\,(\mathbf{G}^i \otimes \mathbf{G}^j)\ ,$$

$$\mathbf{C}^{-1} := (C^{-1})^{ij}\,\mathbf{G}_i \otimes \mathbf{G}_j = \mathbf{g}^i \cdot \mathbf{g}^j\,(\mathbf{G}_i \otimes \mathbf{G}_j)\ , \tag{2.4.45}$$

and

$$\mathbf{b} := b^{ij}\,\mathbf{g}_i \otimes \mathbf{g}_j = \mathbf{G}^i \cdot \mathbf{G}^j\,\mathbf{g}_i \otimes \mathbf{g}_j\ ,$$

$$\mathbf{b}^{-1} := (b^{-1})_{ij}\,\mathbf{g}^i \otimes \mathbf{g}^j = \mathbf{G}_i \cdot \mathbf{G}_j\,\mathbf{g}^i \otimes \mathbf{g}^j\ . \tag{2.4.46}$$

Application. Prove the following identities

$$(\mathbf{F}^m)^T = (\mathbf{F}^T)^m\ ,\quad \mathrm{tr}\,(\mathbf{F}^T)^m = \mathrm{tr}\,\mathbf{F}^m\ ,\quad \det\mathbf{F} = \det\mathbf{F}^T\ .$$

We first have

$$(\mathbf{F}^m)^T = (\mathbf{F}\,\mathbf{F}\,\ldots\,\mathbf{F})^T = \mathbf{F}^T\,\mathbf{F}^T\,\ldots\,\mathbf{F}^T = (\mathbf{F}^T)^m\ . \tag{2.4.47}$$

Using this result and the identity $\mathrm{tr}\,\mathbf{A} = \mathrm{tr}\,\mathbf{A}^T$ by (1.3.51) we obtain

$$\mathrm{tr}\,(\mathbf{F}^T)^m = \mathrm{tr}\,(\mathbf{F}^m)^T = \mathrm{tr}\,\mathbf{F}^m\ . \tag{2.4.48}$$

Recall that the relation (1.8.6) holds for any second-order tensor $\mathbf{A}$. If we specify this equation for $\mathbf{F}$ and $\mathbf{F}^T$ and compare the corresponding results with each other we see in view of (2.4.48) that

$$\mathrm{III}_{\mathbf{F}} = \det\mathbf{F} = \mathrm{III}_{\mathbf{F}^T} = \det\mathbf{F}^T\ , \tag{2.4.49}$$

which has already been considered in the derivation of (2.4.42). Note that the relations (2.4.47) to (2.4.49) hold for arbitrary second-order tensors.

Application. Let $\Psi = \Psi\,(\mathrm{I}_\mathbf{C}, \mathrm{II}_\mathbf{C}, \mathrm{III}_\mathbf{C})$ be a scalar-valued function of the invariants $\mathrm{I}_\mathbf{C}$, $\mathrm{II}_\mathbf{C}$ and $\mathrm{III}_\mathbf{C}$ of the right CAUCHY-GREEN tensor $\mathbf{C}$. Construct the partial derivatives of the function Ψ with respect to $\mathbf{C}$ and $\mathbf{b}$ and show that $\Psi_{,\mathbf{b}}\,\mathbf{b} = \mathbf{F}\,\Psi_{,\mathbf{C}}\,\mathbf{F}^T$.

By means of the chain rule we may write

$$\Psi_{,\mathbf{C}} = \frac{\partial\Psi}{\partial\mathbf{C}} = \frac{\partial\Psi}{\partial\mathrm{I}_\mathbf{C}}\frac{\partial\mathrm{I}_\mathbf{C}}{\partial\mathbf{C}} + \frac{\partial\Psi}{\partial\mathrm{II}_\mathbf{C}}\frac{\partial\mathrm{II}_\mathbf{C}}{\partial\mathbf{C}} + \frac{\partial\Psi}{\partial\mathrm{III}_\mathbf{C}}\frac{\partial\mathrm{III}_\mathbf{C}}{\partial\mathbf{C}}\ ,$$

which, by considering (1.8.7), (1.8.8) and (1.8.10), becomes:

$$\Psi_{,\mathbf{C}} = \left(\frac{\partial\Psi}{\partial\mathrm{I}_\mathbf{C}} + \frac{\partial\Psi}{\partial\mathrm{II}_\mathbf{C}}\mathrm{I}_\mathbf{C}\right)\mathbf{G} - \frac{\partial\Psi}{\partial\mathrm{II}_\mathbf{C}}\mathbf{C} + \frac{\partial\Psi}{\partial\mathrm{III}_\mathbf{C}}\mathrm{III}_\mathbf{C}\,\mathbf{C}^{-1}\ . \tag{2.4.50}$$

By means of (2.4.41) the function $\Psi = \Psi\,(\mathrm{I}_\mathbf{b}, \mathrm{II}_\mathbf{b}, \mathrm{III}_\mathbf{b})$ remains unchanged if the invariants of $\mathbf{C}$ are replaced by those of $\mathbf{b}$. Consequently, an expression similar to (2.4.50)

$$\Psi_{,\mathbf{b}} = \left(\frac{\partial\Psi}{\partial\mathrm{I}_\mathbf{b}} + \frac{\partial\Psi}{\partial\mathrm{II}_\mathbf{b}}\mathrm{I}_\mathbf{b}\right)\mathbf{g} - \frac{\partial\Psi}{\partial\mathrm{II}_\mathbf{b}}\mathbf{b} + \frac{\partial\Psi}{\partial\mathrm{III}_\mathbf{b}}\mathrm{III}_\mathbf{b}\,\mathbf{b}^{-1} \tag{2.4.51}$$

holds also for $\Psi,_{\mathbf{b}}$. If we now pre-contract (2.4.50) by $\mathbf{F}$, post-contract by $\mathbf{F}^T$ and consider that, according to (2.4.34) and (2.4.35),

$$\mathbf{F}\,\mathbf{G}\,\mathbf{F}^T = \mathbf{b}\ ,\quad \mathbf{F}\,\mathbf{C}\,\mathbf{F}^T = \mathbf{b}^2\ ,\quad \mathbf{F}\,\mathbf{C}^{-1}\,\mathbf{F}^T = \mathbf{g} \tag{2.4.52}$$

we end up in view of (2.4.41), (2.4.51) with the identity

$$\Psi,_{\mathbf{b}}\,\mathbf{b} = \mathbf{F}\,\Psi,_{\mathbf{C}}\,\mathbf{F}^T\ . \tag{2.4.53}$$

From (2.4.51) we also see that

$$\Psi,_{\mathbf{b}}\,\mathbf{b} = \mathbf{b}\,\Psi,_{\mathbf{b}} \tag{2.4.54}$$

indicating $\Psi,_{\mathbf{b}}$ and $\mathbf{b}$ to be coaxial. Finally, we note that the expressions

$$\mathbf{S} = 2\rho_0 \frac{\partial\Psi}{\partial\mathbf{C}}\ ,\quad \boldsymbol{\tau} = 2\rho_0\,\mathbf{b}\frac{\partial\Psi}{\partial\mathbf{b}} = 2\rho_0 \frac{\partial\Psi}{\partial\mathbf{b}}\,\mathbf{b} \tag{2.4.55}$$

with the density ρ_0 in the undeformed configuration define the second PIOLA-KIRCHHOFF stress tensor $\mathbf{S}$ and the KIRCHHOFF stress tensor $\boldsymbol{\tau}$, respectively, if the function Ψ is identified as the specific elastic strain energy of the considered body. This feature will be confirmed in chapter 6.

2.5 GREEN-LAGRANGE strain tensor, ALMANSI strain tensor

In (2.4.34), (2.4.35) we have introduced the symmetric tensors $\mathbf{C}$ and $\mathbf{b}$ which may be expressed alternatively in terms of the deformation gradient $\mathbf{F}$ or the stretch tensors $\mathbf{U}$ and $\mathbf{v}$ as follows:

$$\mathbf{C} = \mathbf{F}^T\,\mathbf{F} = \mathbf{U}^2\ , \tag{2.5.1}$$

$$\mathbf{b} = \mathbf{F}\,\mathbf{F}^T = \mathbf{v}^2\ . \tag{2.5.2}$$

The above expressions will be employed in the following to define objective strain measures. We further recall the relations (2.1.21) and (2.1.22)

$$\mathbf{g}_i = \mathbf{G}_i + \mathbf{u},_i = (G_{ji} + U_j|_i)\,\mathbf{G}^j = F_{ji}\,\mathbf{G}^j\ , \tag{2.5.3}$$

$$\mathbf{G}_i = \mathbf{g}_i - \mathbf{u},_i = (g_{ji} - u_j||_i)\,\mathbf{g}^j = (f^{-1})_{ji}\,\mathbf{g}^j \tag{2.5.4}$$

permitting to express the deformed basis $\mathbf{g}_i$ and the undeformed basis $\mathbf{G}_i$ in terms of the displacement vector $\mathbf{u}$.

We now consider the infinitesimal material vector $d\mathbf{X} = \mathbf{G}_i\, d\Theta^i$ at $\mathbf{X}$ which is transformed after deformation into $d\mathbf{x} = \mathbf{g}_i\, d\Theta^i$ at $\mathbf{x}$. The transformations between $d\mathbf{X}$ and $d\mathbf{x}$ are given according to (2.2.10) by

$$d\mathbf{x} = \mathbf{F}\, d\mathbf{X}\ ,\quad d\mathbf{X} = \mathbf{F}^{-1}\, d\mathbf{x}\ . \tag{2.5.5}$$

Let dS be the length of $d\mathbf{X}$ and let ds be the length of $d\mathbf{x}$. Thus we may write by considering (2.5.1) and (2.5.2):

$$ds^2 = d\mathbf{x}\cdot d\mathbf{x} = (d\mathbf{X}\,\mathbf{F}^T)\cdot(\mathbf{F}\,d\mathbf{X})$$
$$= d\mathbf{X}\,(\mathbf{F}^T\mathbf{F})\,d\mathbf{X} = d\mathbf{X}\,\mathbf{C}\,d\mathbf{X}\ , \tag{2.5.6}$$

$$dS^2 = d\mathbf{X}\cdot d\mathbf{X} = (d\mathbf{x}\,\mathbf{F}^{-T})\cdot(\mathbf{F}^{-1}\,d\mathbf{x})$$
$$= d\mathbf{x}\,\mathbf{F}^{-T}\,\mathbf{F}^{-1}\,d\mathbf{x} = d\mathbf{x}\,\mathbf{b}^{-1}\,d\mathbf{x}\ . \tag{2.5.7}$$

From the above expressions we see that the deformation tensor $\mathbf{C}$, which presents a material tensor, gives the squared length ds^2 of the line-element $d\mathbf{x}$ into which the given line-element $d\mathbf{X}$ is deformed. On the contrary, the deformation tensor $\mathbf{b}^{-1}$ being a spatial variable describes the initial squared length dS^2 of a line-element $d\mathbf{x}$ identified in the deformed configuration.

GREEN-LAGRANGE strain tensor E. For its definition we form by considering (2.5.6) and (2.5.7)

$$ds^2 - dS^2 = d\mathbf{X}\,(\mathbf{F}^T\mathbf{F} - \mathbf{G})\,d\mathbf{X} = d\mathbf{X}\,(\mathbf{C}-\mathbf{G})\,d\mathbf{X} = d\mathbf{X}\,(2\mathbf{E})\,d\mathbf{X} \tag{2.5.8}$$

from which we receive for $\mathbf{E}$ the following expression

$$\mathbf{E} := E_{ij}\,\mathbf{G}^i\otimes\mathbf{G}^j = \frac{1}{2}(\mathbf{F}^T\mathbf{F}-\mathbf{G}) = \frac{1}{2}(\mathbf{C}-\mathbf{G}) \tag{2.5.9}$$

in terms of the right CAUCHY-GREEN tensor $\mathbf{C}$ and the identity tensor $\mathbf{G}$:

$$\mathbf{G} := G_{ij}\,\mathbf{G}^i\otimes\mathbf{G}^j = \mathbf{G}_i\otimes\mathbf{G}^i = \mathbf{I}\ . \tag{2.5.10}$$

Thus, the GREEN-LAGRANGE strain tensor $\mathbf{E}$ turns out to be a tensor which gives the change in the squared length of the material vector $d\mathbf{X}$ identified in the undeformed state B_o. In other words, $\mathbf{E}$ enables to evaluate changes of length for an observer related to the undeformed state. This justifies to call $\mathbf{E}$ a "material strain tensor".

Accordingly, the strain tensor $\mathbf{E}$ will be used with components E_{ij} defined with respect to the undeformed state. For their calculation we introduce (2.5.10) together with the expression (2.2.3):

$$\mathbf{F} = \mathbf{g}_i\otimes\mathbf{G}^i \tag{2.5.11}$$

into (2.5.9)

$$\mathbf{E} := E_{ij}\,\mathbf{G}^i\otimes\mathbf{G}^j = \frac{1}{2}\Big[(\mathbf{G}^i\otimes\mathbf{g}_i)(\mathbf{g}_j\otimes\mathbf{G}^j) - G_{ij}\,(\mathbf{G}^i\otimes\mathbf{G}^j)\Big]$$

to obtain by considering (2.5.3) the following results

$$E_{ij} = \frac{1}{2}(\mathbf{g}_i\cdot\mathbf{g}_j - \mathbf{G}_i\cdot\mathbf{G}_j) = \frac{1}{2}(g_{ij} - G_{ij})$$
$$= \frac{1}{2}(\mathbf{G}_i\cdot\mathbf{u}_{,j} + \mathbf{G}_j\cdot\mathbf{u}_{,i} + \mathbf{u}_{,i}\cdot\mathbf{u}_{,j}) = \frac{1}{2}(U_i|_j + U_j|_i + U_r|_i\,U^r|_j)\ . \tag{2.5.12}$$

The first expression in terms of $\mathbf{g}_i = \mathbf{x},_i$ and $\mathbf{G}_i = \mathbf{X},_i$ is of particular interest for the isoparametric finite element formulation. If LAGRANGIAN coordinates $\Theta^i \to X^i$ are used to describe the deformation state, then

$$\mathbf{G}_i = \mathbf{i}_i \ , \quad \mathbf{g}_i = \frac{\partial \mathbf{x}}{\partial X^i} = \frac{\partial x^k}{\partial X^i}\,\mathbf{i}_k$$

so that (2.5.12) becomes:

$$E_{ij} = \frac{1}{2}\left[\frac{\partial x^k}{\partial X^i}\frac{\partial x^m}{\partial X^j}\,\delta_{km} - \delta_{ij}\right] . \qquad (2.5.13)$$

The GREEN strain tensor $\mathbf{E}$ is an *objective* strain measure. If the body in its initial state B_o is subjected to a translation, then $\mathbf{F} = \mathbf{G}$. For a pure rotation we have $\mathbf{F} = \mathbf{R}$. In both cases it can easily confirmed by virtue of (2.5.9) that $\mathbf{E} = \mathbf{0}$.

ALMANSI strain tensor e. For its definition the difference $ds^2 - dS^2$ is now to be expressed, in contrast to (2.5.8), in terms of the spatial vector $d\mathbf{x}$. Using (2.5.6) and (2.5.7) we obtain

$$ds^2 - dS^2 = d\mathbf{x}\,(\mathbf{g} - \mathbf{F}^{-T}\,\mathbf{F}^{-1})\,d\mathbf{x} = d\mathbf{x}\,(\mathbf{g} - \mathbf{b}^{-1})\,d\mathbf{x} = d\mathbf{x}\,(2\,\mathbf{e})\,d\mathbf{x} \ , \qquad (2.5.14)$$

where

$$\mathbf{e} := e_{ij}\,\mathbf{g}^i \otimes \mathbf{g}^j = \frac{1}{2}(\mathbf{g} - \mathbf{F}^{-T}\,\mathbf{F}^{-1}) = \frac{1}{2}(\mathbf{g} - \mathbf{b}^{-1}) \qquad (2.5.15)$$

is the so-called *ALMANSI strain tensor* and

$$\mathbf{g} := g_{ij}\,\mathbf{g}^i \otimes \mathbf{g}^j = \mathbf{g}_i \otimes \mathbf{g}^i = \mathbf{I} \qquad (2.5.16)$$

abbreviates the identity tensor related to the deformed basis $\mathbf{g}_i$. In contrast to $\mathbf{E}$, the ALMANSI tensor $\mathbf{e}$ permits to evaluate changes of length if a material element $d\mathbf{x}$ is identified in the actual deformed state. Accordingly, $\mathbf{e}$ is a *spatial* tensor. In this context we recall that the components of all spatial tensors will by defined with respect to the deformed basis and denoted systematically by lower case letters, e.g. for $\mathbf{e}$ by e_{ij}.

The components of $\mathbf{e}$ can be evaluated by the same procedure as has been used above for the derivation of the expressions (2.5.12). In view of (2.4.46) and (2.5.4), (2.5.15), the corresponding results are of the form:

$$\begin{aligned} e_{ij} &= \frac{1}{2}(\mathbf{g}_i \cdot \mathbf{g}_j - \mathbf{G}_i \cdot \mathbf{G}_j) &&= \frac{1}{2}(g_{ij} - G_{ij}) \\ &= \frac{1}{2}(\mathbf{g}_i \cdot \mathbf{u},_j + \mathbf{g}_j \cdot \mathbf{u},_i - \mathbf{u},_i \cdot \mathbf{u},_j) &&= \frac{1}{2}(u_i\|_j + u_j\|_i - u_r\|_i\, u^r\|_j) \ , \end{aligned} \qquad (2.5.17)$$

where the covariant derivatives – denoted by double stroke $(\ldots)\|_i$ – refer to the deformed basis. If spatial coordinates $\Theta^i \to x^i$ are used:

$$\mathbf{g}_i = \mathbf{i}_i \,, \quad \mathbf{G}_i = \frac{\partial \mathbf{X}}{\partial x^i} = \frac{\partial X^m}{\partial x^i}\mathbf{i}_m \tag{2.5.18}$$

the above result takes the form

$$e_{ij} = \frac{1}{2}\left(\delta_{ij} - \frac{\partial X^m}{\partial x^i}\frac{\partial X^n}{\partial x^j}\delta_{mn}\right) . \tag{2.5.19}$$

The comparison of equations (2.5.12) and (2.5.17) shows that the pure covariant components e_{ij} and E_{ij} are identical,

$$e_{ij} = E_{ij} = \frac{1}{2}(g_{ij} - G_{ij}) \,, \quad e^j_i \neq E^j_i \,, \quad e^{ij} \neq E^{ij} \tag{2.5.20}$$

a similar statement being however not valid for other types of components.

Finally we add the relations

$$2\,E_{ij} = C_{ij} - G_{ij} \,, \tag{2.5.21}$$

$$2\,e_{ij} = g_{ij} - (b^{-1})_{ij} \tag{2.5.22}$$

obtained from (2.4.45), (2.4.46) and (2.5.9), (2.5.15).

Remark. Since both **C** and **E** are symmetric tensors, they each have three real principal values connected with three principal directions. Equation (2.5.21) indicates that the principal axes of the two tensors **C** and **E** coincide, since if the coordinate axes are selected such that the components C_{ij} for $i \neq j$ vanish, then the components E_{ij} will be also zero for $i \neq j$, referring to the same axes.

A similar conclusion holds for the two tensors $\mathbf{b}^{-1}$ and **e** whose principal axes coincide at the spatial point **x**, but which are in general not parallel to the principal axes of **C** and **E** at **X**. In section 2.6 we will observe that the rotation tensor **R** used in the polar decomposition theorem (2.4.4) rotates the principal axes of **C** at **X** into the principal axes of $\mathbf{b}^{-1}$ at **x**.

To complete this section we will deal briefly with the mechanical interpretation of the strain tensors introduced above, particularly of their components. For this purpose we first define the stretch in a given direction of the undeformed body.

Stretch. The *stretch* λ of a material line-element d**X** transformed after deformation into d**x** is defined as ratio $\lambda = ds/dS$, dS and ds being the length of d**X** and d**x**, respectively. To evaluate λ we use for the unit vector in the direction of d**X** the notation

$$\mathbf{N} := N^i\,\mathbf{G}_i = \frac{d\mathbf{X}}{dS} \; . \tag{2.5.23}$$

Considering, in addition, the BIOT strain tensor **H** (2.4.30)

$$\mathbf{H} = \mathbf{U} - \mathbf{G} \quad \rightarrow \quad H_{ij} = U_{ij} - G_{ij} \tag{2.5.24}$$

and using equations (2.5.1), (2.5.6), (2.5.9) we then receive

$$\lambda = \frac{ds}{dS} = \sqrt{\frac{d\mathbf{X}}{dS}\,\mathbf{C}\,\frac{d\mathbf{X}}{dS}} = \sqrt{\mathbf{N}\,\mathbf{C}\,\mathbf{N}} = \sqrt{\mathbf{N}\,(2\,\mathbf{E} + \mathbf{G})\,\mathbf{N}}$$

$$= \sqrt{\mathbf{N}\,\mathbf{U}^2\,\mathbf{N}} = \sqrt{\mathbf{N}\,(\mathbf{H} + \mathbf{G})^2\,\mathbf{N}} \ , \tag{2.5.25}$$

or in component form

$$\lambda = \sqrt{N^i\,C_{ij}\,N^j} = \sqrt{N^i\,(2\,E_{ij} + G_{ij})\,N^j}$$

$$= \sqrt{N^i\,U_{ik}\,U^k_j\,N^j} = \sqrt{N^i\,(H_{ik}\,H^k_j + 2\,H_{ij} + G_{ij})\,N^j} \ . \tag{2.5.26}$$

Similarly, the ratio $1/\lambda$ can also be determined. By using (2.5.7) and the unit vector

$$\mathbf{n} := n^i\,\mathbf{g}_i = \frac{d\mathbf{x}}{ds} \tag{2.5.27}$$

in direction of $d\mathbf{x}$ as abbreviation we obtain

$$\frac{1}{\lambda} = \frac{dS}{ds} = \sqrt{\frac{d\mathbf{x}}{ds}\,\mathbf{b}^{-1}\,\frac{d\mathbf{x}}{ds}} = \sqrt{\mathbf{n}\,\mathbf{b}^{-1}\,\mathbf{n}} = \sqrt{\mathbf{n}\,(\mathbf{g} - 2\,\mathbf{e})\,\mathbf{n}}$$

$$= \sqrt{\mathbf{n}\,(\mathbf{v}^2)^{-1}\,\mathbf{n}} = \sqrt{\mathbf{n}\,(\mathbf{g} - \mathbf{h})^2\,\mathbf{n}} \tag{2.5.28}$$

or in component form

$$\frac{1}{\lambda} = \sqrt{n^i\,(b^{-1})_{ij}\,n^j} = \sqrt{n^i\,(g_{ij} - 2\,e_{ij})n^j} \tag{2.5.29}$$

$$= \sqrt{n^i\,(v^{-1})_{ik}\,(v^{-1})^k_j\,n^j} = \sqrt{n^i(h_{ik}\,h^k_j - 2h_{ij} + g_{ij})\,n^j} \ ,$$

where relations (2.5.2), (2.5.15) and (2.4.32) have been also considered. We recall that all tensor components appearing in (2.5.29) refer to the deformed basis and are denoted therefore by lower case letters while those occuring in (2.5.26) are defined with respect to the undeformed basis.

Unit extension. The unit extension ε of an undeformed element $d\mathbf{X}$ occupying the position $d\mathbf{x}$ in the deformed state is defined by

$$\varepsilon = \frac{ds - dS}{dS} = \frac{ds}{dS} - 1 = \lambda - 1 \ , \tag{2.5.30}$$

where λ denotes the stretch introduced in (2.5.25). To gain more insight into the geometrical meaning of the expressions (2.5.26) we evaluate ε first for the case where the unit vector $\mathbf{N}$ shows in the direction of the base vector $\mathbf{G}_i$

$$\mathbf{N} := N^i\,\mathbf{G}_i = \mathbf{G}_{<i>} = \frac{\mathbf{G}_i}{\sqrt{G_{ii}}} \quad \rightarrow \quad N^i = \frac{1}{\sqrt{G_{ii}}} \ . \tag{2.5.31}$$

Thus we find from (2.5.26) and (2.5.30)

$$\varepsilon_{<i>} = \lambda_{<i>} - 1 = \sqrt{\frac{C_{ii}}{G_{ii}}} - 1 = \sqrt{2\frac{E_{ii}}{G_{ii}} + 1} - 1$$

$$= \sqrt{\frac{U_{ik}\,U_i^k}{G_{ii}}} - 1 = \sqrt{\left(\frac{H_{ik}\,H_i^k + 2\,H_{ii} + G_{ii}}{G_{ii}}\right)} - 1 \qquad (2.5.32)$$

where $\varepsilon_{<i>}$ and $\lambda_{<i>}$ are the values of ε and λ for $\mathbf{N} = \mathbf{G}_{<i>}$. The above result (2.5.32) can be easily specified to the case where the base vectors $\mathbf{G}_i$ coincide with the unit eigenvectors $\mathbf{N}_i$ of the tensor $\mathbf{C}$. Denoting the components of the *coaxial* tensors $\mathbf{C}$, $\mathbf{E}$, $\mathbf{U}$ and $\mathbf{H}$ with respect to $\mathbf{N}_i \otimes \mathbf{N}_i$ by Λ_i, E_i, λ_i and H_i:

$$\mathbf{C} = \sum_{i=1}^{3} \Lambda_i\,\mathbf{N}_i \otimes \mathbf{N}_i\ , \quad \mathbf{E} = \sum_{i=1}^{3} E_i\,\mathbf{N}_i \otimes \mathbf{N}_i\ ,$$

$$\mathbf{U} = \sum_{i=1}^{3} \lambda_i\,\mathbf{N}_i \otimes \mathbf{N}_i\ , \quad \mathbf{H} = \sum_{i=1}^{3} H_i\,\mathbf{N}_i \otimes \mathbf{N}_i \qquad (2.5.33)$$

and considering that, in this case, $G_{ii} = 1$ we obtain

$$\varepsilon_i = \sqrt{\Lambda_i} - 1 = \sqrt{2\,E_i + 1} - 1 = \lambda_i - 1 = H_i\ , \qquad (2.5.34)$$

or

$$E_i = \varepsilon_i + \frac{1}{2}(\varepsilon_i)^2 = H_i + \frac{1}{2}(H_i)^2\ , \qquad (2.5.35)$$

where ε_i is the unit extension in direction of $\mathbf{N}_i$. If the deformations are infinitesimal then the quadratic terms in ε_i and H_i can be neglected in (2.5.35) showing that for this special case the distinction between the tensors $\mathbf{H}$ and $\mathbf{E}$ is irrelevant. From (2.5.34) we also deduce that the eigenvalues of $\mathbf{H}$ give the extensions in direction of the eigenvectors $\mathbf{N}_i$ while the eigenvalues λ_i of $\mathbf{U}$ correspond to the stretches in the same directions.

2.6 Eigenvectors and eigenvalues of deformation variables

As has been discussed in section 1.9 in a general form any symmetric second-order tensor $\mathbf{T}$ possesses three real eigenvalues λ_i and three orthogonal principal axes which are determined by the eigenvectors $\mathbf{N}_i$ of the associated eigenvalue problem. If the tensor $\mathbf{T}$ refers to the orthonormal basis $\mathbf{N}_i$

$$\mathbf{T} = \hat{T}^{ij}\,\mathbf{N}_i \otimes \mathbf{N}_j\ , \qquad (2.6.1)$$

it possesses only three nonvanishing components

$$\hat{T}^{ii} = \hat{T}_{ii} = \lambda_i\ , \quad \hat{T}^{ij} = \hat{T}_{ij} = 0 \quad \text{for } i \neq j \qquad (2.6.2)$$

being identical with the eigenvalues λ_i. This permits to represent **T** by the so-called *spectral decomposition*

$$\mathbf{T} = \sum_{i=1}^{3} \lambda_i \, \mathbf{N}_i \otimes \mathbf{N}_i \ . \tag{2.6.3}$$

This way, **T** is completely determined by its real eigenvalues λ_i and the eigenvectors $\mathbf{N}_i$ which have together, in view of the constraints $\mathbf{N}_i \cdot \mathbf{N}_j = \delta_{ij}$, three independent components.

In this section we deal with the eigenvalue problems of the right and left stretch tensors **U** and **v**

$$\mathbf{F} = \mathbf{R}\,\mathbf{U} \ , \quad \mathbf{F} = \mathbf{v}\,\mathbf{R} \tag{2.6.4}$$

introduced by means of the polar decomposition theorem (2.4.4) as well as the eigenvalue problems of the deformation variables **C** and **b** defined in (2.4.34) and (2.4.35)

$$\mathbf{C} = \mathbf{F}^T \mathbf{F} = \mathbf{U}^2 \ , \tag{2.6.5}$$

$$\mathbf{b} = \mathbf{F}\,\mathbf{F}^T = \mathbf{v}^2 \ . \tag{2.6.6}$$

To give a geometrical interpretation of the related eigenvalues we will make use of the relation (2.5.25) rewritten in the form:

$$\lambda^2 = \left(\frac{ds}{dS}\right)^2 = \mathbf{N}\,\mathbf{C}\,\mathbf{N} = N^i \, C_{ij} \, N^j \ . \tag{2.6.7}$$

In this context we recall that λ describes the stretch of a material element $d\mathbf{X}$ having the direction **N** in the undeformed state. We also remember that **U** and **C** are *material* tensors while **v** and **b** present their *spatial* counterparts.

We first consider the right *stretch* tensor **U**. The corresponding eigenvalue problem is given by*:

$$|\,\mathbf{U} - \lambda\,\mathbf{I}\,| = 0 \ , \quad \mathbf{I} = \mathbf{G}_i \otimes \mathbf{G}^i \ . \tag{2.6.8}$$

If the above determinant is expanded in terms of the components

$$\mathbf{U} = U_{ij}\, \mathbf{G}^i \otimes \mathbf{G}^j = \tilde{\mathbf{G}}_i \cdot \mathbf{g}_j \, (\mathbf{G}^i \otimes \mathbf{G}^j) \ , \tag{2.6.9}$$

defined in (2.4.26), the *characteristic polynomial* reads similar to (1.9.7) as

$$\lambda^3 - I_U \, \lambda^2 + II_U \, \lambda - III_U = 0 \ . \tag{2.6.10}$$

Here,

$$I_U = \operatorname{tr} \mathbf{U} = U^j_j = \lambda_1 + \lambda_2 + \lambda_3 \quad , \tag{2.6.11}$$

---

* We will show soon that λ is identical with the stretch defined in (2.6.7).

$$\mathrm{II}_U = \frac{1}{2}\left[(\mathrm{tr}\,\mathbf{U})^2 - \mathrm{tr}\,\mathbf{U}^2\right] = \frac{1}{2}(U^i_j U^k_k - U^i_j U^j_i) = \lambda_1 \lambda_2 + \lambda_2 \lambda_3 + \lambda_3 \lambda_1 \,, \tag{2.6.12}$$

$$\mathrm{III}_U = \det \mathbf{U} = |U^i_j| = e_{ijk} U^i_1 U^j_2 U^k_3 = \lambda_1 \lambda_2 \lambda_3 \tag{2.6.13}$$

are the *invariants* of the symmetric tensor **U** defined in a general form in (1.8.1) to (1.8.3) and for a symmetric tensor $\boldsymbol{\sigma}$ with three real eigenvalues in (1.9.8) and (1.9.17).

The roots λ_i (i = 1, 2, 3) of the characteristic equation (2.6.10) are the eigenvalues of the tensor **U** and the unit vectors $\mathbf{N}_i$ satisfying the homogeneous equation

$$(\mathbf{U} - \lambda_i \mathbf{I})\, \mathbf{N}_i = \mathbf{0} \tag{2.6.14}$$

determine the mutually orthogonal *principal axes* of **U**. If the tensor **U** is referred to the orthonormal basis formed by the vectors $\mathbf{N}_i$

$$\mathbf{U} = \sum_{i=1}^{3} \lambda_i \mathbf{N}_i \otimes \mathbf{N}_i \,, \tag{2.6.15}$$

its components consist uniquely of the diagonal elements λ_i, the off-diagonal elements being in this case zero. Equation (2.6.15) is called the *spectral decomposition* of the tensor **U**. We note that the position of the indices in (2.6.15) is irrelevant since $\mathbf{N}_j$ (j = 1, 2, 3) form a rectangular Cartesian reference frame.

Inserting (2.6.15) into (2.6.5)

$$\mathbf{C} = \left(\sum_{i=1}^{3} \lambda_i \mathbf{N}_i \otimes \mathbf{N}_i\right)\left(\sum_{j=1}^{3} \lambda_j \mathbf{N}_j \otimes \mathbf{N}_j\right) = \sum_{i=1}^{3}\sum_{j=1}^{3} \lambda_i \lambda_j \delta_{ij} \mathbf{N}_i \otimes \mathbf{N}_j \tag{2.6.16}$$

yields for **C** the following spectral decomposition

$$\mathbf{C} := \sum_{i=1}^{3} \Lambda_i \mathbf{N}_i \otimes \mathbf{N}_i = \sum_{i=1}^{3} \lambda_i^2 \mathbf{N}_i \otimes \mathbf{N}_i \,, \quad \Lambda_i = \lambda_i^2 \,, \tag{2.6.17}$$

according to which we may state:

Remark. The principal axes of **C** are identical with those of the stretch tensor **U** and determined by the eigenvectors $\mathbf{N}_i$. Consequently, **C** and **U** are *coaxial* tensors. The eigenvalues Λ_i of **C** are identical with the square of the eigenvalues of **U** such that $\Lambda_i = \lambda_i^2$. Evidently it is possible to determine the eigenvalues Λ_i directly through the condition

$$\det(\mathbf{C} - \Lambda\, \mathbf{I}) = 0 \tag{2.6.18}$$

corresponding to the eigenvalue problem of **C**.

Remark. By means of (2.6.15) and the equality $\Lambda_i = \lambda_i^2$ the stretch tensor **U** can be constructed through the eigenvalue problem (2.6.18) of **C** without an explicit use of the polar decomposition theorem (2.6.4).

We now discuss the mechanical significance of the eigenvalues $\Lambda_i = \lambda_i^2$ of the right CAUCHY-GREEN **C**. For this, we refer to the relation (2.6.7) in order to evaluate the values of the vector **N** for which λ takes extreme values. In view of the constraint

$$\mathbf{N} \cdot \mathbf{N} = 1 \tag{2.6.19}$$

this problem can be solved by determining the extreme values of the function

$$F = \mathbf{N\,C\,N} - \Lambda\,(\mathbf{N{\cdot}N} - 1)\ , \tag{2.6.20}$$

where Λ is the LAGRANGE-multiplier. The necessary condition for an extremum is

$$\frac{\partial F}{\partial \mathbf{N}} = 2\,(\mathbf{C} - \Lambda\,\mathbf{I})\,\mathbf{N} = \mathbf{0}\ . \tag{2.6.21}$$

Accordingly, a nontrivial solution exists if the condition (2.6.18) is satisfied corresponding exactly to the eigenvalue problem of the tensor **C**. Thus we deduce:

Remark. The eigenvectors $\mathbf{N}_i$ of **C** or **U** determine the directions in the undeformed body for which the stretch λ receives extreme values. These are identical with the eigenvalues of **U**. If we consider a volume element of the undeformed body whose edges are defined by the unit vectors $\mathbf{N}_i$ (i = 1, 2, 3), this element receives no angle changes but solely stretches λ_i.

We now turn our attention to the left stretch tensor **v**. We shall show that the eigenvalue problem of **v** is closely related to that of **U** in the sense that principal values of both tensors are identical. In this context we will discover an essential property of the rotation tensor **R** according to which its original definition (2.4.10)

$$\mathbf{R} = \tilde{\mathbf{G}}_i \otimes \mathbf{G}^i \quad \rightarrow \quad \tilde{\mathbf{G}}_i = \mathbf{R}\,\mathbf{G}_i \tag{2.6.22}$$

may be replaced by a new one which uses the principal axes of the tensor **U** at **X** and those of the tensor **v** at **x**. For the derivation we start from the equation

$$(\mathbf{U} - \lambda_i\,\mathbf{I})\,\mathbf{N}_i = \mathbf{0} \tag{2.6.23}$$

indicating that λ_i are the principal values of **U** and $\mathbf{N}_i$ are the associated principal directions. We now express **U** in accordance with (2.6.4) by

$$\mathbf{U} = \mathbf{R}^T\,\mathbf{v}\,\mathbf{R} \tag{2.6.24}$$

and contract the above equation from the left side by **R**. This delivers

$$(\mathbf{R}\,\mathbf{R}^T\,\mathbf{v} - \lambda_i\,\mathbf{I})\,\mathbf{R}\mathbf{N}_i = \mathbf{0} \tag{2.6.25}$$

which, by using the unit vectors

$$\mathbf{n}_i = \mathbf{R}\,\mathbf{N}_i\ , \quad \mathbf{N}_i = \mathbf{R}^T\,\mathbf{n}_i \tag{2.6.26}$$

as abbreviations and taking the orthogonality condition $\mathbf{R}\,\mathbf{R}^T = \mathbf{I}$ into account, becomes

$$(\mathbf{v} - \lambda_i \mathbf{I})\, \mathbf{n}_i = \mathbf{0} \ . \tag{2.6.27}$$

This equality implies that λ_i are the roots of the equation

$$\det(\mathbf{v} - \lambda \mathbf{I}) = 0 \ , \tag{2.6.28}$$

i.e. the principal values of the left stretch tensor **v**. The unit vectors $\mathbf{n}_i$ (2.6.26) give the corresponding principal directions. Accordingly, we may set for **v**

$$\mathbf{v} = \sum_{i=1}^{3} \lambda_i \, \mathbf{n}_i \otimes \mathbf{n}_i \ . \tag{2.6.29}$$

This result, in turn, can be used to obtain the spectral decomposition for the left CAUCHY-GREEN tensor **b**. According to (2.6.6) we have

$$\mathbf{b} := \sum_{i=1}^{3} \Lambda_i \, \mathbf{n}_i \otimes \mathbf{n}_i = \sum_{i=1}^{3} \lambda_i^2 \, \mathbf{n}_i \otimes \mathbf{n}_i \ . \tag{2.6.30}$$

Note that this relation is similar to that given in (2.6.17) for **C**. The results obtained above can now be summarized as follows:

Remark. The principal values of the stretch tensor **U** and **v** are identical. Equation (2.6.26) shows that the principal directions $\mathbf{n}_i$ of the tensor **v** are obtained from the principal directions $\mathbf{N}_i$ of the tensor **U** by application of the rotation tensor **R**. A similar conclusion holds for the deformation variables **C** and **b** possessing the same principal values $\Lambda_i = \lambda_i^2$ associated with the principal directions $\mathbf{N}_i$ and $\mathbf{n}_i$.

According to section 1.9 this implies the equalities

$$\mathrm{I}_{\mathbf{U}} = \mathrm{I}_{\mathbf{v}} \ , \quad \mathrm{II}_{\mathbf{U}} = \mathrm{II}_{\mathbf{v}} \ , \quad \mathrm{III}_{\mathbf{U}} = \mathrm{III}_{\mathbf{v}} \tag{2.6.31}$$

and

$$\mathrm{I}_{\mathbf{C}} = \mathrm{I}_{\mathbf{b}} \ , \quad \mathrm{II}_{\mathbf{C}} = \mathrm{II}_{\mathbf{b}} \ , \quad \mathrm{III}_{\mathbf{C}} = \mathrm{III}_{\mathbf{b}} \tag{2.6.32}$$

for the invariants connected with **u**, **v** and **C**, **b**, in agreement with the previous results (2.4.41).

Remark. In view of (2.6.26), **R** turns out to be a tensor rotating the principal axes of **U** at the material point **X** into the principal axes of **v** at the spatial point **x** so that

$$\mathbf{R} = \mathbf{n}_i \otimes \mathbf{N}^i = \mathbf{n}^i \otimes \mathbf{N}_i \tag{2.6.33}$$

as an alternative formulation to (2.6.22).

The next question is to determine the vectors at **x** into which the eigenvectors of **U** at **X** are transformed after deformation. To solve this problem we recall that the deformation gradient **F** is a tensor, which transforms any arbitrary basis $\mathbf{G}_i$ into its deformed counterpart $\mathbf{g}_i$ through $\mathbf{g}_i = \mathbf{F}\,\mathbf{G}_i$. Accordingly, if we apply this procedure to determine $\mathbf{N}_i$ after deformation and consider (2.6.4), (2.6.15)

$$\mathbf{F}\,\mathbf{N}_k = \mathbf{R}\,\mathbf{U}\,\mathbf{N}_k = \mathbf{R}\left(\sum_{i=1}^{3} \lambda_i\,\mathbf{N}_i \otimes \mathbf{N}_i\right)\mathbf{N}_k \tag{2.6.34}$$

we receive by virtue of (2.6.26)

$$\mathbf{F}\,\mathbf{N}_k = \lambda_k\,\mathbf{n}_k\,,\quad \mathbf{F}^{-1}\,\mathbf{n}_k = \frac{1}{\lambda_k}\mathbf{N}_k \quad \text{or} \quad \mathbf{F} = \sum_{i=1}^{3} \lambda_i\,\mathbf{n}_i \otimes \mathbf{N}_i\,,\quad \mathbf{F}^{-1} = \sum_{i=1}^{3} \mathbf{N}_i \otimes \frac{1}{\lambda_i}\mathbf{n}_i \tag{2.6.35}$$

permitting to deduce the following:

Remark. When the principal values λ_i (Λ_i) of $\mathbf{U}(\mathbf{C})$ are distinct then there is a unique set of three orthogonal principal directions $\mathbf{N}_i$ at $\mathbf{X}$ and the deformation carries them into a unique set of mutually orthogonal directions described by the unit vectors $\mathbf{n}_i$, which are the principal directions of $\mathbf{v}(\mathbf{b})$ at $\mathbf{x}$. If the values λ_i are not distinct, the vectors $\mathbf{N}_i$ are not uniquely determined, but it is always possible to select an orthonormal set of vectors $\mathbf{N}_i$, which are the principal directions of $\mathbf{U}(\mathbf{C})$ at $\mathbf{X}$ and which are carried by the deformation into an orthogonal set of three principal directions $\mathbf{n}_i$ of $\mathbf{v}(\mathbf{b})$ at $\mathbf{x}$ (see Malvern 1969).

If we rewrite (2.6.14) in the form

$$\mathbf{U}\,\mathbf{N}_i = \lambda_i\,\mathbf{N}_i\,, \tag{2.6.36}$$

which holds also for $\mathbf{U} \to \mathbf{v}$ and $\mathbf{N}_i \to \mathbf{n}_i$, we may add a further important result:

Remark. If the stretch tensor $\mathbf{U}(\mathbf{v})$ operates on the set of all vectors at a point $\mathbf{X}(\mathbf{x})$, it produces length changes (stretches) of the vectors and also provides additional rotation of all vectors except those showing in the principal directions $\mathbf{N}_i$ ($\mathbf{n}_i$).

The essential results discussed in this section are illustrated in Fig. 2.9 for a 2D deformation. The first figure illustrates the main characteristics of the deformation gradient $\mathbf{F}$, which transforms any arbitrary set of base vectors $\mathbf{G}_i$ selected at $\mathbf{X}$ into its deformed counterpart $\mathbf{g}_i$ at $\mathbf{x}$. The second figure demonstrates the orthogonal tensor $\mathbf{R}$ to be a tensor transforming the principal directions $\mathbf{N}_i$ of $\mathbf{U}$ into the principal directions $\mathbf{n}_i$ of $\mathbf{v}$. In the third figure the deformation of a rectangular element is illustrated which is spanned in the initial state by the unit eigenvectors $\mathbf{N}_i$ of $\mathbf{U}$. Such an element receives during deformation only stretches described by λ_i. Its shape in the deformed state is determined by the principal directions $\mathbf{n}_i$ of $\mathbf{v}$ at $\mathbf{x}$. Conversely, in the last figure a rectangular element of the deformed body with edges determined by the eigenvectors $\mathbf{n}_i$ of $\mathbf{v}$ is shown. In the undeformed configuration B_0 the shape of this element is determined by the principal directions $\mathbf{N}_i$ of $\mathbf{U}$ and the corresponding edge lengths are given by $1/\lambda_i$.

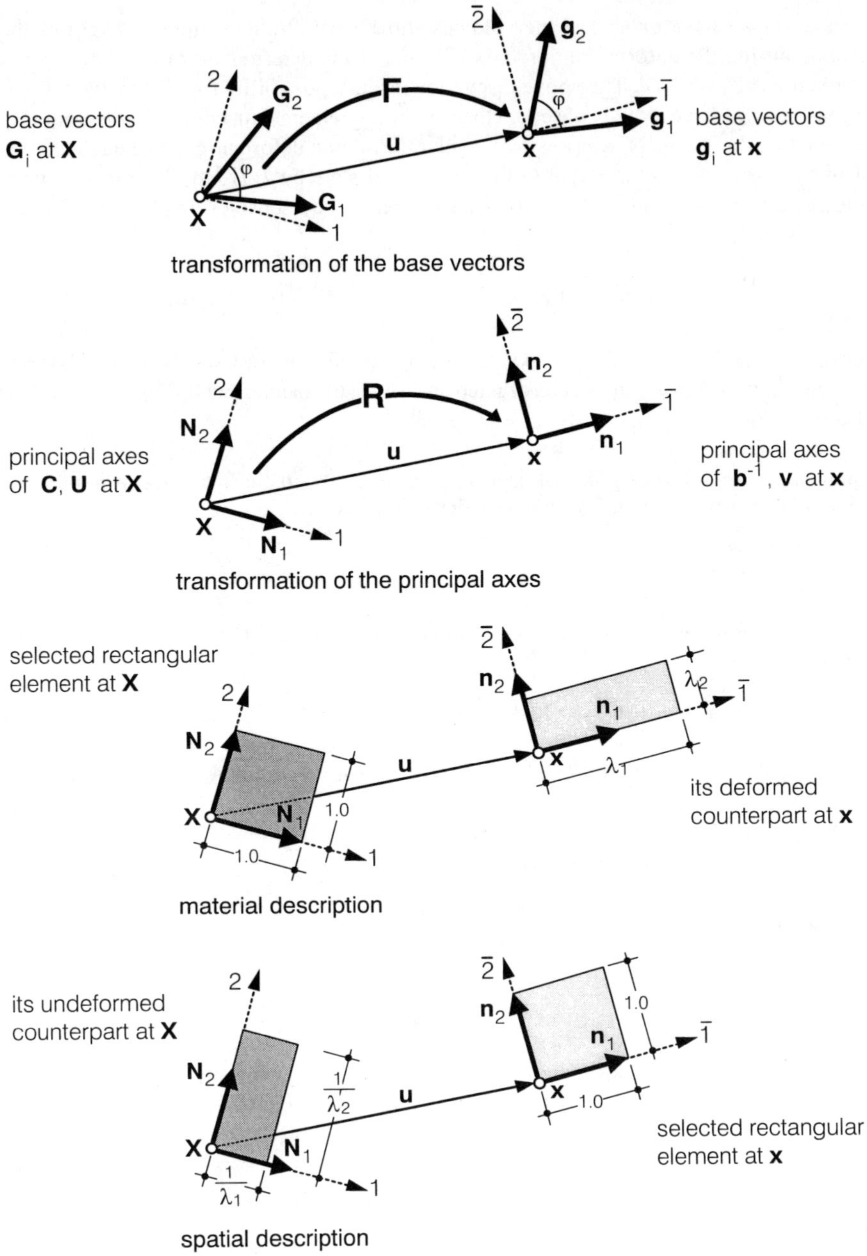

Fig. 2.9. Deformation of a volume element with edges showing in principal directions $\mathbf{N}_i$ or $\mathbf{n}_i$ as a 2D illustration

In (2.2.31) we have expressed the volume ratio $J = dV/dV_0$ describing changes of the volume during the deformation process in terms of the determinant of the deformation gradient $\mathbf{F} = F^{i}_{\cdot j}\, \mathbf{G}_i \otimes \mathbf{G}^j$. The above discussion offers a possibility for alternative presentations of J. We have already observed that a volume element with edges showing initially in principal directions $\mathbf{N}_i$ remains rectangular during the deformation. The edge lengths of this volume element are given in the deformed state by λ_i (i = 1, 2, 3) if they were selected originally as unity (Fig.2.9). Thus we may set by considering (2.6.13), (2.6.17) and (2.6.31)

$$J = \frac{dV}{dV_0} = \lambda_1 \lambda_2 \lambda_3 = III_U = III_{\mathbf{v}} = \sqrt{\Lambda_1 \Lambda_2 \Lambda_3} = \sqrt{III_C} = \sqrt{III_{\mathbf{b}}}, \qquad (2.6.37)$$

confirming the earlier results (2.4.43) from geometrical point of view. Note that the third invariants in (2.6.37) can be constructed in a unified manner according to (1.8.3) or (1.8.6).

Application. Show that the principal values of the inverse tensor $\mathbf{b}^{-1}$ are reciprocal to the principal values of the tensor **b** and that the principal directions coincide.

To prove this we set

$$(\mathbf{b} - \Lambda_i \mathbf{I})\, \mathbf{n}_i = \mathbf{0} \qquad (2.6.38)$$

with Λ_i as the principal values of **b** and $\mathbf{n}_i$ as its principal directions. The contraction of this relation by $\mathbf{b}^{-1}$ yields

$$(\mathbf{b}^{-1}\mathbf{b} - \Lambda_i \mathbf{b}^{-1})\, \mathbf{n}_i = \mathbf{0} \quad \rightarrow \quad \left(\mathbf{b}^{-1} - \frac{1}{\Lambda_i}\mathbf{I}\right) \mathbf{n}_i = \mathbf{0} \qquad (2.6.39)$$

confirming the above statement. This can be shown alternatively as follows. Since the equality

$$\mathbf{b}^{-1}\mathbf{b} = \mathbf{b}\,\mathbf{b}^{-1} = \mathbf{I} \qquad (2.6.40)$$

holds, the tensors $\mathbf{b}^{-1}$ and **b** are coaxial. Hence

$$\mathbf{b} = \sum_{i=1}^{3} \Lambda_i\, \mathbf{n}_i \otimes \mathbf{n}_i\,, \quad \mathbf{b}^{-1} = \sum_{i=1}^{3} \frac{1}{\Lambda_i}\, \mathbf{n}_i \otimes \mathbf{n}_i\,. \qquad (2.6.41)$$

2.7 Unified definitions of strain tensors

The spectral decompositions (2.6.15), (2.6.17), (2.6.29) and (2.6.30) offer a new possibility to introduce various strain tensors by a unified procedure on the basis of the stretch tensors **U** and **v**. The selection of the stretches **U** and **v** as starting point is somewhat arbitrary, but suitable as the condition **U** = **v** = **I** corresponds to a rigid body motion being free of deformations. *Material strain* variables may be introduced by

$$\mathbf{E}^{(m)}(\mathbf{U}) := \begin{cases} \frac{1}{m}(\mathbf{U}^m - \mathbf{G}) & \text{for } m \neq 0\,, \\ \ln \mathbf{U} & \text{for } m = 0\,, \end{cases} \qquad (2.7.1)$$

where m denotes positive integers. The strain variables defined above are *coaxial* to **U**, that means, they have the same principal directions as **U**. Their principal values are given in terms of those of **U** denoted by λ_i (i = 1, 2, 3) by

$$\mathbf{E}^{(m)}(\lambda_i) := \begin{cases} \dfrac{1}{m}\sum\limits_{i=1}^{3} [(\lambda_i)^m - 1]\, \mathbf{N}_i \otimes \mathbf{N}_i & \text{for } m \neq 0 \,, \\ \sum\limits_{i=1}^{3} \ln \lambda_i\, \mathbf{N}_i \otimes \mathbf{N}_i & \text{for } m = 0 \,. \end{cases} \tag{2.7.2}$$

Similarly, *spatial strain* tensors can be defined on the basis of **v** by the rule:

$$\mathbf{e}^{(m)}(\mathbf{v}) := \begin{cases} \dfrac{1}{m}(\mathbf{v}^m - \mathbf{g}) & \text{for } m \neq 0 \,, \\ \ln \mathbf{v} & \text{for } m = 0 \end{cases} \tag{2.7.3}$$

with negative integers m. Since **U** and **v** have the same eigenvalues, the spectral decompositions for $\mathbf{e}^{(m)}$ may be given similar to (2.7.2) in terms of λ_i. The definitions (2.7.1) and (2.7.3) may be extended to arbitrary real powers m (Ogden 1984, Morman 1986).* Furthermore it is possible to define strain measures which are not coaxial to **U** or **v**.

Strain tensors introduced in the previous sections may be now classified as follows. We have

- *GREEN-LAGRANGE strain tensor* : $\mathbf{E}^{(2)} = \mathbf{E} = \frac{1}{2}(\mathbf{U}^2 - \mathbf{G})$ for m = 2 ,
- *BIOT strain tensor* : $\mathbf{E}^{(1)} = \mathbf{H} = \mathbf{U} - \mathbf{G}$ for m = 1 ,
- *HENCKY strain tensor* : $\mathbf{E}^{(0)} = \ln \mathbf{U}$ for m = 0

as *material* tensors and

- *ALMANSI strain tensor* : $\mathbf{e}^{(-2)} = \mathbf{e} = \frac{1}{2}(\mathbf{g} - \mathbf{v}^{-2})$ for m = −2 ,
- *spatial BIOT strain tensor* : $\mathbf{e}^{(-1)} = \mathbf{h} = \mathbf{g} - \mathbf{v}^{-1}$ for m = −1 ,
- *spatial HENCKY strain tensor* : $\mathbf{e}^{(0)} = \ln \mathbf{v}$ for m = 0

as *spatial* tensors.

The HENCKY strain tensor ln **U** can be expressed by an infinite power series in **U** which, according to (1.12.2), is given by

$$\ln \mathbf{U} = \ln [\mathbf{G} + (\mathbf{U} - \mathbf{G})] = (\mathbf{U} - \mathbf{G}) - \frac{1}{2}(\mathbf{U} - \mathbf{G})^2 + \frac{1}{3}(\mathbf{U} - \mathbf{G})^3 \ldots \,. \tag{2.7.4}$$

A similar series expansion in **v** holds also for ln **v**.

---

* A detailed discussion of this aspect is given in section 2.11.

Table 2.2. Definition of material and spatial deformation tensors

notations	definitions	components
material deformation gradient	$\mathbf{F} = \mathbf{g}_i \otimes \mathbf{G}^i$	$\mathbf{F} = F_{ij}\, \mathbf{G}^i \otimes \mathbf{G}^j = \mathbf{G}_i \cdot \mathbf{g}_j\, (\mathbf{G}^i \otimes \mathbf{G}^j)$
	$\mathbf{F}^T = \mathbf{G}^i \otimes \mathbf{g}_i$	$\mathbf{F}^T = (F^T)_{ij}\, \mathbf{G}^i \otimes \mathbf{G}^j = \mathbf{G}_j \cdot \mathbf{g}_i\, (\mathbf{G}^i \otimes \mathbf{G}^j)$
spatial deformation gradient	$\mathbf{F}^{-1} = \mathbf{G}_i \otimes \mathbf{g}^i$	$\mathbf{F}^{-1} = (f^{-1})_{ij} (\mathbf{g}^i \otimes \mathbf{g}^j) = \mathbf{G}_j \cdot \mathbf{g}_i\, (\mathbf{g}^i \otimes \mathbf{g}^j)$
	$\mathbf{F}^{-T} = \mathbf{g}^i \otimes \mathbf{G}_i$	$\mathbf{F}^{-T} = (f^{-T})_{ij} (\mathbf{g}^i \otimes \mathbf{g}^j) = \mathbf{G}_i \cdot \mathbf{g}_j\, (\mathbf{g}^i \otimes \mathbf{g}^j)$
right stretch tensor	$\mathbf{U} = \mathbf{R}^T\mathbf{F} = (\tilde{\mathbf{G}}_j \otimes \mathbf{G}^j)^T\, \mathbf{F}$	$\mathbf{U} = U_{ij}\, \mathbf{G}^i \otimes \mathbf{G}^j = \tilde{\mathbf{G}}_i \cdot \mathbf{g}_j\, (\mathbf{G}^i \otimes \mathbf{G}^j)$
	$\mathbf{U}^{-1} = \mathbf{F}^{-1}\mathbf{R} = \mathbf{F}^{-1}\, (\tilde{\mathbf{G}}^j \otimes \mathbf{G}_j)$	$\mathbf{U}^{-1} = (U^{-1})^{ij} (\mathbf{G}_i \otimes \mathbf{G}_j) = \mathbf{g}^i \cdot \tilde{\mathbf{G}}^j (\mathbf{G}_i \otimes \mathbf{G}_j)$
left stretch tensor	$\mathbf{v} = \mathbf{F}\,\mathbf{R}^T = \mathbf{F}\, (\hat{\mathbf{g}}^j \otimes \mathbf{g}_j)$	$\mathbf{v} = v^{ij}\, \mathbf{g}_i \otimes \mathbf{g}_j = \mathbf{G}^i \cdot \hat{\mathbf{g}}^j\, (\mathbf{g}_i \otimes \mathbf{g}_j)$
	$\mathbf{v}^{-1} = \mathbf{R}\,\mathbf{F}^{-1} = (\mathbf{g}^j \otimes \hat{\mathbf{g}}_j)\, \mathbf{F}^{-1}$	$\mathbf{v}^{-1} = (v^{-1})_{ij} (\mathbf{g}^i \otimes \mathbf{g}^j) = \hat{\mathbf{g}}_i \cdot \mathbf{G}_j\, (\mathbf{g}^i \otimes \mathbf{g}^j)$
right CAUCHY-GREEN tensor	$\mathbf{C} = \mathbf{F}^T\mathbf{F} = \mathbf{U}^2$	$\mathbf{C} = C_{ij}\, \mathbf{G}^i \otimes \mathbf{G}^j = \mathbf{g}_i \cdot \mathbf{g}_j\, (\mathbf{G}^i \otimes \mathbf{G}^j)$
	$\mathbf{C}^{-1} = \mathbf{F}^{-1}\mathbf{F}^{-T} = \mathbf{U}^{-2}$	$\mathbf{C}^{-1} = (C^{-1})^{ij} (\mathbf{G}_i \otimes \mathbf{G}_j) = \mathbf{g}^i \cdot \mathbf{g}^j\, (\mathbf{G}_i \otimes \mathbf{G}_j)$
left CAUCHY-GREEN tensor	$\mathbf{b} = \mathbf{F}\,\mathbf{F}^T = \mathbf{v}^2$	$\mathbf{b} = b^{ij}\, \mathbf{g}_i \otimes \mathbf{g}_j = \mathbf{G}^i \cdot \mathbf{G}^j\, (\mathbf{g}_i \otimes \mathbf{g}_j)$
	$\mathbf{b}^{-1} = \mathbf{F}^{-T}\mathbf{F}^{-1} = \mathbf{v}^{-2}$	$\mathbf{b}^{-1} = (b^{-1})_{ij}\, (\mathbf{g}^i \otimes \mathbf{g}^j) = \mathbf{G}_i \cdot \mathbf{G}_j\, (\mathbf{g}^i \otimes \mathbf{g}^j)$
GREEN-LAGRANGE strain tensor	$\mathbf{E} = \frac{1}{2}\, (\mathbf{F}^T\mathbf{F} - \mathbf{G}) = \frac{1}{2}\, (\mathbf{C} - \mathbf{G})$	$\mathbf{E} = E_{ij}\, \mathbf{G}^i \otimes \mathbf{G}^j = \frac{1}{2} (\mathbf{g}_i \cdot \mathbf{g}_j - \mathbf{G}_i \cdot \mathbf{G}_j) (\mathbf{G}^i \otimes \mathbf{G}^j)$
ALMANSI strain tensor	$\mathbf{e} = \frac{1}{2} (\mathbf{g} - \mathbf{F}^{-T}\mathbf{F}^{-1}) = \frac{1}{2} (\mathbf{g} - \mathbf{b}^{-1})$	$\mathbf{e} = e_{ij}\, \mathbf{g}^i \otimes \mathbf{g}^j = \frac{1}{2}\, (\mathbf{g}_i \cdot \mathbf{g}_j - \mathbf{G}_i \cdot \mathbf{G}_j) (\mathbf{g}^i \otimes \mathbf{g}^j)$
BIOT strain tensor	$\mathbf{H} = \mathbf{U} - \mathbf{G}$	$\mathbf{H} = H_{ij}\, \mathbf{G}^i \otimes \mathbf{G}^j = (\tilde{\mathbf{G}}_i \cdot \mathbf{g}_j - \mathbf{G}_i \cdot \mathbf{G}_j)\, (\mathbf{G}^i \otimes \mathbf{G}^j)$
spatial BIOT strain tensor	$\mathbf{h} = \mathbf{g} - \mathbf{v}^{-1}$	$\mathbf{h} = h_{ij}\, \mathbf{g}^i \otimes \mathbf{g}^j = (\mathbf{g}_i \cdot \mathbf{g}_j - \hat{\mathbf{g}}_i \cdot \mathbf{G}_j)\, (\mathbf{g}^i \otimes \mathbf{g}^j)$
important equalities	$F_{ij} = (f^{-T})_{ij},\; U_{ij} = (v^{-1})_{ij},\; E_{ij} = e_{ij}$	holding only for pure covariant components

By means of the relation (2.5.1)

$$\mathbf{C} := \mathbf{F}^T\,\mathbf{F} = \mathbf{U}^2$$

the material strain tensors can be expressed in terms of the right CAUCHY-GREEN tensor **C**. Concerning the definition of objective strain measures the spatial counterpart of **C** is the inverse CAUCHY-GREEN tensor $\mathbf{b}^{-1}$ which is, according to (2.5.2), related to **v** by

$$\mathbf{b}^{-1} := \mathbf{F}^{-T}\,\mathbf{F}^{-1} = (\mathbf{v}^{-1})^2 = \mathbf{v}^{-2} \tag{2.7.5}$$

Thus we see that the deformation gradient **F** and its inverse $\mathbf{F}^{-1}$ are closely related to the material and spatial formulation, respectively.

The material strain tensors $\mathbf{E}^{(m)}$ (m = 0, 1, 2) presented above are essentially characterized by two properties:

- they are all used with components defined with respect to the undeformed basis $\mathbf{G}^i \otimes \mathbf{G}^j$.
- their eigenvectors $\mathbf{N}_i$ define in the undeformed state B_o rectangular volume elements which receive no change of shape during deformation (Fig. 2.9).

On the contrary, the main characteristics of spatial strain variables $\mathbf{e}^m$ (m = 0, –1,–2) are:

- they are all used with components defined with respect to the deformed basis $\mathbf{g}^i \otimes \mathbf{g}^j$.
- their eigenvectors $\mathbf{n}_i$ define in the deformed state B rectangular volume elements which were also in the initial state B_o of the same shape (Fig. 2.9).

The definitions of the deformation tensors and their components are summarized in Table 2.2.

2.8 Isochoric and volumetric deformations

In (2.6.37) the JACOBIAN J which describes changes of volume during deformation is expressed in terms of the principle values λ_i of the right stretch tensor **U**. For the subsequent derivation we rewrite J by considering (2.2.31) and (2.6.13) in the form

$$J = \frac{dV}{dV_o} = \lambda_1 \lambda_2 \lambda_3 = | U^r_{\cdot s} | = \det \mathbf{F} = | F^i_{\cdot j} | \; . \tag{2.8.1}$$

The JACOBIAN J can be used to decompose the deformation gradient into an isochoric and a volumetric part. A deformation is said to be *isochoric* if the volume

$$V_t = \iiint_{B^*} dV \tag{2.8.2}$$

of an arbitrary part B* of the body remains unchanged during deformation. Accordingly, isochoric deformations are *volume-preserving*. A deformation is said to be *volumetric* if the deformation gradient **F** can be described in the form $\mathbf{F} = \alpha\,\mathbf{I}$, where α is a positive scalar-valued function and **I** is the identity tensor.

Isochoric deformations are described by one of the following conditions

$$J = 1 \quad \text{or} \quad \rho = \rho_o \quad \text{or} \quad \operatorname{div} \dot{\mathbf{x}} = 0 \; , \tag{2.8.3}$$

the first one being the consequence of (2.8.1), while the remaining ones are to be proved in section 5.1. Note that ρ and ρ_o are the mass densities of the deformed and undeformed body, while $\dot{\mathbf{x}} = \mathrm{D}\mathbf{x}/\mathrm{Dt}$ is the material time derivative of the spatial coordinates $\mathbf{x}$.

If we now present the deformation gradient $\mathbf{F}$ by using J in the form

$$\mathbf{F} = J^{\frac{1}{3}} \mathbf{F}^* = (J^{\frac{1}{3}} \mathbf{I}) \mathbf{F}^* , \tag{2.8.4}$$

we easily see by considering (2.8.1) that

$$\mathbf{F}^* = J^{-\frac{1}{3}} \mathbf{F} \quad \rightarrow \quad \det \mathbf{F}^* = \det (J^{-\frac{1}{3}} \mathbf{F}) = J^{-1} \det \mathbf{F} = 1 . \tag{2.8.5}$$

Accordingly, $\mathbf{F}^*$ is the part of the deformation gradient describing pure isochoric deformations while the part $J^{1/3}\,\mathbf{I}$ expresses volumetric deformations.

Inserting the polar decomposition $\mathbf{F} = \mathbf{R}\,\mathbf{U}$ into (2.8.5) we obtain

$$\mathbf{F}^* = J^{-\frac{1}{3}} \mathbf{R}\,\mathbf{U} = \mathbf{R}\,(J^{-\frac{1}{3}} \mathbf{U}) = \mathbf{R}\,\mathbf{U}^* . \tag{2.8.6}$$

The eigenvalues λ_i^* of the new stretch tensor $\mathbf{U}^*$ are then given by

$$\mathbf{U}^* = J^{-\frac{1}{3}} \mathbf{U} = \sum_{i=1}^{3} \lambda_i^* \mathbf{N}_i \otimes \mathbf{N}_i = \sum_{i=1}^{3} J^{-\frac{1}{3}} \lambda_i \mathbf{N}_i \otimes \mathbf{N}_i \quad \rightarrow \quad \lambda_i^* = J^{-\frac{1}{3}} \lambda_i \tag{2.8.7}$$

and satisfy in view of (2.8.1) the incompressibility condition

$$\lambda_1^* \lambda_2^* \lambda_3^* = 1 , \tag{2.8.8}$$

indicating that $\mathbf{U}^*$ describes *isochoric* deformations.

For further useful results concerning homogeneous and nonhomogeneous isochoric deformations we refer to Ogden 1984. We also note, that the decomposition (2.8.4) is due to Flory 1961 and presents a suitable tool for modelling incompressible rubber-like materials. For more on this, see e.g. Simo and Taylor 1991.

2.9 Rate of deformation tensor and spin tensor

Important kinematic variables. In elasticity theory the actual process by which a continuum transforms from one configuration to another is mostly considered irrelevant: the initial and final states alone are sufficient to define the strains. However, to describe irreversible processes, e.g. the plastic deformations of a solid it is necessary to specify the process and to follow the history of deformation. The deformation process can in principle be described by the trajectory of every point $\mathbf{X}$

$$\mathbf{x} = \mathbf{x}(\mathbf{X}, t) \quad \rightarrow \quad x^i = x^i (X^1, X^2, X^3, t)$$

as in (2.1.3), although the explicit formulation of such an equation is a very difficult problem even for quite simple deformation processes (Malvern 1969).

The important kinematic tensors to describe path-dependent behaviour, e.g. in visco-plasticity are the *rate of deformation tensor* (also called the *stretching tensor* or *velocity strain*) and the *spin tensor* (also called the *vorticity tensor*), which are to be introduced in this section. The corresponding definitions are based on the *velocity vector* $\mathbf{v} = \dot{\mathbf{x}}$, which follows from the displacement vector $\mathbf{u}$ (2.1.19)

$$\mathbf{u} = \mathbf{x}(\mathbf{X}, t) - \mathbf{X} \tag{2.9.1}$$

by means of the material time derivative to be denoted by $D/Dt = (\dot{\ })$. The material time derivative is understood as a differentiation with respect to time t holding the material coordinates $\mathbf{X}$ fixed.* Thus from (2.9.1)

$$\mathbf{v} = \dot{\mathbf{u}} = \dot{\mathbf{x}} \ , \quad \dot{\mathbf{x}} := \frac{D\mathbf{x}}{Dt} = \frac{\partial \mathbf{x}}{\partial t} \ . \tag{2.9.2}$$

For later use we also need the material time derivatives of the deformed base vectors $\mathbf{g}_i$ and $\mathbf{g}^i$. By virtue of (2.2.6) and (2.2.7) the corresponding results are of the form

$$\dot{\mathbf{g}}_i = \dot{\mathbf{F}}\,\mathbf{G}_i = \dot{\mathbf{F}}\,\mathbf{F}^{-1}\,\mathbf{g}_i = -\mathbf{F}\,\dot{\mathbf{F}}^{-1}\,\mathbf{g}_i \ , \tag{2.9.3}$$

$$\dot{\mathbf{g}}^i = \dot{\mathbf{F}}^{-T}\,\mathbf{G}^i = \dot{\mathbf{F}}^{-T}\,\mathbf{F}^{T}\,\mathbf{g}^i = -\mathbf{F}^{-T}\,\dot{\mathbf{F}}^{T}\,\mathbf{g}^i \ , \tag{2.9.4}$$

where according to (2.2.3) and (2.2.4)

$$\dot{\mathbf{F}} = \dot{\mathbf{g}}_i \otimes \mathbf{G}^i \ , \quad \dot{\mathbf{F}}^{T} = \mathbf{G}^i \otimes \dot{\mathbf{g}}_i \ , \quad \dot{\mathbf{F}}^{-1} = \mathbf{G}_i \otimes \dot{\mathbf{g}}^i \ , \quad \dot{\mathbf{F}}^{-T} = \dot{\mathbf{g}}^i \otimes \mathbf{G}_i \ . \tag{2.9.5}$$

We note that in the derivation of (2.9.3) and (2.9.4) the identities

$$\mathbf{F}\,\mathbf{F}^{-1} = \mathbf{I} \ \rightarrow \ \dot{\mathbf{F}}\,\mathbf{F}^{-1} = -\mathbf{F}\,\dot{\mathbf{F}}^{-1} \ \rightarrow \ \dot{\mathbf{g}}_i \otimes \mathbf{g}^i = -\mathbf{g}_i \otimes \dot{\mathbf{g}}^i \tag{2.9.6}$$

have been used. A further useful result is the partial derivative of (2.9.2) with respect to Θ^i

$$\mathbf{v}_{,i} = \dot{\mathbf{u}}_{,i} = \dot{\mathbf{x}}_{,i} = \dot{\mathbf{g}}_i \ , \tag{2.9.7}$$

whose contraction by $d\Theta^i$

$$d\mathbf{v} = \mathbf{v}_{,i}\, d\Theta^i = \dot{\mathbf{g}}_i\, d\Theta^i = d\dot{\mathbf{x}} \tag{2.9.8}$$

delivers the *relative velocity vector* $d\mathbf{v}$ corresponding to the difference of the velocity vectors of two infinitesimally adjacent space points $\mathbf{x} + d\mathbf{x}$ and $\mathbf{x}$ (Fig. 2.10).

Spatial gradient of velocity. The *spatial gradient of velocity* is denoted by $\mathbf{l}$ and defined by the following expressions

---

* A more detailed explanation of this definition is given in section 4.1

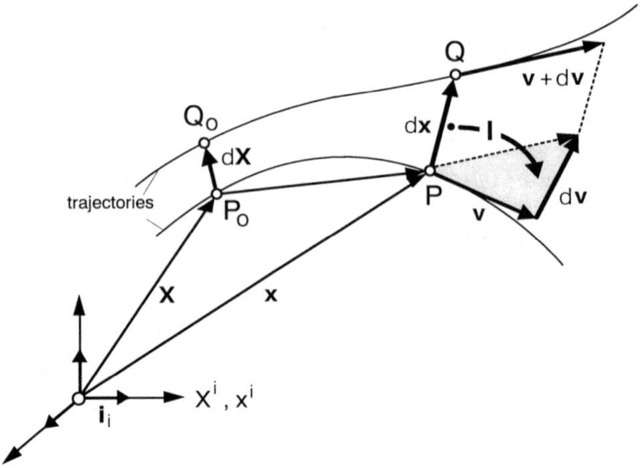

Fig. 2.10. Transformation of the deformed material vector **dx** into the relative velocity vector **dv** by the spatial gradient of velocity **l**

$$\mathbf{l} := \operatorname{grad}\mathbf{v} = \frac{\partial \mathbf{v}}{\partial x^i} \otimes \mathbf{i}^i = \frac{\partial \mathbf{v}}{\partial \Theta^j} \otimes \frac{\partial \Theta^j}{\partial x^i}\mathbf{i}^i = \mathbf{v}_{,j} \otimes \mathbf{g}^j \tag{2.9.9}$$

which are similar to those given in (2.2.12) for grad **X**. In view of the equalities

$$\mathbf{v}_{,i} \otimes \mathbf{g}^i = \dot{\mathbf{x}}_{,i} \otimes \mathbf{g}^i = \dot{\mathbf{g}}_i \otimes \mathbf{g}^i = \dot{\mathbf{F}}\,\mathbf{F}^{-1} = -\mathbf{g}_i \otimes \dot{\mathbf{g}}^i = -\mathbf{F}\,\dot{\mathbf{F}}^{-1} \tag{2.9.10}$$

in accordance with (2.9.3), (2.9.6) and (2.9.7) the above definition can be replaced by

$$\mathbf{l} = \operatorname{grad}\dot{\mathbf{x}} = \dot{\mathbf{x}}_{,i} \otimes \mathbf{g}^i = \dot{\mathbf{g}}_i \otimes \mathbf{g}^i = \dot{\mathbf{F}}\,\mathbf{F}^{-1} = -\mathbf{g}_i \otimes \dot{\mathbf{g}}^i = -\mathbf{F}\,\dot{\mathbf{F}}^{-1} \ , \tag{2.9.11}$$

where $\dot{\mathbf{F}}$ is given in (2.9.5). By virtue of this relation equations (2.9.3) and (2.9.4) become

$$\dot{\mathbf{g}}_i = \mathbf{l}\,\mathbf{g}_i = \mathbf{g}_i\,\mathbf{l}^T \ , \quad \dot{\mathbf{g}}^i = -\mathbf{l}^T\mathbf{g}^i = -\mathbf{g}^i\,\mathbf{l} \tag{2.9.12}$$

showing that **l** is a tensor mapping the basis $\mathbf{g}_i$ and its material time derivative $\dot{\mathbf{g}}_i$ into each other. It can easily be confirmed that the above expressions for $\dot{\mathbf{g}}_i$ and $\dot{\mathbf{g}}^i$ satisfy automatically the following identity:

$$\overline{(\mathbf{g}_i \cdot \mathbf{g}^j)}^{\boldsymbol{\cdot}} = \dot{\delta}_i^j = 0 \quad \rightarrow \quad \dot{\mathbf{g}}_i \cdot \mathbf{g}^j + \mathbf{g}_i \cdot \dot{\mathbf{g}}^j = 0 \ . \tag{2.9.13}$$

By contracting the first relation in (2.9.12) by $d\Theta^i$ and by considering (2.9.8) it can furthermore be deduced that:

$$\dot{\mathbf{g}}_i\, d\Theta^i = \mathbf{l}\,\mathbf{g}_i\, d\Theta^i \quad \rightarrow \quad d\mathbf{v} = d\dot{\mathbf{x}} = \mathbf{l}\, d\mathbf{x} \ . \tag{2.9.14}$$

The above result can be interpreted geometrically as follows. Fig. 2.10 illustrates two neighbouring points P_o and Q_o of the undeformed body moving along their own trajectories to the positions P and Q at time t. The velocity vectors of P_o and Q_o at time t are represented by the vectors **v** and **v**+d**v** which are tangential at P and Q to the corresponding trajectories. Even if an observer is not able to recognize the points P_o and Q_o as well as their trajectories he can record the relative velocity d**v** presenting the difference of the velocity vectors associated with Q and P as well as the relative positions of Q and P represented by d**x**. Thus **l** turns out to be the tensor permitting to transform d**x** into d**v**.

The rate of deformation tensor d and the spin tensor w. As a second-order tensor the spatial velocity gradient **l** can be presented in the form

$$\mathbf{l} := l_{ij}\, \mathbf{g}^i \otimes \mathbf{g}^j = \mathbf{d} + \mathbf{w} \tag{2.9.15}$$

as a sum of a *symmetric* tensor

$$\mathbf{d} := d_{ij}\, \mathbf{g}^i \otimes \mathbf{g}^j = \frac{1}{2}(\mathbf{l} + \mathbf{l}^T) \tag{2.9.16}$$

called the *rate of deformation tensor* (*stretching tensor*) and a *skew-symmetric* one

$$\mathbf{w} := w_{ij}\, \mathbf{g}^i \otimes \mathbf{g}^j = \frac{1}{2}(\mathbf{l} - \mathbf{l}^T) \tag{2.9.17}$$

called the *spin tensor* (*vorticity tensor*). The variables **l**, **d** and **w** are all *spatial* tensors denoted by lower case letters. Their components can be evaluated by the standard procedure (1.2.13) leading in view of (2.9.11) and (2.9.16), (2.9.17) to

$$\mathbf{l} = l_{ij}\, \mathbf{g}^i \otimes \mathbf{g}^j = \mathbf{g}_i \cdot \dot{\mathbf{g}}_j\, (\mathbf{g}^i \otimes \mathbf{g}^j) \ ,$$

$$\mathbf{d} = d_{ij}\, \mathbf{g}^i \otimes \mathbf{g}^j = \frac{1}{2}\, \dot{\overline{\mathbf{g}_i \cdot \mathbf{g}_j}}\, (\mathbf{g}^i \otimes \mathbf{g}^j) \ ,$$

$$\mathbf{w} = w_{ij}\, \mathbf{g}^i \otimes \mathbf{g}^j = \frac{1}{2}(\mathbf{g}_i \cdot \dot{\mathbf{g}}_j - \dot{\mathbf{g}}_i \cdot \mathbf{g}_j)\, (\mathbf{g}^i \otimes \mathbf{g}^j) \ . \tag{2.9.18}$$

The above results together with other important connections for **l**, **d** and **w** are summarized in Table 2.3.

Interpretation of d. To give a geometrical interpretation of **d** we form the material time derivative of the expression $ds^2 = d\mathbf{x} \cdot d\mathbf{x}$ introduced in (2.5.6), ds being the length of the material element d**x** at time t. In view of (2.9.14) and (2.9.15) we receive:

$$\frac{D}{Dt}(ds^2) = 2\, d\mathbf{x} \cdot d\dot{\mathbf{x}} = 2\, d\mathbf{x}\, \mathbf{l}\, d\mathbf{x} = 2\, d\mathbf{x}\, (\mathbf{d} + \mathbf{w})\, d\mathbf{x} \ . \tag{2.9.19}$$

Since the tensor **w** is skew-symmetric, the expression d**x** **w** d**x** vanishes in (2.9.19). Thus the rate of change of the square length ds^2 is entirely described by **d**

$$\frac{D}{Dt}(ds^2) = 2\, d\mathbf{x}\, \mathbf{d}\, d\mathbf{x} \ , \tag{2.9.20}$$

which justifies to call **d** *rate of deformation tensor*.

Table 2.3. Spatial velocity gradient **l**, its symmetric and skew-symmetric part **d** and **w**

notations	definitions	components
the spatial gradient of velocity **l**	$\mathbf{l} = \dot{\mathbf{x}}_{,i} \otimes \mathbf{g}^i = \dot{\mathbf{g}}_i \otimes \mathbf{g}^i = \dot{\mathbf{F}}\,\mathbf{F}^{-1}$ $= \mathbf{d} + \mathbf{w}$	$\mathbf{l} = \mathrm{l}_{ij}\, \mathbf{g}^i \otimes \mathbf{g}^j \quad = \mathbf{g}_i \cdot \dot{\mathbf{g}}_j\, (\mathbf{g}^i \otimes \mathbf{g}^j)$
the rate of deformation tensor **d**	$\mathbf{d} = \mathbf{d}^T = \frac{1}{2}(\mathbf{l} + \mathbf{l}^T)$	$\mathbf{d} = \mathrm{d}_{ij}\, \mathbf{g}^i \otimes \mathbf{g}^j \quad = \frac{1}{2} \overset{\bullet}{\overline{\mathbf{g}_i \cdot \mathbf{g}_j}}\, (\mathbf{g}^i \otimes \mathbf{g}^j)$
the spin tensor **w**	$\mathbf{w} = -\,\mathbf{w}^T = \frac{1}{2}(\mathbf{l} - \mathbf{l}^T)$	$\mathbf{w} = \mathrm{w}_{ij}\, \mathbf{g}^i \otimes \mathbf{g}^j \quad = \frac{1}{2}(\mathbf{g}_i \cdot \dot{\mathbf{g}}_j - \dot{\mathbf{g}}_i \cdot \mathbf{g}_j)(\mathbf{g}^i \otimes \mathbf{g}^j)$
connections	$\mathbf{d} = \frac{1}{2} \overset{\bullet}{\overline{\mathbf{g}_i \cdot \mathbf{g}_j}}\, (\mathbf{g}^i \otimes \mathbf{g}^j) = \mathrm{L}_v\, \mathbf{e} = \mathbf{F}^{-T}\, \dot{\mathbf{E}}\, \mathbf{F}^{-1} \longrightarrow \mathrm{d}_{ij} = \dot{\mathrm{e}}_{ij} = \dot{\mathrm{E}}_{ij} = \frac{1}{2} \overset{\bullet}{\overline{\mathbf{g}_i \cdot \mathbf{g}_j}}$	

Interpretation of w. Accordingly, $\mathbf{d} = \mathbf{0}$ signalizes that the instantaneous motion is a rigid body motion. If we now specify (2.9.14) for this special case $\mathbf{l} = \mathbf{w}$ and remember that the tensor **w** is skew-symmetric then we may set according to (1.3.56)

$$\mathrm{d}\mathbf{v} = \mathbf{w}\, \mathrm{d}\mathbf{x} = \boldsymbol{\omega} \times \mathrm{d}\mathbf{x} \,, \tag{2.9.21}$$

where $\boldsymbol{\omega}$ is the axial vector of **w**. It can be proved that the vector $\boldsymbol{\omega}$ entering in (2.9.21) is expressible in the form

$$\boldsymbol{\omega} = \frac{1}{2} \operatorname{curl} \mathbf{v} \,, \tag{2.9.22}$$

where curl **v** is defined by (Green and Zerna 1968):

$$\operatorname{curl} \mathbf{v} = \mathbf{g}^r \times \mathbf{v}_{,r} = \mathbf{g}^r \times (\mathrm{v}_s|_r\, \mathbf{g}^s) \,. \tag{2.9.23}$$

To this end we insert the expression (2.9.17) together with (2.9.9) into (2.9.21). We then obtain by considering (1.3.53), (1.3.54) and (2.9.23):

$$\mathrm{d}\mathbf{v} = \mathbf{w}\, \mathrm{d}\mathbf{x} = \frac{1}{2}(\mathbf{v}_{,i} \otimes \mathbf{g}^i - \mathbf{g}^i \otimes \mathbf{v}_{,i})\, \mathrm{d}\mathbf{x} = \frac{1}{2}(\mathbf{g}^i \times \mathbf{v}_{,i}) \times \mathrm{d}\mathbf{x} = \frac{1}{2} \operatorname{curl} \mathbf{v} \times \mathrm{d}\mathbf{x} \,. \tag{2.9.24}$$

Thus the comparison with (2.9.21) confirms the equality (2.9.22), in view of the arbitrariness of the vector $\mathrm{d}\mathbf{x}$. With (2.9.22) equation (2.9.21) becomes

$$\mathrm{d}\mathbf{v} = \mathbf{w}\, \mathrm{d}\mathbf{x} = \frac{1}{2} \operatorname{curl} \mathbf{v} \times \mathrm{d}\mathbf{x} \,, \tag{2.9.25}$$

which, in turn, justifies to call **w** *vorticity* tensor.

Important connections. The rate of deformation tensor **d** defined in (2.9.16) and the material time derivatives of the GREEN-LAGRANGE tensor $\dot{\mathbf{E}}$ and the ALMANSI tensor $\dot{\mathbf{e}}$ can be transformed into each other by means of the deformation gradient **F** and the spatial gradient of velocity $\mathbf{l}$. To derive the corresponding relations attention is first focused on equation (2.5.8), which by considering $d\mathbf{X} = \mathbf{F}^{-1}\, d\mathbf{x}$ gives:

$$\frac{D}{Dt}(ds^2 - dS^2) = \frac{D}{Dt}(ds^2) = 2\, d\mathbf{X}\, \dot{\mathbf{E}}\, d\mathbf{X} = 2\, d\mathbf{x}\, (\mathbf{F}^{-T}\, \dot{\mathbf{E}}\, \mathbf{F}^{-1})\, d\mathbf{x} \ . \qquad (2.9.26)$$

By using (2.9.14) we similarly obtain from (2.5.14):

$$\frac{D}{Dt}(ds^2 - dS^2) = \frac{D}{Dt}(ds^2) = 2\, (d\dot{\mathbf{x}}\, \mathbf{e}\, d\mathbf{x} + d\mathbf{x}\, \dot{\mathbf{e}}\, d\mathbf{x} + d\mathbf{x}\, \mathbf{e}\, d\dot{\mathbf{x}})$$

$$= 2\, d\mathbf{x}\, (\mathbf{l}^T\, \mathbf{e} + \dot{\mathbf{e}} + \mathbf{e}\, \mathbf{l})\, d\mathbf{x} \ . \qquad (2.9.27)$$

The relations (2.9.20) and (2.9.26), (2.9.27) hold for arbitrary $d\mathbf{x}$. Accordingly, their comparison delivers by considering the component relation (2.9.18) for $\mathbf{d}$:

$$\mathbf{d} = \frac{1}{2}\, \overline{\mathbf{g}_i \cdot \mathbf{g}_j}^{\,\cdot}\, (\mathbf{g}^i \otimes \mathbf{g}^j) = L_v\, \mathbf{e} = \dot{\mathbf{e}} + \mathbf{l}^T\, \mathbf{e} + \mathbf{e}\, \mathbf{l} = \mathbf{F}^{-T}\, \dot{\mathbf{E}}\, \mathbf{F}^{-1} \ , \qquad (2.9.28)$$

where in accordance with (2.5.15) and (2.5.17)

$$L_v\mathbf{e} = \dot{e}_{ij}\, \mathbf{g}^i \otimes \mathbf{g}^j = \frac{1}{2}\, \overline{\mathbf{g}_i \cdot \mathbf{g}_j}^{\,\cdot}\, (\mathbf{g}^i \otimes \mathbf{g}^j) = \dot{\mathbf{e}} + \mathbf{l}^T\, \mathbf{e} + \mathbf{e}\, \mathbf{l} \ . \qquad (2.9.29)$$

In view of this result and (2.5.9), (2.5.12) we also have

$$d_{ij} = \dot{e}_{ij} = \dot{E}_{ij} = \frac{1}{2}\, \overline{\mathbf{g}_i \cdot \mathbf{g}_j}^{\,\cdot} \qquad (2.9.30)$$

corresponding to the component form of the relation (2.9.28). The operator L_v used in (2.9.29) defines the LIE-derivative which is understood as the material time derivative of an arbitrary spatial tensor holding the deformed basis constant. From (2.9.29) it is apparent that the tensor $\mathbf{l}$ enables to express the LIE-derivative $L_v\, \mathbf{e}$ in absolute notation.

Remark. Concerning the equalities given in (2.9.30) we notice the following:

$$d^{ij} := d_{mn}\, g^{mi}\, g^{nj} = \dot{e}_{mn}\, g^{mi}\, g^{nj} \neq \dot{e}^{ij} := \overline{e_{mn}\, g^{mi}\, g^{nj}}^{\,\cdot} \ , \qquad (2.9.31)$$

$$d^{ij} := d_{mn}\, g^{mi}\, g^{nj} = \dot{E}_{mn}\, g^{mi}\, g^{nj} \neq \dot{E}^{ij} := \dot{E}_{mn}\, G^{mi}\, G^{nj} \ . \qquad (2.9.32)$$

For components defined with respect to the deformed basis the order of forming the material time derivative and the raising of indices may not be interchanged:

$$\overline{e^{ij}}^{\,\cdot} = \overline{e_{mn}\, g^{mi}\, g^{nj}}^{\,\cdot} \neq \dot{e}_{mn}\, g^{mi}\, g^{nj} \ . \qquad (2.9.33)$$

Table 2.3 summarizes important results established in connection with the spatial gradient of velocity.

2.10 Pull-back and push-forward operations

Pull-back and push-forward. *Pull-back* and *push-forward* are operations which transport the components of a tensor from the deformed basis into the undeformed one or vice versa. They will be mainly used for forming the so-called *LIE-derivatives* of spatial tensors. To introduce them we repeat the transformations (2.2.6) and (2.2.7)

$$\mathbf{g}_i = \mathbf{F}\,\mathbf{G}_i\,, \qquad \mathbf{G}_i = \mathbf{F}^{-1}\,\mathbf{g}_i$$

$$\mathbf{g}^i = \mathbf{F}^{-T}\,\mathbf{G}^i\,, \qquad \mathbf{G}^i = \mathbf{F}^T\,\mathbf{g}^i\,, \tag{2.10.1}$$

between the undeformed $\mathbf{G}_i$ and deformed base vectors $\mathbf{g}_i$ and consider as an example a second-order spatial tensor $\mathbf{S}$

$$\mathbf{S} = S_{ij}\,\mathbf{g}^i \otimes \mathbf{g}^j = S^{ij}\,\mathbf{g}_i \otimes \mathbf{g}_j \tag{2.10.2}$$

with components referring to the deformed basis. The rules to be established for pull-back and push-forward operations depend on the type of tensor components. Emphasis is first given to the covariant components S_{ij}.

Pull-back $\mathbf{\Phi}^*$ is an operation which transports the components of a spatial tensor $\mathbf{S}$ into the undeformed basis to obtain a material tensor which is given, in view of (2.10.1), by

$$\mathbf{\Phi}^*(\mathbf{S}) = S_{ij}\,\mathbf{G}^i \otimes \mathbf{G}^j = \mathbf{F}^T\,(S_{ij}\,\mathbf{g}^i \otimes \mathbf{g}^j)\,\mathbf{F} = \mathbf{F}^T\,\mathbf{S}\,\mathbf{F}\,. \tag{2.10.3}$$

Conversely, *push-forward* $\mathbf{\Phi}_*$ is an operation transporting the components of a material tensor $\mathbf{T}$

$$\mathbf{T} = T_{ij}\,\mathbf{G}^i \otimes \mathbf{G}^j = T^{ij}\,\mathbf{G}_i \otimes \mathbf{G}_j \tag{2.10.4}$$

into the deformed basis to obtain a spatial tensor. The corresponding transformation is of the form:

$$\mathbf{\Phi}_*(\mathbf{T}) = T_{ij}\,\mathbf{g}^i \otimes \mathbf{g}^j = \mathbf{F}^{-T}\,(T_{ij}\,\mathbf{G}^i \otimes \mathbf{G}^j)\,\mathbf{F}^{-1} = \mathbf{F}^{-T}\,\mathbf{T}\,\mathbf{F}^{-1}\,. \tag{2.10.5}$$

The above rules hold for covariant components. If the tensors $\mathbf{S}$ and $\mathbf{T}$ are used with the contravariant components S^{ij} and T^{ij} then they are to be replaced by the following ones:

Pull-back $\mathbf{\Phi}^*$:

$$\mathbf{\Phi}^*(\mathbf{S}) = S^{ij}\,\mathbf{G}_i \otimes \mathbf{G}_j = \mathbf{F}^{-1}\,(S^{ij}\,\mathbf{g}_i \otimes \mathbf{g}_j)\,\mathbf{F}^{-T} = \mathbf{F}^{-1}\,\mathbf{S}\,\mathbf{F}^{-T}\,, \tag{2.10.6}$$

Push-forward $\mathbf{\Phi}_*$:

$$\mathbf{\Phi}_*(\mathbf{T}) = T^{ij}\,\mathbf{g}_i \otimes \mathbf{g}_j = \mathbf{F}\,(T^{ij}\,\mathbf{G}_i \otimes \mathbf{G}_j)\,\mathbf{F}^T = \mathbf{F}\,\mathbf{T}\,\mathbf{F}^T\,. \tag{2.10.7}$$

The rules introduced above are illustrated in Fig. 2.11. It is clear that, if pull-back and push-forward are applied successively to a tensor with the same component type then the final result will be the initial tensor. This can be confirmed by means of (2.10.3) and (2.10. 5):

$$\mathbf{\Phi}^* (\mathbf{S}) = \mathbf{F}^T \mathbf{S}\, \mathbf{F} \quad \rightarrow \quad \mathbf{\Phi}_* [\mathbf{\Phi}^* (\mathbf{S})] = \mathbf{F}^{-T} (\mathbf{F}^T \mathbf{S}\, \mathbf{F})\, \mathbf{F}^{-1} = \mathbf{S} \; . \qquad (2.10.8)$$

We again point out that the operators $\mathbf{\Phi}^*$ and $\mathbf{\Phi}_*$ depend upon the type of the tensor components intended to be moved into another basis.

As an example we consider the right CAUCHY-GREEN tensor $\mathbf{C}$ and the identity tensor $\mathbf{g}$ which are, in view of the equality $C_{ij} = g_{ij}$, related to each other by

$$\mathbf{C} = C_{ij}\, \mathbf{G}^i \otimes \mathbf{G}^j = \mathbf{\Phi}^* (\mathbf{g}) = \mathbf{F}^T \mathbf{g}\, \mathbf{F} \,, \quad \mathbf{g} = g_{ij}\, \mathbf{g}^i \otimes \mathbf{g}^j = \mathbf{\Phi}_* (\mathbf{C}) = \mathbf{F}^{-T} \mathbf{C}\, \mathbf{F}^{-1} \; . \qquad (2.10.9)$$

In the case of the GREEN-LAGRANGE strain tensor $\mathbf{E}$ and the ALMANSI strain tensor $\mathbf{e}$ we similarly have

$$\mathbf{E} = E_{ij}\, \mathbf{G}^i \otimes \mathbf{G}^j = \mathbf{\Phi}^* (\mathbf{e}) = \mathbf{F}^T \mathbf{e}\, \mathbf{F} \,, \quad \mathbf{e} = e_{ij}\, \mathbf{g}^i \otimes \mathbf{g}^j = \mathbf{\Phi}_* (\mathbf{E}) = \mathbf{F}^{-T} \mathbf{E}\, \mathbf{F}^{-1} \,, \qquad (2.10.10)$$

again due to the equality of the covariant components, $E_{ij} = e_{ij}$.

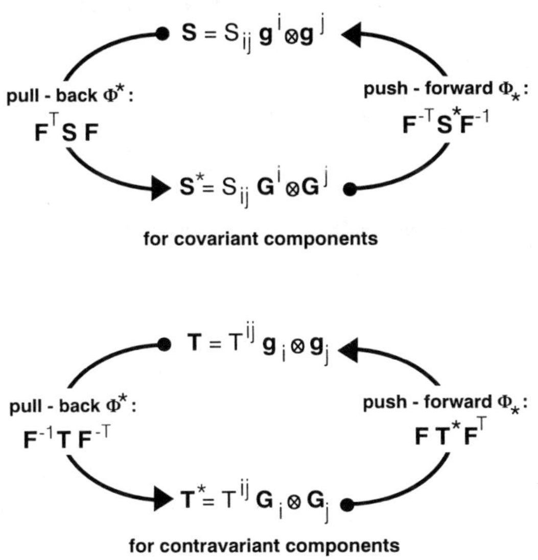

Fig. 2.11. Pull-back and push-forward operations for tensors with covariant and contravariant components

LIE-derivative. Our next aim is to show how the operators $\boldsymbol{\Phi}^*$ and $\boldsymbol{\Phi}_*$ can be used to form the LIE-derivative of a spatial tensor. The first example is the ALMANSI tensor $\mathbf{e} = e_{ij}\, \mathbf{g}^i \otimes \mathbf{g}^j$ with covariant components. Its material time derivative – denoted by $(\dot{\ })$ – is given by

$$\dot{\mathbf{e}} = \frac{D\mathbf{e}}{Dt} = \dot{e}_{ij}\, \mathbf{g}^i \otimes \mathbf{g}^j + e_{ij}\, \dot{\mathbf{g}}^i \otimes \mathbf{g}^j + e_{ij}\, \mathbf{g}^i \otimes \dot{\mathbf{g}}^j \ . \tag{2.10.11}$$

The LIE-derivative of $\mathbf{e}$ corresponds to the first term of the right-hand side of this equation

$$L_v\, \mathbf{e} = \overset{\nabla}{\mathbf{e}} = \dot{e}_{ij}\, \mathbf{g}^i \otimes \mathbf{g}^j \ . \tag{2.10.12}$$

Accordingly, the *LIE-derivative* can be regarded as a material time derivative of a spatial variable to be formed by considering the deformed basis constant. The main advantage of the LIE-derivative is that its application to a strain tensor, e.g. ALMANSI tensor, leads to *objective rates* of this tensor.

By using (2.10.3) and (2.10.5) the LIE-derivative of $\mathbf{e}$ can be constructed by the following calculation steps:

pull-back:

$$e_{ij}\, \mathbf{G}^i \otimes \mathbf{G}^j = \boldsymbol{\Phi}^*(\mathbf{e}) = \mathbf{F}^T \mathbf{e}\, \mathbf{F} \ , \tag{2.10.13}$$

form material time derivative:

$$\dot{e}_{ij}\, \mathbf{G}^i \otimes \mathbf{G}^j = \frac{D}{Dt} \boldsymbol{\Phi}^*(\mathbf{e}) = \dot{\mathbf{F}}^T \mathbf{e}\, \mathbf{F} + \mathbf{F}^T \dot{\mathbf{e}}\, \mathbf{F} + \mathbf{F}^T \mathbf{e}\, \dot{\mathbf{F}} \ , \tag{2.10.14}$$

push-forward:

$$\begin{aligned} \dot{e}_{ij}\, \mathbf{g}^i \otimes \mathbf{g}^j &= \boldsymbol{\Phi}_*\left[\frac{D}{Dt} \boldsymbol{\Phi}^*(\mathbf{e})\right] = \mathbf{F}^{-T}\left[\frac{D}{Dt} \boldsymbol{\Phi}^*(\mathbf{e})\right]\mathbf{F}^{-1} \\ &= \mathbf{F}^{-T}\, \dot{\mathbf{F}}^T \mathbf{e} + \dot{\mathbf{e}} + \mathbf{e}\, \dot{\mathbf{F}}\, \mathbf{F}^{-1} \ . \end{aligned} \tag{2.10.15}$$

The expression (2.10.15) is identical with the LIE-derivative (2.10.12). We may therefore write

$$L_v\, \mathbf{e} = \overset{\nabla}{\mathbf{e}} = \dot{e}_{ij}\, \mathbf{g}^i \otimes \mathbf{g}^j = \dot{\mathbf{e}} + \mathbf{l}^T \mathbf{e} + \mathbf{e}\, \mathbf{l} \ , \tag{2.10.16}$$

where the spatial gradient of velocity $\mathbf{l}$ defined in (2.9.11)

$$\mathbf{l} = \dot{\mathbf{F}}\, \mathbf{F}^{-1} \tag{2.10.17}$$

occurs as abbreviation. We observe that pull-back and push-forward serve to express the LIE-derivative (2.10.12) in absolute notation. As a further example we consider the CAUCHY stress tensor $\boldsymbol{\sigma} = \sigma^{ij}\, \mathbf{g}_i \otimes \mathbf{g}_j$ with contravariant components. In this case the LIE-derivative

$$L_v\, \boldsymbol{\sigma} = \overset{\nabla}{\boldsymbol{\sigma}} = \dot{\sigma}^{ij}\, \mathbf{g}_i \otimes \mathbf{g}_j \tag{2.10.18}$$

is obtained by the following calculation steps:

pull-back:

$$\sigma^{ij}\,\mathbf{G}_i \otimes \mathbf{G}_j \;=\; \boldsymbol{\Phi}^*(\boldsymbol{\sigma}) \;=\; \mathbf{F}^{-1}\,\boldsymbol{\sigma}\,\mathbf{F}^{-T} \;, \tag{2.10.19}$$

form material time derivative:

$$\dot{\sigma}^{ij}\,\mathbf{G}_i \otimes \mathbf{G}_j \;=\; \frac{D}{Dt}\boldsymbol{\Phi}^*(\boldsymbol{\sigma}) \;=\; \dot{\overline{\mathbf{F}^{-1}}}\boldsymbol{\sigma}\,\mathbf{F}^{-T} + \mathbf{F}^{-1}\,\dot{\boldsymbol{\sigma}}\,\mathbf{F}^{-T} + \mathbf{F}^{-1}\,\boldsymbol{\sigma}\,\dot{\overline{\mathbf{F}^{-T}}} \;, \tag{2.10.20}$$

push-forward:

$$\begin{aligned} \dot{\sigma}^{ij}\,\mathbf{g}_i \otimes \mathbf{g}_j \;&=\; \boldsymbol{\Phi}_*\left[\frac{D}{Dt}\boldsymbol{\Phi}^*(\boldsymbol{\sigma})\right] \;=\; \mathbf{F}\left[\frac{D}{Dt}\boldsymbol{\Phi}^*(\boldsymbol{\sigma})\right]\mathbf{F}^T \\ &=\; \mathbf{F}\,\dot{\overline{\mathbf{F}^{-1}}}\,\boldsymbol{\sigma} + \dot{\boldsymbol{\sigma}} + \boldsymbol{\sigma}\,\dot{\overline{\mathbf{F}^{-T}}}\,\mathbf{F}^T \;. \end{aligned} \tag{2.10.21}$$

By considering the identities

$$\begin{aligned} &\mathbf{F}\,\mathbf{F}^{-1} = \mathbf{I} \quad \rightarrow \quad \dot{\mathbf{F}}\,\mathbf{F}^{-1} = -\mathbf{F}\,\dot{\overline{\mathbf{F}^{-1}}} = \mathbf{l} \\ &\mathbf{F}^{-T}\,\mathbf{F}^{T} = \mathbf{I} \quad \rightarrow \quad \dot{\overline{\mathbf{F}^{-T}}}\,\mathbf{F}^{T} = -\mathbf{F}^{-T}\,\dot{\mathbf{F}}^{T} = -\mathbf{l}^T \end{aligned} \tag{2.10.22}$$

in accordance with (2.10.17), equation (2.10.21) can be transformed into

$$L_v\,\boldsymbol{\sigma} \;=\; \overset{\nabla}{\boldsymbol{\sigma}} \;=\; \dot{\sigma}^{ij}\,\mathbf{g}_i \otimes \mathbf{g}_j \;=\; \dot{\boldsymbol{\sigma}} - \mathbf{l}\,\boldsymbol{\sigma} - \boldsymbol{\sigma}\,\mathbf{l}^T \;. \tag{2.10.23}$$

Both results (2.10.16) and (2.10.23) demonstrate that the velocity gradient **l** is a variable permitting to describe the connection between the material time derivative and the LIE-derivative of a spatial tensor. We also notice that the LIE-derivative of any spatial tensor **b** can be presented in a general form as

$$L_v\,\mathbf{b} \;=\; \overset{\nabla}{\mathbf{b}} \;=\; \boldsymbol{\Phi}_*\left[\frac{D}{Dt}\boldsymbol{\Phi}^*(\mathbf{b})\right] \tag{2.10.24}$$

in terms of *pull-back* $\boldsymbol{\Phi}^*$ and *push-forward* operation $\boldsymbol{\Phi}_*$ given in (2.10.3) and (2.10.5) for covariant components and in (2.10.6), (2.10.7) for the contravariant components of the tensor **b**. Finally it should be pointed out that

$$\dot{b}^{ij}\,\mathbf{g}_i \otimes \mathbf{g}_j \;\neq\; \dot{b}_{ij}\,\mathbf{g}^i \otimes \mathbf{g}^j \;, \tag{2.10.25}$$

as can easily be observed from the following derivation:

$$\begin{aligned} \dot{b}_{ij}\,\mathbf{g}^i \otimes \mathbf{g}^j \;&=\; \dot{\overline{(b^{mn}\,g_{mi}\,g_{nj})}}\,\mathbf{g}^i \otimes \mathbf{g}^j \\ &=\; (\dot{b}^{mn}\,g_{mi}\,g_{nj} + b^{mn}\,\dot{g}_{mi}\,g_{nj} + b^{mn}\,g_{mi}\,\dot{g}_{nj})\,\mathbf{g}^i \otimes \mathbf{g}^j \;. \end{aligned} \tag{2.10.26}$$

For a comprehensive discussion of pull-back and push-forward operations we refer to Marsden and Hughes 1983 and for specific applications to Wriggers 1988.

The LIE-derivatives of an arbitrary spatial tensor **a** are summarized below for components of different types.

$$
\begin{aligned}
L_v^{b}\,\mathbf{a} &= L_v\,(a_{ij}\,\mathbf{g}^i \otimes \mathbf{g}^j) = \dot{a}_{ij}\,\mathbf{g}^i \otimes \mathbf{g}^j = \dot{\mathbf{a}} + \mathbf{l}^T\mathbf{a} + \mathbf{a}\mathbf{l} \ ,\\
L_v^{\#}\,\mathbf{a} &= L_v\,(a^{ij}\,\mathbf{g}_i \otimes \mathbf{g}_j) = \dot{a}^{ij}\,\mathbf{g}_i \otimes \mathbf{g}_j = \dot{\mathbf{a}} - \mathbf{l}\mathbf{a} - \mathbf{a}\mathbf{l}^T \ ,\\
L_v^{\backslash}\,\mathbf{a} &= L_v\,(a^{i}_{\cdot j}\,\mathbf{g}_i \otimes \mathbf{g}^j) = \dot{a}^{i}_{\cdot j}\,\mathbf{g}_i \otimes \mathbf{g}^j = \dot{\mathbf{a}} - \mathbf{l}\mathbf{a} + \mathbf{a}\mathbf{l} \ ,\\
L_v^{/}\,\mathbf{a} &= L_v\,(a_{i}^{\cdot j}\,\mathbf{g}^i \otimes \mathbf{g}_j) = \dot{a}_{i}^{\cdot j}\,\mathbf{g}^i \otimes \mathbf{g}_j = \dot{\mathbf{a}} + \mathbf{l}^T\mathbf{a} - \mathbf{a}\mathbf{l}^T \ .
\end{aligned} \tag{2.10.27}
$$

2.11 Isotropic tensor functions of C and b

Rotation of the initial and deformed state. We consider a body occupying the position B_o in the initial unstrained state and the position B after deformation. A point P_o in B_o is determined with respect to a global reference frame by the position vector $\mathbf{X}$. After deformation P_o takes the position P and its position vector referring to the same origin 0 is denoted by $\mathbf{x}$. Attention is now focused on the right and left CAUCHY-GREEN tensors $\mathbf{C}$ and $\mathbf{b}$ which have been used in section 2.5 to introduce various strain measures. Now, our aim is first to examine how the tensors $\mathbf{C}$ and $\mathbf{b}$ are transformed if one of the states, the initial state B_o or the actual state B, is subjected to a rotation. In a further step tensor functions will be defined which are said to be isotropic to $\mathbf{C}$ and $\mathbf{b}$. We recall that any rotation may be described by an orthogonal tensor $\mathbf{Q}$ as has been shown in section 1.4.

We first assume the initial state B_o to be rotated by means of an orthogonal tensor $\mathbf{Q}$ into the position $\overset{\blacklozenge}{B}_o$ (Fig. 2.12). This rotation transforms the vectors $\mathbf{G}_i$ associated with the state B_o into $\overset{\blacklozenge}{\mathbf{G}}_i$ in $\overset{\blacklozenge}{B}_o$. We make use of the notation $(\overset{\blacklozenge}{\ldots})$ to characterize variables which are determined by using $\overset{\blacklozenge}{\mathbf{G}}_i$ in place of $\mathbf{G}_i$. Some useful results are recorded below referring to the material description:

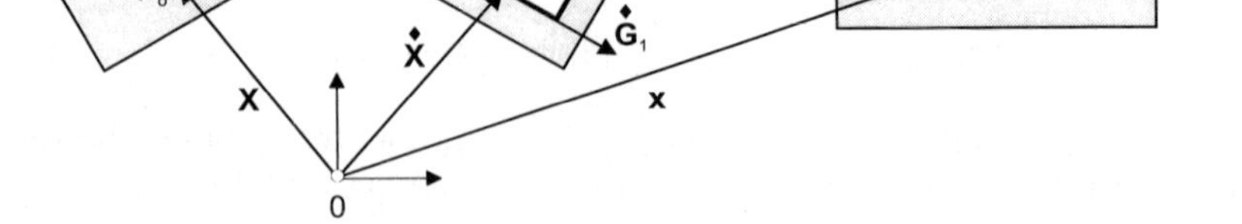

Fig. 2.12. Rotation of the initial state B_o into $\overset{\blacklozenge}{B}_o$ as a 2D illustration

base vectors in $\overset{\diamond}{\mathrm{B}}_o$ *:*

$$\overset{\diamond}{\mathbf{G}}_i = \mathbf{Q}\,\mathbf{G}_i \ , \quad \mathbf{G}_i = \mathbf{Q}^T \overset{\diamond}{\mathbf{G}}_i \ , \tag{2.11.1}$$

deformation gradient:

$$\overset{\diamond}{\mathbf{F}} = \mathbf{g}_i \otimes \overset{\diamond}{\mathbf{G}}{}^i = (\mathbf{g}_i \otimes \mathbf{G}^i)\,\mathbf{Q}^T = \mathbf{F}\,\mathbf{Q}^T \ , \quad \mathbf{F} = \overset{\diamond}{\mathbf{F}}\,\mathbf{Q} \ , \tag{2.11.2}$$

right CAUCHY-GREEN tensor:

$$\overset{\diamond}{\mathbf{C}} = \overset{\diamond}{\mathbf{F}}{}^T \overset{\diamond}{\mathbf{F}} = \mathbf{Q}\,\mathbf{F}^T\,\mathbf{F}\,\mathbf{Q}^T = \mathbf{Q}\,\mathbf{C}\,\mathbf{Q}^T \ , \qquad \mathbf{C} = \mathbf{Q}^T\,\overset{\diamond}{\mathbf{C}}\,\mathbf{Q} \ . \tag{2.11.3}$$

The above results are obtained starting from the definitions given in Table 2.2. From (2.11.3) we see that the transformation $\mathrm{B}_o \rightarrow \overset{\diamond}{\mathrm{B}}_o$ induces the change of the material tensor $\mathbf{C}$ into $\overset{\diamond}{\mathbf{C}}$. But the components of the tensors $\mathbf{C} = C_{ij}\,\mathbf{G}^i \otimes \mathbf{G}^j$ and $\overset{\diamond}{\mathbf{C}} = \overset{\diamond}{C}_{ij}\,\overset{\diamond}{\mathbf{G}}{}^i \otimes \overset{\diamond}{\mathbf{G}}{}^j$ are identical as can be verified by means of (2.11.1) and (2.11.3):

$$\overset{\diamond}{C}_{ij} = \overset{\diamond}{\mathbf{G}}_i\,\overset{\diamond}{\mathbf{C}}\,\overset{\diamond}{\mathbf{G}}_j = \mathbf{G}_i\,\mathbf{Q}^T\,\overset{\diamond}{\mathbf{C}}\,\mathbf{Q}\,\mathbf{G}_j = \mathbf{G}_i\,\mathbf{C}\,\mathbf{G}_j = C_{ij} \ . \tag{2.11.4}$$

A further important result is that the rotation $\mathrm{B}_o \rightarrow \overset{\diamond}{\mathrm{B}}_o$ leaves the spatial variable $\mathbf{b}$, in contrast to $\mathbf{C}$, unchanged. In fact, by considering (2.11.2) we have

$$\overset{\diamond}{\mathbf{b}} = \overset{\diamond}{\mathbf{F}}\,\overset{\diamond}{\mathbf{F}}{}^T = \mathbf{F}\,\mathbf{Q}^T\,\mathbf{Q}\,\mathbf{F}^T = \mathbf{F}\,\mathbf{I}\,\mathbf{F}^T = \mathbf{b} \ . \tag{2.11.5}$$

If we now rotate the deformed state B into $\overset{*}{\mathrm{B}}$ through an arbitrary orthogonal tensor $\mathbf{Q}$ preserving the initial state B_o unchanged (Fig. 2.13) and use the notation $(\overset{*}{\ldots})$ for the tensors which are evaluated by replacing the basis $\mathbf{g}_i$ by the rotated one $\overset{*}{\mathbf{g}}_i = \mathbf{Q}\,\mathbf{g}_i$, we obtain:

base vectors associated with $\overset{*}{\mathrm{B}}$*:*

$$\overset{*}{\mathbf{g}}_i = \mathbf{Q}\,\mathbf{g}_i \ , \quad \mathbf{g}_i = \mathbf{Q}^T\,\overset{*}{\mathbf{g}}_i \ , \tag{2.11.6}$$

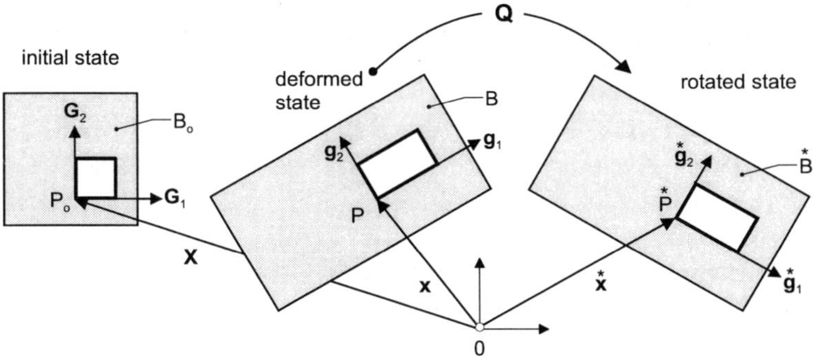

Fig. 2.13. Rotation of the deformed state B into $\overset{*}{\mathrm{B}}$ as a 2D illustration

deformation gradient:

$$\overset{*}{\mathbf{F}} = \overset{*}{\mathbf{g}}_i \otimes \mathbf{G}^i = \mathbf{Q}\,\mathbf{g}_i \otimes \mathbf{G}^i = \mathbf{Q}\,\mathbf{F}\ , \qquad \mathbf{F} = \mathbf{Q}^T\,\overset{*}{\mathbf{F}}\ , \tag{2.11.7}$$

left CAUCHY-GREEN tensor:

$$\overset{*}{\mathbf{b}} = \overset{*}{\mathbf{F}}\,\overset{*}{\mathbf{F}}{}^T = \mathbf{Q}\,\mathbf{F}\,\mathbf{F}^T\,\mathbf{Q}^T = \mathbf{Q}\,\mathbf{b}\,\mathbf{Q}^T\ , \quad \mathbf{b} = \mathbf{Q}^T\,\overset{*}{\mathbf{b}}\,\mathbf{Q}\ . \tag{2.11.8}$$

From (2.11.3) and (2.11.8) it is apparent that $\mathbf{b}$ is transformed into $\overset{*}{\mathbf{b}}$ in the same way as $\mathbf{C}$ into $\overset{\bullet}{\mathbf{C}}$. By means of (2.11.6), (2.11.8) we furthermore confirm that

$$\overset{*}{b}_{ij} = \overset{*}{\mathbf{g}}_i\,\overset{*}{\mathbf{b}}\,\overset{*}{\mathbf{g}}_j = \mathbf{g}_i\,\mathbf{Q}^T\,\overset{*}{\mathbf{b}}\,\mathbf{Q}\,\mathbf{g}_j = \mathbf{g}_i\,\mathbf{b}\,\mathbf{g}_j = b_{ij}\ , \tag{2.11.9}$$

and

$$\overset{*}{\mathbf{C}} = \overset{*}{\mathbf{F}}{}^T\,\overset{*}{\mathbf{F}} = \mathbf{F}^T\,\mathbf{Q}^T\,\mathbf{Q}\,\mathbf{F} = \mathbf{F}^T\,\mathbf{I}\,\mathbf{F} = \mathbf{C}\ . \tag{2.11.10}$$

The above relations are also similar to those given in (2.11.4) and (2.11.5) in connection with the transformation $B_o \rightarrow \overset{\bullet}{B}_o$. Accordingly, we see that the rotation of the initial state $B_o \rightarrow \overset{\bullet}{B}_o$ induces similar changes on the *material* variable $\mathbf{C}$ as the rotation of the actual state $B \rightarrow \overset{*}{B}$ on the spatial variable $\mathbf{b}$. The above results will be used below to give a geometrical interpretation of the so-called *isotropic* tensor functions.

Isotropic tensor functions. A tensor-valued function $\mathbf{G}\,(\mathbf{C})$ is said to be an isotropic tensor function of $\mathbf{C}$ if the transformation

$$\mathbf{G}\,(\overset{\bullet}{\mathbf{C}}) := \mathbf{G}\,(\mathbf{Q}\,\mathbf{C}\,\mathbf{Q}^T) = \mathbf{Q}\,\mathbf{G}\,(\mathbf{C})\,\mathbf{Q}^T \tag{2.11.11}$$

with $\overset{\bullet}{\mathbf{C}} = \mathbf{Q}\,\mathbf{C}\,\mathbf{Q}^T$ holds for all orthogonal tensors $\mathbf{Q}$. By comparing the above transformation with (2.11.3) we see that $\mathbf{G}\,(\overset{\bullet}{\mathbf{C}})$ is obtained from $\mathbf{G}\,(\mathbf{C})$ in the same way as the tensor $\overset{\bullet}{\mathbf{C}}$ from $\mathbf{C}$. Remembering the definition of $\overset{\bullet}{\mathbf{C}}$ we see that $\mathbf{G}\,(\overset{\bullet}{\mathbf{C}})$ is the value of the function $\mathbf{G}\,(\mathbf{C})$ if the initial state B_o is rotated into a new position $\overset{\bullet}{B}_o$ which induces the change of $\mathbf{C}$ into $\overset{\bullet}{\mathbf{C}}$.

By considering the orthogonality condition $\mathbf{Q}^{-1} = \mathbf{Q}^T$ it is not difficult to show that the tensor function $\mathbf{C}^m$, where m is an integer, is isotropic to $\mathbf{C}$. Consequently, tensor power series in $\mathbf{C}$, e.g. of the form

$$e^{\mathbf{C}} = \mathbf{I} + \mathbf{C} + \frac{1}{2!}\,\mathbf{C}^2 + \ldots$$

$$\ln\,(\mathbf{I} + \mathbf{C}) = \mathbf{C} - \frac{1}{2}\,\mathbf{C}^2 + \frac{1}{3}\,\mathbf{C}^3 \ldots \tag{2.11.12}$$

are isotropic tensor functions. It can also be shown that $\mathbf{C}^{1/m}$, where m is an integer, is isotropic*. To accomplish this we use the relation

* $\mathbf{C}^{1/m}$ is defined to be coaxial with $\mathbf{C}$ so that $\mathbf{C}^{1/m} = \sum_{i=1}^{3} (\Lambda_i)^{1/m}\,\mathbf{N}_i \otimes \mathbf{N}_i$, where Λ_i and $\mathbf{N}_i$, respectively, are the eigenvalues and the eigenvectors of $\mathbf{C}$.

$$(\mathbf{Q}\,\mathbf{C}^{1/m}\,\mathbf{Q}^T)^m = (\mathbf{Q}\,\mathbf{C}^{1/m}\,\mathbf{Q}^T)\,(\mathbf{Q}\,\mathbf{C}^{1/m}\,\mathbf{Q}^T)\ldots(\mathbf{Q}\,\mathbf{C}^{1/m}\,\mathbf{Q}^T) = \mathbf{Q}\,\mathbf{C}\,\mathbf{Q}^T \qquad (2.11.13)$$

leading to

$$\mathbf{Q}\,\mathbf{C}^{1/m}\,\mathbf{Q}^T = (\mathbf{Q}\,\mathbf{C}\,\mathbf{Q}^T)^{1/m}\ . \qquad (2.11.14)$$

Hence $\mathbf{C}^{1/m}$ is an *isotropic tensor* function. Notice that, if $\mathbf{G}\,(\mathbf{C})$ is isotropic, so is $\mathbf{G}^{-1}\,(\mathbf{C})$. Therefore, $\mathbf{C}^{-1/m}$ is isotropic. Hence, $\mathbf{U} = \mathbf{C}^{1/2}$ and $\mathbf{U}^{-1} = \mathbf{C}^{-1/2}$ are isotropic tensor functions of $\mathbf{C}$ (Ting 1985).

Isotropic tensor functions $\mathbf{g}\,(\mathbf{b})$ of $\mathbf{b}$ are defined similarly by the requirement

$$\mathbf{g}\,(\overset{*}{\mathbf{b}}) := \mathbf{g}\,(\mathbf{Q}\,\mathbf{b}\,\mathbf{Q}^T) = \mathbf{Q}\,\mathbf{g}\,(\mathbf{b})\,\mathbf{Q}^T \qquad (2.11.15)$$

which can be used to show that $\mathbf{b}^m$ and $\mathbf{b}^{1/m}$ with integer values of m are isotropic to $\mathbf{b}$. Note that the transformation (2.11.15) is similar to (2.11.8) and describes in this sense the change of $\mathbf{g}\,(\mathbf{b})$ caused by the rotation of the deformed state B into $\overset{*}{\mathrm{B}}$, e.i. the transformation $\mathbf{b}$ into $\overset{*}{\mathbf{b}}$.

The tensor functions $\mathbf{C}^{m/2}$ and $\mathbf{b}^{m/2}$ with an arbitrary real number m have been used by Seth 1964 to introduce the following generalized strain measures in *material* description

$$\mathbf{E}^{(m)} = \begin{cases} \dfrac{1}{m}\,(\mathbf{C}^{m/2} - \mathbf{I}) & m \neq 0\ , \\[2mm] \dfrac{1}{2}\ln \mathbf{C} & m = 0\ , \end{cases} \qquad (2.11.16)$$

and in *spatial* description

$$\mathbf{e}^{(m)} = \begin{cases} \dfrac{1}{m}\,(\mathbf{I} - \mathbf{b}^{-m/2}) & m \neq 0\ , \\[2mm] \dfrac{1}{2}\ln \mathbf{b} & m = 0\ . \end{cases} \qquad (2.11.17)$$

For integer values of m, the above equations reduce to the strain measures of GREEN-LAGRANGE ($\mathbf{E}$), ALMANSI ($\mathbf{e}$), HENCKY ($\ln \mathbf{U}$), etc. In fact the validity of the relations (2.11.16) and (2.11.17) for noninteger real values of m is assured if $1/m\,(\mathbf{C}^{m/2}\!-\!\mathbf{I})$ and $1/m\,(\mathbf{I}\!-\!\mathbf{b}^{-m/2})$, respectively, are defined to be coaxial with the right and left stretch tensors $\mathbf{U} = \mathbf{C}^{1/2}$ and $\mathbf{v} = \mathbf{b}^{1/2}$ and having principal values $1/m\,[(\lambda_i)^m\!-\!1]$ and $1/m\,[1\!-\!(\lambda_i)^{-m}]$ ($i = 1, 2, 3$), λ_i being the eigenvalues common to $\mathbf{U}$ and $\mathbf{v}$ (Morman 1986). As pointed out by Seth 1964, by introducing a degree of freedom in the exponent m, it is possible to condense the nonlinear effects of deformation into the definition of strain, and thus rely less on representing the nonlinear behaviour in the constitutive equations.

Application of the representation theorem. The representation theorem by Truesdell and Noll 1965 states that any isotropic function $\mathbf{f}\,(\mathbf{b})$ may be expressed by a quadratic polynomial in $\mathbf{b}$

$$\mathbf{f}\,(\mathbf{b}) = \delta_0\,\mathbf{I} + \delta_1\,\mathbf{b} + \delta_2\,\mathbf{b}^2\ , \qquad (2.11.18)$$

where δ_0, δ_1 and δ_2 are functions of the invariants of **b** given according to (1.8.1) to (1.8.3) by:

$$\mathrm{I}_{\mathbf{b}} = \mathrm{tr}\,\mathbf{b}\ , \quad \mathrm{II}_{\mathbf{b}} = \frac{1}{2}\left[(\mathrm{I}_{\mathbf{b}})^2 - \mathrm{tr}\,\mathbf{b}^2\right]\ , \quad \mathrm{III}_{\mathbf{b}} = \det\mathbf{b}\ . \tag{2.11.19}$$

The representation theorem (2.11.18) provides, by using the eigenvalues $(\lambda_i)^2$ and the invariants of **b**, useful representations for $\mathbf{b}^{m/2}$ without recourse to eigenvector calculations.

To show this we recall that all the tensors occuring in (2.11.18) are coaxial. If we replace them by their spectral decompositions we receive for $\mathbf{f}(\mathbf{b}) = \mathbf{b}^{m/2}$

$$\begin{aligned} f\left[(\lambda_1)^2\right] &:= (\lambda_1)^m = \delta_0 + \delta_1(\lambda_1)^2 + \delta_2(\lambda_1)^4\ , \\ f\left[(\lambda_2)^2\right] &:= (\lambda_2)^m = \delta_0 + \delta_1(\lambda_2)^2 + \delta_2(\lambda_2)^4\ , \\ f\left[(\lambda_3)^2\right] &:= (\lambda_3)^m = \delta_0 + \delta_1(\lambda_3)^2 + \delta_2(\lambda_3)^4\ , \end{aligned} \tag{2.11.20}$$

where $(\lambda_i)^2$ (i = 1, 2, 3) are the principal values of **b**. When the λ_i are distinct, the system of equations (2.11.20) has a unique solution for δ_0, δ_1 and δ_2 given by

$$\begin{aligned} \delta_0 &= \sum_{i=1}^{3} \frac{f\,[(\lambda_i)^2]\,\mathrm{III}_{\mathbf{b}}\,(\lambda_i)^{-2}}{[2\,(\lambda_i)^4 - \mathrm{I}_{\mathbf{b}}\,(\lambda_i)^2 + \mathrm{III}_{\mathbf{b}}\,(\lambda_i)^{-2}]}\ , \\ \delta_1 &= \sum_{i=1}^{3} \frac{f\,[(\lambda_i)^2]\,[(\lambda_i)^2 - \mathrm{I}_{\mathbf{b}}]}{[2\,(\lambda_i)^4 - \mathrm{I}_{\mathbf{b}}\,(\lambda_i)^2 + \mathrm{III}_{\mathbf{b}}\,(\lambda_i)^{-2}]}\ , \\ \delta_2 &= \sum_{i=1}^{3} \frac{f\,[(\lambda_i)^2]}{[2\,(\lambda_i)^4 - \mathrm{I}_{\mathbf{b}}\,(\lambda_i)^2 + \mathrm{III}_{\mathbf{b}}\,(\lambda_i)^{-2}]}\ . \end{aligned} \tag{2.11.21}$$

Inserting the expressions (2.11.21) into (2.11.18) yields

$$\mathbf{b}^{m/2} = \sum_{i=1}^{3} (\lambda_i)^m \left[\frac{\mathbf{b}^2 - [\mathrm{I}_{\mathbf{b}} - (\lambda_i)^2]\,\mathbf{b} + \mathrm{III}_{\mathbf{b}}\,(\lambda_i)^{-2}\,\mathbf{I}}{2\,(\lambda_i)^4 - \mathrm{I}_{\mathbf{b}}\,(\lambda_i)^2 + \mathrm{III}_{\mathbf{b}}\,(\lambda_i)^{-2}}\right] \tag{2.11.22}$$

with the invariants $\mathrm{I}_{\mathbf{b}}$, $\mathrm{III}_{\mathbf{b}}$ defined in (2.11.19).The eigenvalues λ_i can be determined from the characteristic equation

$$\lambda^6 - \mathrm{I}_{\mathbf{b}}\,\lambda^4 + \mathrm{II}_{\mathbf{b}}\,\lambda^2 - \mathrm{III}_{\mathbf{b}} = 0\ , \tag{2.11.23}$$

which yields (Morman 1986)

$$(\lambda_j)^2 = \frac{1}{3}\left[\mathrm{I}_{\mathbf{b}} + 2\,(\mathrm{I}_{\mathbf{b}}^2 - 3\,\mathrm{II}_{\mathbf{b}})^{1/2}\cos\frac{1}{3}(\Theta + 2\pi j)\right]\ , \quad (j = 1, 2, 3) \tag{2.11.24}$$

where

$$\Theta = \arccos \left[\frac{2\, I_b^3 - 9\, I_b\, II_b + 27\, III_b}{2\, (I_b^2 - 3\, II_b)^{3/2}} \right] . \tag{2.11.25}$$

Comparing (2.11.22) with the spectral representation for f (**b**)

$$f(\mathbf{b}) := \mathbf{b}^{m/2} = \sum_{i=1}^{3} (\lambda_i)^m\, \mathbf{n}_i \otimes \mathbf{n}_i \ , \tag{2.11.26}$$

where $\mathbf{n}_i$ are the eigenvectors of **b** we obtain

$$\mathbf{n}_i \otimes \mathbf{n}_i = \frac{\mathbf{b}^2 - [I_b - (\lambda_i)^2]\, \mathbf{b} + III_b\, (\lambda_i)^{-2}\, \mathbf{I}}{[2\, (\lambda_i)^4 - I_b\, (\lambda_i)^2 + III_b\, (\lambda_i)^{-2}]} . \tag{2.11.27}$$

If $\lambda_1 = \lambda_2 \neq \lambda_3$ it is easily deduced from equations (2.11.26) and (2.11.27) that

$$\mathbf{b}^{m/2} = (\lambda_1)^m\, \mathbf{I} + [(\lambda_3)^m - (\lambda_1)^m] \left[\frac{\mathbf{b}^2 - [I_b - (\lambda_3)^2]\, \mathbf{b} + III_b\, (\lambda_3)^{-2}\, \mathbf{I}}{2\, (\lambda_3)^4 - I_b\, (\lambda_3)^2 + III_b\, (\lambda_3)^{-2}} \right] \tag{2.11.28}$$

while for $\lambda_1 = \lambda_2 = \lambda_3$ we have

$$\mathbf{b}^{m/2} = (\lambda_1)^m\, \mathbf{I} . \tag{2.11.29}$$

An analogous representation for $\mathbf{C}^{m/2}$ can be given by a similar procedure.

Exercises

2.1. A unit vector **N** is given in a point P_0 of a body in its undeformed state B_0. Express the unit vector **n** which determines the direction of **N** in the actual deformed state in terms of the deformation gradient **F** and **N**.

2.2. Evaluate the stretch in a given direction **N** of an undeformed body in terms of the deformation gradient **F** and the left stretch tensor **v**.

2.3. Fig. 2.14 shows the deformed configuration of an infinitesimal quadrilateral volume element dV_0 subjected to in-plane deformations so that $\mathbf{G}_3 = \mathbf{g}_3$. Evaluate the principal stretches λ_i as well as the components of the rotation tensor **R** involved in the polar decomposition theorem with respect to the basis $\mathbf{i}_i \otimes \mathbf{i}_j$.

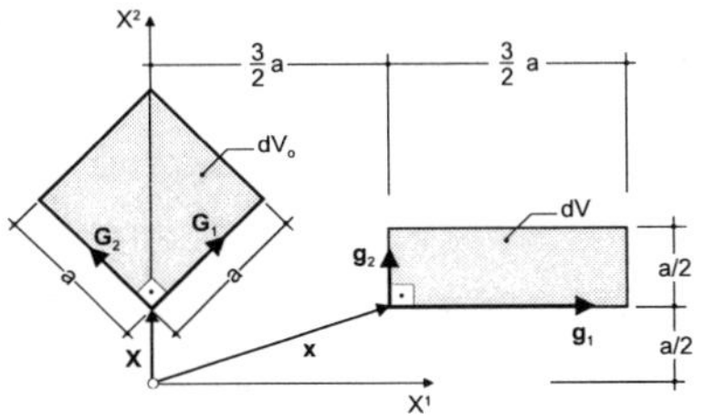

Fig. 2.14. Illustration to exercise 2.3

2.4. Form the first and the second variation of the invariants of the right CAUCHY-GREEN tensor **C**.

2.5. Show that a tensor **A** and its deviatoric part dev **A** are coaxial.

2.6. During a deformation the nodal point 3 of a finite element selected in a mid-surface of a plate is moved in the position $\overline{3}$ indicated in Fig. 2.15 while the other ones remain unchanged. A plane deformation is assumed so that $\mathbf{G}_3 = \mathbf{g}_3$. Approximating the displacement field by a bilinear interpolation polynomial within the finite element area evaluate the following variables in terms of isoparametric coordinates ξ^i: the base vectors $\mathbf{g}_i$; the deformation gradient **F** and its components with respect to $\mathbf{G}_i \otimes \mathbf{G}_j$; the components of the GREEN-LAGRANGE strain tensor **E** with respect to $\mathbf{G}_i \otimes \mathbf{G}_j$; the components of the ALMANSI strain tensor **e** with respect to $\mathbf{g}_i \otimes \mathbf{g}_j$; the components of the right-stretch tensor **U** in principal directions at the nodal point 3; the right-stretch tensor **U** by means of the polar decomposition theorem (select a suitable rotation vector to express the rotation tensor).

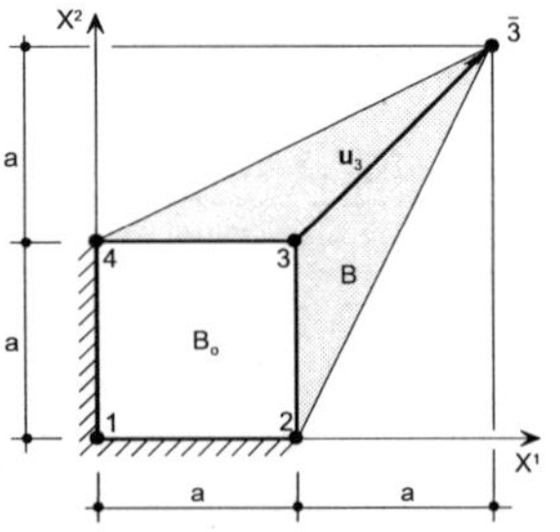

Fig. 2.15. Illustration to exercise 2.6

2.7. Show that det $\bar{\mathbf{b}} = 1$ if $\bar{\mathbf{b}} = J^{-2/3}\,\mathbf{b}$.

2.8. Show that $\mathbf{n}\,\mathbf{l}^T\,\mathbf{n} = \mathbf{n}\,\mathbf{d}\,\mathbf{n}$ for an arbitrary vector **n**.

2.9. Show that the real eigenvalue of a skew-symmetric tensor $\mathbf{A} = -\mathbf{A}^T$ is zero and that the corresponding eigenvector shows in direction of the axial vector of **A**.

2.10. Let $\overset{*}{\mathrm{B}}$ a new configuration of a body obtained by the application of an orthogonal tensor **Q** to the position vector **x** of the deformed configuration B. Evaluate the values of $\mathbf{C}^2$, $\mathbf{b}^2$ and $\mathrm{III}_\mathbf{C}$, $\mathrm{III}_\mathbf{b}$ in the new position $\overset{*}{\mathrm{B}}$.

2.11. The tensor functions $\Psi\,(\mathbf{E})$ and $\bar{\Psi}\,(\mathbf{C})$ are related by

$$\Psi\,(\mathbf{E}) = \Psi\left(\frac{1}{2}\,(\mathbf{C}-\mathbf{G})\right) = \bar{\Psi}\,(\mathbf{C})\ .$$

Establish the relation between the partial derivatives $\Psi_{,\mathbf{E}}$ and $\bar{\Psi}_{,\mathbf{C}}$.

2.12. $\Psi\,(\mathbf{U})$ and $\bar{\Psi}\,(\mathbf{C})$ are tensor functions of the right stretch tensor **U** and the right CAUCHY-GREEN tensor **C**, respectively such that $\bar{\Psi}(\mathbf{C}) = \bar{\Psi}(\mathbf{U}^2) = \Psi(\mathbf{U})$. Establish the relation between the partial derivatives $\Psi_{,\mathbf{U}}$ and $\bar{\Psi}_{,\mathbf{C}}$.

2.13. Express the following tensors in terms of the eigenvalues Λ_i and the eigenvectors $\mathbf{n}_i$ of **b**: $\mathbf{b}^{-1}$, ln **b**, tr **b**, $\mathbf{b}^m$.

2.14. Show that $\mathrm{I}_\mathbf{C} = 3$, $\mathrm{II}_\mathbf{C} = 3$ and $\mathrm{III}_\mathbf{C} = 1$ for the unstrained state B_o of the body.

2.15. Show that $\mathrm{I}_\mathbf{b} = G^{rs}\,g_{rs}$, $\mathrm{II}_\mathbf{b} = \frac{1}{2}\left(\mathrm{I}_\mathbf{b}^2 - G^{rm}\,G^{sn}\,g_{rs}\,g_{mn}\right)$, $\mathrm{III}_\mathbf{b} = \left|G^{rm}\,g_{ms}\right| = g/G$.

3 Stresses

After a detailed discussion of the CAUCHY stress vector and stress tensor this section introduces various stress tensors, shows systematically their connections and finally relates them to the deformation measures of chapter 2 as energy conjugate quantities. Particular attention is given to the physical interpretation of the CAUCHY stress tensor presenting the starting point of the derivations.

3.1 CAUCHY stress tensor

We consider a body with the initial position B_0 taking under external force actions the position B at time t. External forces induce in a body internal forces which can be described by the stress tensors to be defined in this section.

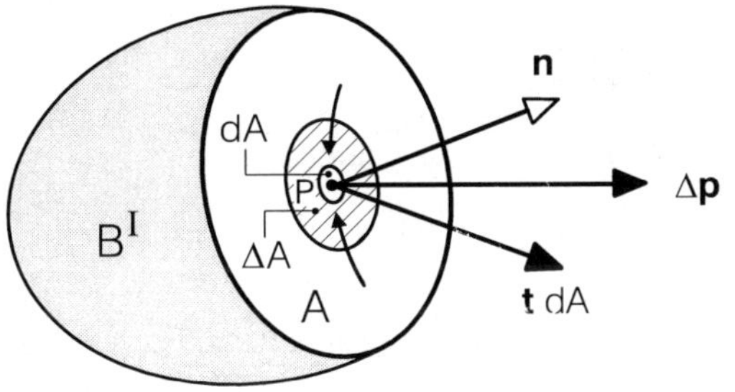

Fig. 3.1. Definition of the CAUCHY stress vector **t**

Be P an arbitrary characteristic point inside the body in its deformed state B. We separate B into two parts B^I and B^{II} by a smooth surface passing through P. Then, one can define ΔA as surface element in the vicinity of P with $\mathbf{n}$ as outward unit normal to ΔA at P. The forces exerted on the part B^I of the body across the area ΔA is equivalent to a force $\Delta\mathbf{p}$ at P and a couple $\Delta\mathbf{m}$. We now imagine that the area ΔA tends to zero keeping P as inner point. A basic postulate of continuum mechanics is that the vector $\Delta\mathbf{p}/\Delta A$ tends to a defined limit

$$\mathbf{t} = \lim_{\Delta A \to 0} \frac{\Delta\mathbf{p}}{\Delta A} \tag{3.1.1}$$

with ΔA approaching zero. On the contrary, we assume that the vector $\Delta \mathbf{m}/\Delta A$ vanishes in this limit case. This excludes the possibility that there may exist continuously distributed couples, which would lead in the limit case to so-called *couple stresses*. A basic assumption of classical continuum mechanics is that the action of one body on another across an infinitesimal surface area dA is adequately represented by a stress vector as defined in (3.1.1).

Stress vector. The stress vector **t** refers to a deformed surface element whose unit normal vector is **n** and represents a force per unit area of the deformed surface A. The vector **t** is invariant since its definition does not depend on the selection of coordinates.

The *CAUCHY postulate* states that the vector **t** remains unchanged for all surfaces having at P the same normal vector **n**. This means that the limit value **t** is independent of the surfaces chosen as long as they all have at P the same normal vector. This feature is illustrated in Fig. 3.2.

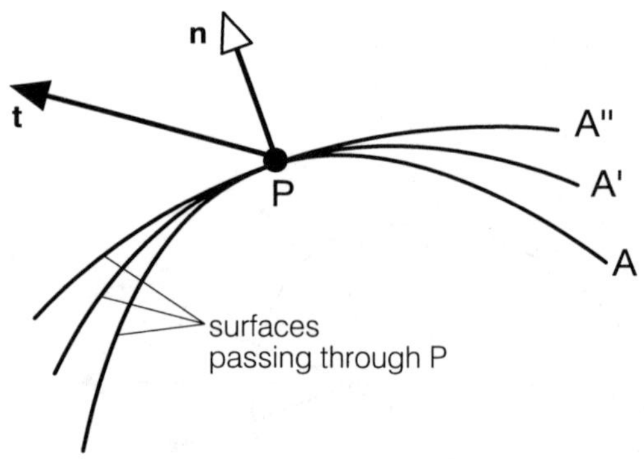

Fig. 3.2. CAUCHY stress vector **t** of surfaces having at **P** the same normal vector **n**

To introduce suitable variables for the description of the stress state we consider at a point P of the deformed body an infinitesimal parallelepiped bounded by the faces Θ^i = constant and $\Theta^i + d\Theta^i$ = constant (Fig. 3.3). In the limit, the edges of this volume element are determined by the vectorial elements

$$d\mathbf{s}_{<i>} = \mathbf{g}_i \, d\Theta^i \; . \quad \text{(no summation over i)}$$

The area of a face spanned by the vectors $\mathbf{g}_j \, d\Theta^j$ and $\mathbf{g}_k \, d\Theta^k$ (no summation over j, k) is denoted by $dA_{<i>}$ and has the value

$$dA_{<i>} = \|\mathbf{g}_j \, d\Theta^j \times \mathbf{g}_k \, d\Theta^k\| = \|\varepsilon_{jki} \, d\Theta^j \, d\Theta^k \, \mathbf{g}^i\| = \sqrt{g \, g^{ii}} \, d\Theta^j \, d\Theta^k \quad (i \neq j \neq k) \, . \quad (3.1.2)$$

Surface elements $dA_{<i>}$ lying in the surfaces $\Theta^i + d\Theta^i$ = constant are called *positive faces* and are characterized by the fact that their outward normal vectors $\mathbf{n}^{<i>}$ are identical with the unit base vectors $\mathbf{g}^{<i>}$ such that $\mathbf{n}^{<i>} = \mathbf{g}^{<i>} = \mathbf{g}^i/\sqrt{g^{ii}}$. On the contrary, $\mathbf{n}^{<i>} = -\mathbf{g}^{<i>}$ characterizes the negative faces $dA_{<i>}$ of the parallelepiped.

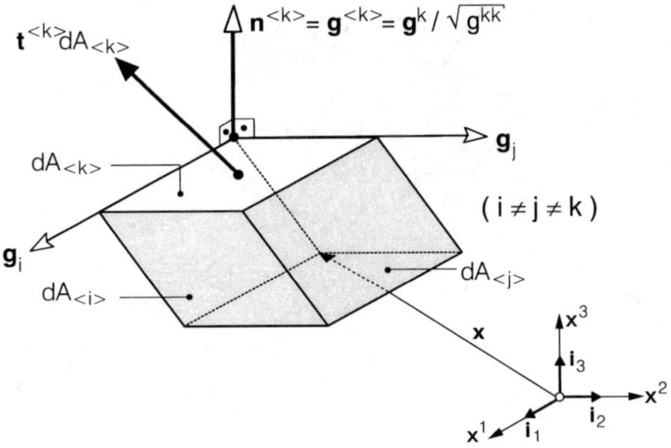

Fig. 3.3. Stress resultant vectors acting upon the surface elements $dA_{<i>}$

Physical stress components. Let $\mathbf{t}^{<i>}$ be the stress vector per unit area acting on the positive surface element $dA_{<i>}$ (i = 1, 2, 3) with the unit normal vector $\mathbf{n}^{<i>} = \mathbf{g}^{<i>}$. As is shown in Fig. 3.3, $\mathbf{t}^{<i>}dA_{<i>}$ (no summation over i) presents the resultant force acting upon the surface element $dA_{<i>}$. The components $\sigma^{<ji>}$ of $\mathbf{t}^{<i>}$ defined by

$$\mathbf{t}^{<i>} = \sigma^{<ji>}\,\mathbf{g}_{<j>} = \sigma^{<ji>}\frac{1}{\sqrt{g_{jj}}}\,\mathbf{g}_j \qquad \text{(summation over j)} \tag{3.1.3}$$

with respect to the unit base vectors $\mathbf{g}_{<j>}$ present therefore forces per unit area of the surface element $dA_{<i>}$ and are thus called *physical stress components* (Fig. 3.4). The indices in $\sigma^{<ji>}$ are selected such that the first index gives the direction $\mathbf{g}_j$ of $\sigma^{<ji>}$, while the second one indicates the element $dA_{<i>}$ upon which $\sigma^{<ji>}$ acts.

Tensorial stress components. By considering (3.1.2), the stress resultant vector of the surface element $dA_{<i>}$ can be expressed as

$$\mathbf{t}^{<i>}\,dA_{<i>} = \mathbf{t}^{<i>}\sqrt{g\,g^{ii}}\,d\Theta^j\,d\Theta^k = \sqrt{g}\,\mathbf{t}^i\,d\Theta^j\,d\Theta^k \quad \text{(no summation over i, } i \neq j \neq k) \tag{3.1.4}$$

where as abbreviation the vector

$$\mathbf{t}^i = \mathbf{t}^{<i>}\sqrt{g^{ii}} \qquad \text{(no summation over i)} \tag{3.1.5}$$

has been introduced which will be shown to be a *contravariant* vector. On the contrary, $\mathbf{t}^{<i>}$ does not satisfy the contravariant transformation rule as is pointed out by putting the index i in pointed brackets. The decomposition of the contravariant vector $\mathbf{t}^i$

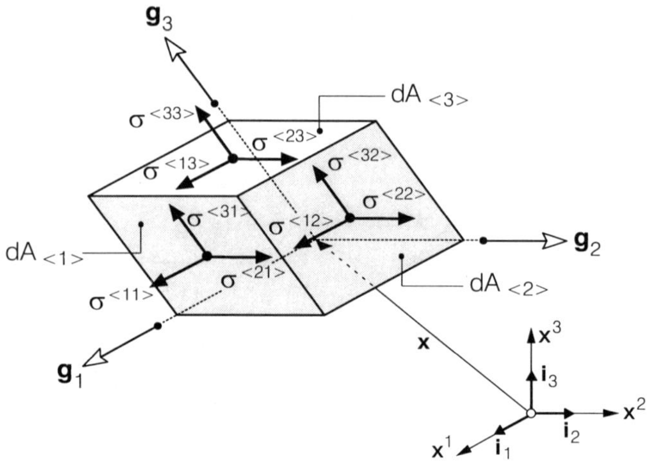

Fig. 3.4. Physical components of the CAUCHY stress tensor $\boldsymbol{\sigma}$

$$\mathbf{t}^i = \sigma^{ji}\,\mathbf{g}_j \tag{3.1.6}$$

with respect to the covariant basis $\mathbf{g}_j$ defines tensorial stress components σ^{ji}. The relation between tensorial σ^{ji} and physical components $\sigma^{<ji>}$ can be obtained by inserting (3.1.3) and (3.1.6) into (3.1.5). The result reads as:

$$\sigma^{<ji>} = \sqrt{g_{jj} / g^{ii}}\;\sigma^{ji}\ . \tag{3.1.7}$$

The indices in σ^{ji} have a similar meaning as those of the physical components $\sigma^{<ji>}$. The first index j indicates the direction $\mathbf{g}_j$ of the component σ^{ji} and the second one i the surface element $dA_{<i>}$. In section 5.3 we will show that the components $\sigma^{ij} = \sigma^{ji}$ are symmetric with respect to the indices i and j. This property will be used in the subsequent derivations.

It remains to show that $\mathbf{t}^\alpha$ are *contravariant* vectors. To achieve this we separate at a point P an infinitesimal tetrahedron (Fig. 3.5) from the deformed body bounded by the surfaces Θ^i = constant (i = 1, 2, 3) and a given surface whose unit normal vector is

$$\mathbf{n} = n^i\,\mathbf{g}_i = n_i\,\mathbf{g}^i\ . \tag{3.1.8}$$

The faces Θ^i = constant of the tetrahedron have the areas 1/2 $dA_{<i>}$ and the surface with the unit normal $\mathbf{n}$ the area 1/2 dA. The areas $dA_{<i>}$ and dA are related in vectorial form by (Green and Zerna 1968).

$$\mathbf{n}\,dA = \sum_{i=1}^{3} \mathbf{n}^{<i>}\,dA_{<i>} = \sum_{i=1}^{3} \frac{\mathbf{g}^i}{\sqrt{g^{ii}}}\,dA_{<i>}\ , \tag{3.1.9}$$

leading with (3.1.8) in component form to

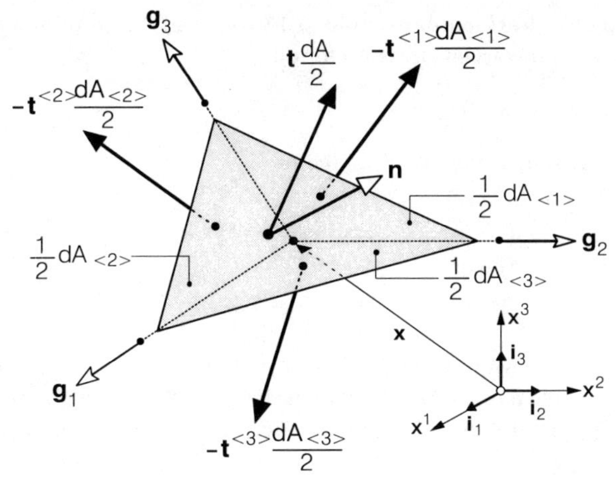

Fig. 3.5. Resultant stress vectors acting upon the faces of the tetrahedron

$$n_i \sqrt{g^{ii}}\, dA = dA_{<i>} \; . \tag{3.1.10}$$

Relation (3.1.9) expresses that the area dA is vectorially equivalent to the sum of the areas $dA_{<i>}$.

The resultant stress vectors acting on the faces of the considered tetrahedron are given in Fig. 3.5 according to the definition of **t** and $\mathbf{t}^{<i>}$. Thus the equation of motion of the infinitesimal tetrahedron is described by

$$\mathbf{t}\, dA = \mathbf{t}^{<i>}\, dA_{<i>} \; , \qquad \text{(summation over i)} \tag{3.1.11}$$

where volume and inertia forces do not occur since they are of higher order of magnitude than surface forces. The transformation of (3.1.11) by means of (3.1.5) and (3.1.10)

$$\mathbf{t}\, dA = \mathbf{t}^{<i>} \sqrt{g^{ii}}\, n_i\, dA = \mathbf{t}^i\, n_i\, dA \qquad \text{(summation over i)} \tag{3.1.12}$$

leads after division by dA and with (3.1.6) to

$$\mathbf{t} = \mathbf{t}^i\, n_i = \sigma^{ji}\, n_i\, \mathbf{g}_j \; . \tag{3.1.13}$$

Since **t** is an invariant vector and the components n_i transform according to the covariant rule the above result demonstrates by virtue of the quotient rule that $\mathbf{t}^i$ is a contravariant vector.

CAUCHY stress tensor σ. Using the tensorial components σ^{ij} introduced in (3.1.6) we form by the well-known procedure

$$\boldsymbol{\sigma} = \boldsymbol{\sigma}^T = \sigma^{ij}\, \mathbf{g}_i \otimes \mathbf{g}_j = \sigma^{ji}\, \mathbf{g}_i \otimes \mathbf{g}_j \tag{3.1.14}$$

the so-called CAUCHY *stress tensor* $\boldsymbol{\sigma}$. From (3.1.6) and (3.1.14) it follows that the tensor $\boldsymbol{\sigma}$ is related to the contravariant stress vector $\mathbf{t}^i$ by

$$\boldsymbol{\sigma} = \mathbf{t}^i \otimes \mathbf{g}_i \ . \tag{3.1.15}$$

If we rewrite (3.1.13) by using (3.1.8) in the form

$$\mathbf{t} = (\mathbf{t}^i \otimes \mathbf{g}_i)\,\mathbf{n} \ , \tag{3.1.16}$$

we see in view of (3.1.15) that

$$\mathbf{t} = \boldsymbol{\sigma}\,\mathbf{n} \ . \tag{3.1.17}$$

This result which is known as *CAUCHY theorem* states that $\boldsymbol{\sigma}$ is a tensor transforming the unit normal vector field $\mathbf{n}$ of a surface A into the field of the *CAUCHY stress vector* $\mathbf{t}$ acting upon A. The geometrical illustration of this statement is given in Fig. 3.6.

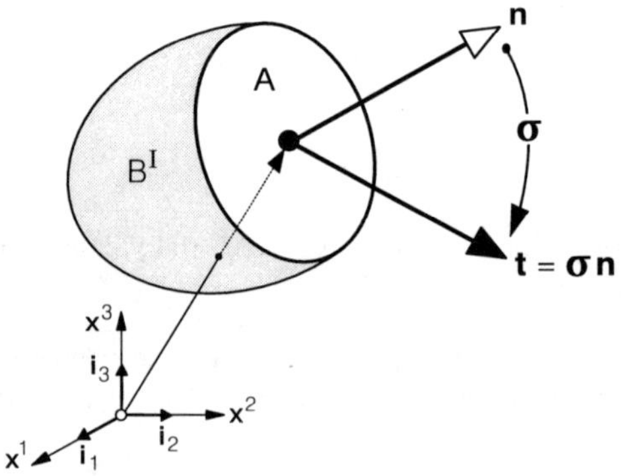

Fig. 3.6. Geometrical interpretation of the CAUCHY stress tensor $\boldsymbol{\sigma}$

3.2 Stress tensors

The CAUCHY stress tensor defined in (3.1.15) presents one possibility among many others to describe the stress state in a deformed body. In this section we will introduce other stress tensors and then establish the connections between them. At the beginning we summarize some formulae needed for the subsequent derivation.

The undeformed surface element dA_o is related to its deformed counterpart dA by (2.2.35):

$$\mathbf{n}\,dA = \det \mathbf{F}\,\mathbf{F}^{-T}\,\mathbf{N}\,dA_o \ , \tag{3.2.1}$$

where $\mathbf{n}$ and $\mathbf{N}$ are the unit normal vectors referring to dA_o and dA respectively and $\mathbf{F}$ denotes the deformation gradient (2.2.3). We also recall the expressions

$$J = \det \mathbf{F} = | F^{i}_{\cdot j} | = \frac{dV}{dV_o} = \sqrt{\frac{g}{G}} , \tag{3.2.2}$$

established in (2.2.31) for the invariant J. It is apparent that J = 1 describes incompressible deformations which are said to be *isochoric*. We finally remember the definition of the *CAUCHY stress tensor* (3.1.17):

$$\boldsymbol{\sigma} = \sigma^{ij} \, \mathbf{g}_i \otimes \mathbf{g}_j : \quad \mathbf{t} = \boldsymbol{\sigma} \, \mathbf{n} \tag{3.2.3}$$

transforming the field of the unit normal vector **n** to the CAUCHY *stress vector* **t** which describes in the form **t** dA the force acting upon a deformed surface element dA.

We now focus our attention on the definition of new stress tensors. Any stress tensor can be defined by a transformation similar to (3.2.3) in terms of a suitable stress vector and the unit normal vector **n** or **N**. We first introduce the following stress vectors starting from the CAUCHY stress vector **t**:

$$\mathbf{t}_o = \frac{dA}{dA_o} \mathbf{t} , \tag{3.2.4}$$

$$\hat{\mathbf{t}}_o = \mathbf{F}^{-1} \, \mathbf{t}_o \qquad \textit{similar to} \qquad \mathbf{G}_i = \mathbf{F}^{-1} \, \mathbf{g}_i , \tag{3.2.5}$$

$$\tilde{\mathbf{t}}_o = \mathbf{R}^T \, \mathbf{t}_o \qquad \textit{similar to} \qquad \mathbf{G}_i = \mathbf{R}^T \, \tilde{\mathbf{G}}_i , \tag{3.2.6}$$

$$\mathbf{t}^* = J \, \mathbf{t} . \tag{3.2.7}$$

Writing equation (3.2.4) in the form

$$\mathbf{t}_o \, dA_o = \mathbf{t} \, dA = d\mathbf{p} \tag{3.2.8}$$

we see that $\mathbf{t}_o$ describes after multiplication with dA_o the actual force d**p** acting upon the deformed surface element dA. In contrast to **t**, the vector $\mathbf{t}_o$ is a force per unit area of the undeformed surface A_o and is called therefore *pseudo stress vector*. Both vectors **t** and $\mathbf{t}_o$ are geometrically interpretable and can be regarded in this sense as *real* stress vectors. This is however not true for the stress vectors $\hat{\mathbf{t}}_o$, $\tilde{\mathbf{t}}_o$ and $\mathbf{t}^*$, which are introduced by purely mathematical transformations. The stress vector $\hat{\mathbf{t}}_o$ is obtained from $\mathbf{t}_o$ in the same manner as the undeformed basis $\mathbf{G}_i$ from its deformed counterpart $\mathbf{g}_i$. In (3.2.6), the rotation tensor **R** from the polar decomposition theorem (2.4.4) is used to rotate $\mathbf{t}_o$ into $\tilde{\mathbf{t}}_o$. This relation is similar to (2.4.9) transforming the rotated basis $\tilde{\mathbf{G}}_i$ into its initial position $\mathbf{G}_i$. The properties of different stress vectors are illustrated in Fig. 3.7. By means of (3.2.4) to (3.2.7), the following relations are obtained permitting to transform different stress vectors into each other:

$$\mathbf{t}_o = \frac{dA}{dA_o} \mathbf{t} = \mathbf{F} \, \hat{\mathbf{t}}_o \qquad = \mathbf{R} \, \tilde{\mathbf{t}}_o , \tag{3.2.9}$$

$$\hat{\mathbf{t}}_o = \mathbf{F}^{-1} \, \mathbf{t}_o = \mathbf{F}^{-1} \, \mathbf{R} \, \tilde{\mathbf{t}}_o = \frac{dA}{dA_o} \mathbf{F}^{-1} \, \mathbf{t} , \tag{3.2.10}$$

$$\tilde{\mathbf{t}}_o = \mathbf{R}^T \mathbf{t}_o = \mathbf{R}^T \mathbf{F} \hat{\mathbf{t}}_o = \frac{dA}{dA_o} \mathbf{R}^T \mathbf{t} \, . \tag{3.2.11}$$

The above stress vectors will be used for the definition of the *material* stress tensors **P**, **S** and **T** and the vector $\mathbf{t}^* = J\,\mathbf{t}$ for the definition of the *spatial* tensor $\boldsymbol{\tau}$ (Fig. 3.7).

The *KIRCHHOFF stress tensor* $\boldsymbol{\tau}$ is related to the stress vector $\mathbf{t}^*$ by

$$\boldsymbol{\tau} = \tau^{ij} \mathbf{g}_i \otimes \mathbf{g}_j : \quad \mathbf{t}^* = J\,\mathbf{t} = \boldsymbol{\tau}\,\mathbf{n} \, , \tag{3.2.12}$$

and satisfies according to (3.2.2) and (3.2.3) the relations

$$\boldsymbol{\tau} = J\,\boldsymbol{\sigma} \quad \rightarrow \quad \tau^{ij} = J\,\sigma^{ij} = \sqrt{\frac{g}{G}}\,\sigma^{ij} \, . \tag{3.2.13}$$

For incompressible deformations J = 1 there is therefore no distinction between $\boldsymbol{\sigma}$ and $\boldsymbol{\tau}$. We now deal with material stress tensors characterized by components defined with respect to the undeformed basis $\mathbf{G}_i$.

The first *PIOLA-KIRCHHOFF stress tensor* **P** is related to the pseudo stress vector $\mathbf{t}_o$ by

$$\mathbf{P} = P^{ij} \mathbf{G}_i \otimes \mathbf{G}_j : \quad \mathbf{t}_o = \mathbf{P}\,\mathbf{N} \, , \tag{3.2.14}$$

where **N** is the unit normal vector of the undeformed surface element dA_o. To establish useful relations for **P** we express $\mathbf{t}_o$, by using (3.2.1), (3.2.3) and (3.2.4), in the form

$$\mathbf{t}_o = \frac{dA}{dA_o}\mathbf{t} = \frac{dA}{dA_o}\boldsymbol{\sigma}\,\mathbf{n} = \det \mathbf{F}\,\boldsymbol{\sigma}\,\mathbf{F}^{-T}\,\mathbf{N} \, . \tag{3.2.15}$$

Since both results (3.2.14) and (3.2.15) hold for an arbitrary vector field **N** we obtain

$$\mathbf{P} = \det \mathbf{F}\,\boldsymbol{\sigma}\,\mathbf{F}^{-T} = \boldsymbol{\tau}\,\mathbf{F}^{-T} \, , \tag{3.2.16}$$

where also (3.2.2) and (3.2.13) have been considered. To transform (3.2.16) into component form we replace $\boldsymbol{\sigma}$ and **P** by the component relations (3.2.3), (3.2.14) and consider the transformations (2.2.6) and (2.2.22)

$$\mathbf{G}_i = \mathbf{F}^{-1}\,\mathbf{g}_i \, , \quad \mathbf{g}_k = F^{i}_{\cdot k}\,\mathbf{G}_i = (\mathbf{G}^i \cdot \mathbf{g}_k)\,\mathbf{G}_i \tag{3.2.17}$$

between the bases $\mathbf{g}_i$ and $\mathbf{G}_i$. This procedure yields

$$\begin{aligned} P^{ij}\,\mathbf{G}_i \otimes \mathbf{G}_j &= \det \mathbf{F}\,(\sigma^{kj}\,\mathbf{g}_k \otimes \mathbf{g}_j)\,\mathbf{F}^{-T} = \det \mathbf{F}\,\sigma^{kj}\,\mathbf{g}_k \otimes \mathbf{G}_j \\ &= \det \mathbf{F}\,\sigma^{kj}\,F^{i}_{\cdot k}\,\mathbf{G}_i \otimes \mathbf{G}_j \, , \end{aligned} \tag{3.2.18}$$

from which we obtain with (3.2.2)

$$P^{ij} = J\,F^{i}_{\cdot k}\,\sigma^{kj} = F^{i}_{\cdot k}\,\tau^{kj} \tag{3.2.19}$$

as final result.

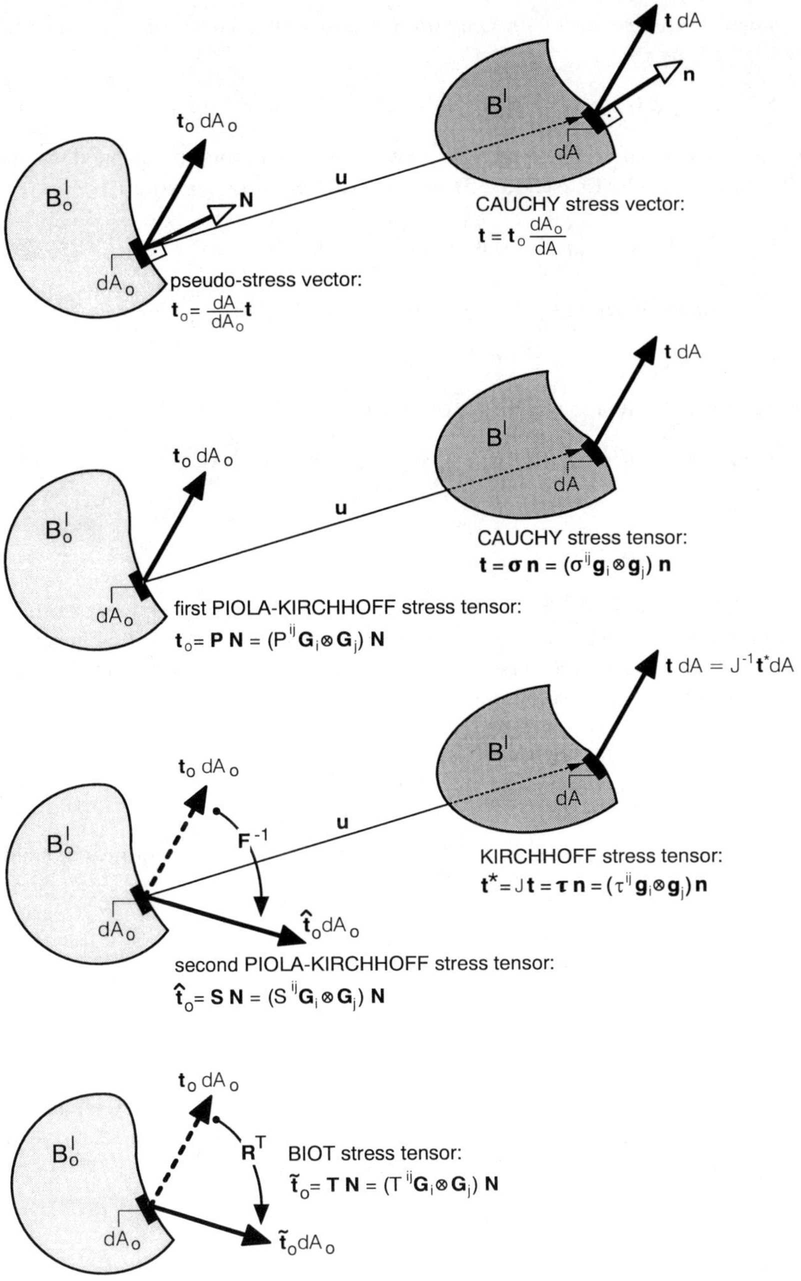

Fig. 3.7. Definition of material and spatial stress tensors

The second *PIOLA-KIRCHHOFF stress tensor* **S** is also a material tensor and related to the stress vector $\hat{\mathbf{t}}_o$ by

$$\mathbf{S} = S^{ij}\,\mathbf{G}_i \otimes \mathbf{G}_j \;:\quad \hat{\mathbf{t}}_o = \mathbf{S}\,\mathbf{N} \;, \tag{3.2.20}$$

with **N** as unit normal vector to dA_o. To relate the stress tensor **S** to the previous ones $\boldsymbol{\sigma}$ and **P** we first consider (3.2.1), (3.2.3) and (3.2.10) permitting to write

$$\hat{\mathbf{t}}_o = \frac{dA}{dA_o}\,\mathbf{F}^{-1}\,\mathbf{t} = \frac{dA}{dA_o}\,\mathbf{F}^{-1}\,\boldsymbol{\sigma}\,\mathbf{n} = \det\mathbf{F}\,\mathbf{F}^{-1}\,\boldsymbol{\sigma}\,\mathbf{F}^{-T}\,\mathbf{N} \;. \tag{3.2.21}$$

Thus the comparison with (3.2.20) yields in view of (3.2.16)

$$\mathbf{S} = \det\mathbf{F}\,\mathbf{F}^{-1}\,\boldsymbol{\sigma}\,\mathbf{F}^{-T} = \mathbf{F}^{-1}\,\mathbf{P} \;. \tag{3.2.22}$$

If we transform this relation by means of (3.2.3), (3.2.17) and (3.2.20)

$$S^{ij}\,\mathbf{G}_i \otimes \mathbf{G}_j = \det\mathbf{F}\,\mathbf{F}^{-1}\,(\sigma^{ij}\,\mathbf{g}_i \otimes \mathbf{g}_j)\,\mathbf{F}^{-T} = \det\mathbf{F}\,\sigma^{ij}\,\mathbf{G}_i \otimes \mathbf{G}_j \tag{3.2.23}$$

we also obtain

$$S^{ij} = J\,\sigma^{ij} = \tau^{ij} \;. \tag{3.2.24}$$

Accordingly, there is no distinction between S^{ij} and τ^{ij}; this result is however valid only for the pure contravariant components of **S** and $\boldsymbol{\tau}$. From (3.2.22) and (3.2.24) it can easily be deduced that the tensors **S** and $\boldsymbol{\tau}$ are symmetric whenever $\boldsymbol{\sigma}$ is symmetric. As a further material stress tensor we also introduce:

The *BIOT* or *JAUMANN stress tensor* is defined by

$$\mathbf{T} = T^{ij}\,\mathbf{G}_i \otimes \mathbf{G}_j \;:\quad \tilde{\mathbf{t}}_o = \mathbf{T}\,\mathbf{N} \;, \tag{3.2.25}$$

where, in view of (3.2.1), (3.2.3) and (3.2.11), the following transformations hold for $\tilde{\mathbf{t}}_o$

$$\tilde{\mathbf{t}}_o = \frac{dA}{dA_o}\,\mathbf{R}^T\,\mathbf{t} = \frac{dA}{dA_o}\,\mathbf{R}^T\,\boldsymbol{\sigma}\,\mathbf{n} = \det\mathbf{F}\,\mathbf{R}^T\,\boldsymbol{\sigma}\,\mathbf{F}^{-T}\,\mathbf{N} \;. \tag{3.2.26}$$

The comparison of this result with (3.2.25) yields

$$\mathbf{T} = \det\mathbf{F}\,\mathbf{R}^T\,\boldsymbol{\sigma}\,\mathbf{F}^{-T} = \mathbf{R}^T\,\mathbf{P} = \mathbf{R}^T\,\mathbf{F}\,\mathbf{F}^{-1}\,\mathbf{P} = \mathbf{U}\,\mathbf{S} \;, \tag{3.2.27}$$

if, in addition, equations (3.2.16), (3.2.22) and the definition (2.4.17) of the right stretch tensor $\mathbf{U} = \mathbf{U}^T$ are considered. By using (3.2.17) and the component relations of the participant tensors, equation (3.2.27) can be transformed into:

$$T^{ij} = U^i_{\;k}\,S^{kj} = R_{k\cdot}^{\;\cdot i}\,P^{kj} = J\,R_{k\cdot}^{\;\cdot i}\,F^k_{\cdot m}\,\sigma^{mj} \;, \tag{3.2.28}$$

where the components of $\mathbf{R} = R^{ij}\,\mathbf{G}_i \otimes \mathbf{G}_j$ and $\mathbf{F} = F^{ij}\,\mathbf{G}_i \otimes \mathbf{G}_j$ refer to the undeformed basis.

Results obtained in (3.2.13), (3.2.16), (3.2.22) and (3.2.27) can now be summarized as follows:

$$
\begin{aligned}
\mathbf{P} &= \mathbf{F}\,\mathbf{S} &&= \mathbf{R}\,\mathbf{T} &&= \boldsymbol{\tau}\,\mathbf{F}^{-T} \;, \\
\mathbf{S} &= \mathbf{F}^{-1}\,\mathbf{P} &&= \mathbf{U}^{-1}\,\mathbf{T} &&= \mathbf{F}^{-1}\,\boldsymbol{\tau}\,\mathbf{F}^{-T} \;, \\
\mathbf{T} &= \mathbf{R}^{T}\,\mathbf{P} &&= \mathbf{U}\,\mathbf{S} &&= \mathbf{R}^{T}\,\boldsymbol{\tau}\,\mathbf{F}^{-T} \;, \\
\boldsymbol{\tau} &= J\,\boldsymbol{\sigma} &&= \mathbf{P}\,\mathbf{F}^{T} &&= \mathbf{F}\,\mathbf{S}\,\mathbf{F}^{T} = \mathbf{R}\,\mathbf{T}\,\mathbf{F}^{T} \;.
\end{aligned}
\qquad (3.2.29)
$$

3.3 Energy conjugate stress and strain variables

The so-called energy conjugate stress and strain variables play an important role to formulate the internal energy of deformable bodies. In section 3.2 we have already introduced various stress measures to describe the stresses in a body under external loads. Any of these stress measures can be used to express the internal energy of a body. However, if a decision is made about the use of a certain strain measure, the stress variable to be used in combination can not be selected arbitrarily. In this section we will show that each stress tensor is related through the *rate of the internal energy* $\rho_o\dot{e}$ to a well defined strain tensor.* Strain and stress variables with this property are referred to as *energy conjugate*.

For the definition of energy conjugate variables we will make use of the material time derivative of the strain tensors.** Thus we suppose that the body in the initial configuration B_o is subjected to the velocity field

$$\dot{\mathbf{u}} = \dot{\mathbf{x}} \quad \rightarrow \quad \dot{\mathbf{u}}_{,i} = \dot{\mathbf{x}}_{,i} = \dot{\mathbf{g}}_i \;. \qquad (3.3.1)$$

Then, we form the material time derivative of the expressions (2.2.3), (2.4.4) and (2.5.9):

$$\dot{\mathbf{F}} = \dot{\mathbf{g}}_i \otimes \mathbf{G}^i \qquad (3.3.2)$$

$$\dot{\mathbf{F}} = \dot{\mathbf{R}}\,\mathbf{U} + \mathbf{R}\,\dot{\mathbf{U}} \;, \qquad (3.3.3)$$

$$\dot{\mathbf{E}} = \frac{1}{2}\,\dot{\mathbf{C}} = \frac{1}{2}\,(\dot{\mathbf{F}}^{T}\,\mathbf{F} + \mathbf{F}^{T}\,\dot{\mathbf{F}}) \;. \qquad (3.3.4)$$

We also recall the equalities (2.9.28)

$$L_v\,\mathbf{e} = \mathbf{d} = \mathbf{F}^{-T}\,\dot{\mathbf{E}}\,\mathbf{F}^{-1} \qquad (3.3.5)$$

relating the LAGRANGE-GREEN strain tensor **E** to the ALMANSI strain tensor **e** and the rate of deformation tensor **d**, where L_v denotes the LIE-derivative.

---

* For a detailed explanation of the rate of the internal energy $\rho_o\dot{e}$ we refer to section 5.5.

** For a comprehensive definition of the material time derivative we refer to section 4.1.

For the derivation we start from the first PIOLA-KIRCHHOFF stress tensor **P** which is a material tensor. According to (1.2.25) and (1.6.11) we may write

$$\mathrm{DIV}\,\mathbf{P} = \mathbf{P}_{,k}\,\mathbf{G}^k = \mathbf{P}|_k\,\mathbf{G}^k \ . \tag{3.3.6}$$

DIV **P** is a force variable. Its scalar product with the velocity field $\dot{\mathbf{u}}$ defines therefore a power which, by means of the identity $\mathbf{G}_i|_k = 0$, can be given as:

$$\begin{aligned}\dot{\mathbf{u}} \cdot \mathrm{DIV}\,\mathbf{P} &= \dot{\mathbf{u}} \cdot (\mathbf{P}|_k\,\mathbf{G}^k) = (\dot{\mathbf{u}} \cdot (\mathbf{P}\,\mathbf{G}^k))|_k - \dot{\mathbf{u}}|_k \cdot (\mathbf{P}\,\mathbf{G}^k) \\ &= (\dot{\mathbf{u}}\,\mathbf{P})|_k \cdot \mathbf{G}^k - \dot{\mathbf{u}}|_k \cdot (\mathbf{P}\,\mathbf{G}^k) \ . \end{aligned} \tag{3.3.7}$$

We integrate the above expression over the volume V_o. Since the first term on the right hand side forms according to (1.6.10) the divergence of the vector $\dot{\mathbf{u}}\,\mathbf{P}$ it can be transformed by means of the GAUSS-GREEN theorem (1.6.15) which, by considering (3.2.14), leads to

$$\iiint_{V_o} (\dot{\mathbf{u}}\,\mathbf{P})|_k \cdot \mathbf{G}^k\, dV_o = \iint_{A_o} (\dot{\mathbf{u}}\,\mathbf{P}) \cdot \mathbf{N}\, dA_o = \iint_{A_o} \dot{\mathbf{u}} \cdot (\mathbf{P}\mathbf{N})\, dA_o = \iint_{A_o} \dot{\mathbf{u}} \cdot \mathbf{t}_o\, dA_o \ . \tag{3.3.8}$$

Thus the final result is

$$\iiint_{V_o} \dot{\mathbf{u}} \cdot \mathrm{DIV}\,\mathbf{P}\, dV_o = \iint_{A_o} \dot{\mathbf{u}} \cdot \mathbf{t}_o\, dA_o - \iiint_{V_o} \dot{\mathbf{u}}|_k \cdot (\mathbf{P}\,\mathbf{G}^k)\, dV_o \ . \tag{3.3.9}$$

The first integral on the right-hand side expresses the power of external forces $\mathbf{t}_o$ acting on the boundary surface of the body. The attention is now focused on the integrand of the second integral

$$\rho_o \dot{e} = \dot{\mathbf{u}}|_k \cdot (\mathbf{P}\,\mathbf{G}^k) = \dot{\mathbf{u}}|_k\,\mathbf{P}\,\mathbf{G}^k \ , \tag{3.3.10}$$

for which we use the notation $\rho_o \dot{e}$, where ρ_o is the mass density of the undeformed body B_o. The expression (3.3.10) describes in the case of isothermal processes the *rate of the internal energy* (per unit undeformed volume) and can be, in view of (1.1.21), (3.3.1) and (3.3.2), transformed into the form

$$\rho_o \dot{e} = \mathbf{P} : (\dot{\mathbf{u}}|_k \otimes \mathbf{G}^k) = \mathbf{P} : (\dot{\mathbf{g}}_k \otimes \mathbf{G}^k) = \mathbf{P} : \dot{\mathbf{F}} \ , \tag{3.3.11}$$

demonstrating the rate of the deformation gradient $\dot{\mathbf{F}}$ and the first PIOLA-KIRCHHOFF stress tensor **P** to be conjugate quantities. In general, a stress tensor and the rate of a deformation measure are said to be *energy conjugate*, if their double contraction is equal to the rate of the internal energy $\rho_o \dot{e}$. The expression $\mathbf{P} : \dot{\mathbf{F}}$ is also called *stress power*.

Equation (3.3.11) presents the starting point for the definition of other pairs of conjugate variables. We replace **P** according to (3.2.29) by $\mathbf{P} = \mathbf{F}\,\mathbf{S}$. By considering the identity (1.3.24)

$$\mathbf{A} : (\mathbf{B}\,\mathbf{C}) = (\mathbf{B}^T\,\mathbf{A}) : \mathbf{C} = (\mathbf{A}\,\mathbf{C}^T) : \mathbf{B} \tag{3.3.12}$$

and the symmetry $\mathbf{S} = \mathbf{S}^{\mathrm{T}}$, the corresponding result can then be transformed as follows:

$$\rho_o \dot{e} = (\mathbf{F}\,\mathbf{S}) : \dot{\mathbf{F}} = \mathbf{S} : (\mathbf{F}^{\mathrm{T}}\,\dot{\mathbf{F}}) = \frac{1}{2}(\mathbf{S} + \mathbf{S}^{\mathrm{T}}) : (\mathbf{F}^{\mathrm{T}}\,\dot{\mathbf{F}})$$

$$= \mathbf{S} : \frac{1}{2}(\mathbf{F}^{\mathrm{T}}\,\dot{\mathbf{F}}) + \mathbf{S} : \frac{1}{2}(\dot{\mathbf{F}}^{\mathrm{T}}\,\mathbf{F}) = \mathbf{S} : \frac{1}{2}(\mathbf{F}^{\mathrm{T}}\,\dot{\mathbf{F}} + \dot{\mathbf{F}}^{\mathrm{T}}\,\mathbf{F}) \ , \tag{3.3.13}$$

leading in view of (3.3.4) to

$$\rho_o \dot{e} = \mathbf{S} : \dot{\mathbf{E}} = \frac{1}{2}\mathbf{S} : \dot{\mathbf{C}} \ . \tag{3.3.14}$$

Thus the rate of the LAGRANGE-GREEN strain tensor $\dot{\mathbf{E}}$ turns out to be energy conjugate to the second KIRCHHOFF-PIOLA stress tensor $\mathbf{S}$.

Similarly it can be shown that the BIOT stress tensor $\mathbf{T}$ and the rate of the right stretch tensor $\dot{\mathbf{U}}$ are conjugate quantities. By using (3.2.29), (3.3.3) and (3.3.12) we first obtain from (3.3.11)

$$\rho_o \dot{e} = (\mathbf{R}\,\mathbf{T}) : \dot{\mathbf{F}} = \mathbf{T} : (\mathbf{R}^{\mathrm{T}}\,\dot{\mathbf{F}}) = \mathbf{T} : (\mathbf{R}^{\mathrm{T}}\,\dot{\mathbf{R}}\,\mathbf{U}) + \mathbf{T} : (\mathbf{R}^{\mathrm{T}}\,\mathbf{R}\,\dot{\mathbf{U}})$$

$$= \mathbf{T} : (\mathbf{R}^{\mathrm{T}}\,\dot{\mathbf{R}}\,\mathbf{U}) + \mathbf{T} : \dot{\mathbf{U}} \ , \tag{3.3.15}$$

where in view of the relation $\mathbf{T} = \mathbf{U}\,\mathbf{S}$ given in (3.2.29)

$$\mathbf{T} : (\mathbf{R}^{\mathrm{T}}\,\dot{\mathbf{R}}\,\mathbf{U}) = (\mathbf{T}\,\mathbf{U}^{\mathrm{T}}) : (\mathbf{R}^{\mathrm{T}}\,\dot{\mathbf{R}}) = (\mathbf{U}\,\mathbf{S}\,\mathbf{U}^{\mathrm{T}}) : (\mathbf{R}^{\mathrm{T}}\,\dot{\mathbf{R}}) = 0 \ . \tag{3.3.16}$$

Due to $\mathbf{S} = \mathbf{S}^{\mathrm{T}}$ the tensor $\mathbf{U}\,\mathbf{S}\,\mathbf{U}^{\mathrm{T}}$ is symmetric, while $\mathbf{R}^{\mathrm{T}}\,\dot{\mathbf{R}}$ is a skew-symmetric tensor since $\mathbf{R}$ is orthogonal:

$$\mathbf{R}^{\mathrm{T}}\,\mathbf{R} = \mathbf{I} \quad \rightarrow \quad \mathbf{R}^{\mathrm{T}}\,\dot{\mathbf{R}} = -\dot{\mathbf{R}}^{\mathrm{T}}\,\mathbf{R} \ . \tag{3.3.17}$$

Thus the double contraction (3.3.16) vanishes and equation (3.3.15) reduces to

$$\rho_o \dot{e} = \mathbf{T} : \dot{\mathbf{U}} = \mathbf{T} : \dot{\mathbf{H}} \ , \tag{3.3.18}$$

where $\mathbf{H}$ is the BIOT strain tensor related to $\mathbf{U}$ by (2.4.30).

The rate of the internal energy $\rho_o \dot{e}$ can be also expressed in terms of the spatial variables $\boldsymbol{\tau}$ and $\mathrm{L_v}\,\mathbf{e}$. Starting from (3.3.14) we find by using (3.2.22), (3.3.12)

$$\rho_o \dot{e} = (\det \mathbf{F}\,\mathbf{F}^{-1}\boldsymbol{\sigma}\,\mathbf{F}^{-\mathrm{T}}) : \dot{\mathbf{E}} = (\det \mathbf{F}\,\boldsymbol{\sigma}\,\mathbf{F}^{-\mathrm{T}}) : (\mathbf{F}^{-\mathrm{T}}\,\dot{\mathbf{E}})$$

$$= \det \mathbf{F}\ \boldsymbol{\sigma} : (\mathbf{F}^{-\mathrm{T}}\,\dot{\mathbf{E}}\,\mathbf{F}^{-1}) \ . \tag{3.3.19}$$

By means of the equalities (3.3.5) and $\det \mathbf{F} = \mathrm{J}$ this result takes the form

$$\rho_o \dot{e} = \mathrm{J}\,\boldsymbol{\sigma} : \mathrm{L_v}\,\mathbf{e} = \boldsymbol{\tau} : \mathrm{L_v}\,\mathbf{e} = \boldsymbol{\tau} : \frac{1}{2}\mathrm{L_v}\,\mathbf{g} = \boldsymbol{\tau} : \mathbf{d} \tag{3.3.20}$$

showing the duality between the KIRCHHOFF stress tensor $\boldsymbol{\tau}$ and the LIE-derivative $\mathrm{L_v}\,\mathbf{e}$ of the ALMANSI strain tensor which, in turn, is related to the LIE-derivative of the identity tensor $\mathbf{g}$ by

$$L_v \mathbf{e} = \frac{1}{2} L_v \mathbf{g} = \frac{1}{2} \overline{(\mathbf{g}_i \cdot \mathbf{g}_j)}^{\cdot} \, \mathbf{g}^i \otimes \mathbf{g}^j \qquad (3.3.21)$$

as can be deduced from (2.9.29).

Concluding this section it is useful to summarize the important results. The following pairs of variables are found to be *energy conjugate*:

$$\textit{material description}: \ \rho_o \dot{e} = \mathbf{P} : \dot{\mathbf{F}} = \mathbf{S} : \dot{\mathbf{E}} = \mathbf{T} : \dot{\mathbf{U}} = \mathbf{T} : \dot{\mathbf{H}}$$

$$\textit{spatial description}: \ \rho_o \dot{e} = J\, \boldsymbol{\sigma} : L_v \mathbf{e} = \boldsymbol{\tau} : L_v \mathbf{e} = \boldsymbol{\tau} : \frac{1}{2} L_v \mathbf{g} = \boldsymbol{\tau} : \mathbf{d} \ . \qquad (3.3.22)$$

In component form the above equalities read as:

$$\textit{material description}: \ \rho_o \dot{e} = P^{ij} \dot{F}_{ij} = S^{ij} \dot{E}_{ij} = T^{ij} \dot{U}_{ij} = T^{ij} \dot{H}_{ij}$$

$$\textit{spatial description}: \ \rho_o \dot{e} = J\, \sigma^{ij} \dot{e}_{ij} = \tau^{ij} \dot{e}_{ij} = \tau^{ij} \frac{1}{2} \dot{g}_{ij} = \tau^{ij} d_{ij} \ , \qquad (3.3.23)$$

where, according to Table 2.2, the following kinematic relations hold for the deformation variables:

$$\dot{F}_{ij} = \overline{\mathbf{G}_i \cdot \mathbf{g}_j}^{\cdot} \ , \quad \dot{U}_{ij} = \dot{H}_{ij} = \overline{\tilde{\mathbf{G}}_i \cdot \mathbf{g}_j}^{\cdot} \ , \quad \dot{E}_{ij} = \dot{e}_{ij} = d_{ij} = \frac{1}{2} \overline{(\mathbf{g}_i \cdot \mathbf{g}_j)}^{\cdot} \ . \qquad (3.3.24)$$

As long as the strain components $\dot{E}_{ij}$, $\dot{e}_{ij}$ and d_{ij} are used with pure covariant indices, there exists no distinction between them. The same holds for the stress components S^{ij} and τ^{ij} which are, in view of (3.2.24), equal as pure contravariant components. Accordingly, the distinction between the following spatial and material formulations for $\rho_o \dot{e}$

$$\rho_o \dot{e} = S^{ij} \dot{E}_{ij} = \tau^{ij} \dot{e}_{ij} = \tau^{ij} d_{ij} \qquad (3.3.25)$$

is only formal, if the position of the indices of the participant tensor components remains unchanged.

3.4 Summary of important definitions

The definition of various stress tensors introduced in this section and the relations existing between them are summarized in Table 3.1.

Table 3.1. Definition of material and spatial stress tensors

notations	definitions	relations
CAUCHY stress tensor	$\boldsymbol{\sigma} = \boldsymbol{\sigma}^T = \sigma^{ij}\, \mathbf{g}_i \otimes \mathbf{g}_j$	$\boldsymbol{\sigma} = J^{-1}\, \boldsymbol{\tau} = J^{-1}\, \mathbf{P}\, \mathbf{F}^T = J^{-1}\, \mathbf{F}\, \mathbf{S}\, \mathbf{F}^T = J^{-1}\, \mathbf{R}\, \mathbf{T}\, \mathbf{F}^T$
KIRCHHOFF stress tensor	$\boldsymbol{\tau} = \boldsymbol{\tau}^T = \tau^{ij}\, \mathbf{g}_i \otimes \mathbf{g}_j$	$\boldsymbol{\tau} = J\, \boldsymbol{\sigma} = \mathbf{P}\, \mathbf{F}^T = \mathbf{F}\, \mathbf{S}\, \mathbf{F}^T = \mathbf{R}\, \mathbf{T}\, \mathbf{F}^T$
first PIOLA-KIRCHHOFF stress tensor	$\mathbf{P} = P^{ij}\, \mathbf{G}_i \otimes \mathbf{G}_j$	$\mathbf{P} = \mathbf{F}\, \mathbf{S} = \mathbf{R}\, \mathbf{T} = \boldsymbol{\tau}\, \mathbf{F}^{-T} = J\, \boldsymbol{\sigma}\, \mathbf{F}^{-T}$
second PIOLA-KIRCHHOFF stress tensor	$\mathbf{S} = \mathbf{S}^T = S^{ij}\, \mathbf{G}_i \otimes \mathbf{G}_j$	$\mathbf{S} = \mathbf{F}^{-1}\, \mathbf{P} = \mathbf{U}^{-1}\, \mathbf{T} = \mathbf{F}^{-1}\, \boldsymbol{\tau}\, \mathbf{F}^{-T} = J\, \mathbf{F}^{-1}\, \boldsymbol{\sigma}\, \mathbf{F}^{-T}$
BIOT stress tensor	$\mathbf{T} = T^{ij}\, \mathbf{G}_i \otimes \mathbf{G}_j$	$\mathbf{T} = \mathbf{R}^T\, \mathbf{P} = \mathbf{U}\, \mathbf{S} = \mathbf{R}^T\, \boldsymbol{\tau}\, \mathbf{F}^{-T} = J\, \mathbf{R}^T\, \boldsymbol{\sigma}\, \mathbf{F}^{-T}$
important equalities	$J = \det \mathbf{F}\,,\quad S^{ij} = J\sigma^{ij} = \tau^{ij}$ (valid for contravariant components)	

Exercises

3.1. Construct the partial derivative of the function $\Psi = \sqrt{\boldsymbol{\tau} : \boldsymbol{\tau}}$ with respect to the KIRCHHOFF stress tensor $\boldsymbol{\tau} = \boldsymbol{\tau}^T$ and show that the result is a tensor which is coaxial to $\boldsymbol{\tau}$. For the derivation, use the definition

$$\Psi_{,\tau} = \frac{\partial \Psi}{\partial \tau^i_j}\, \mathbf{g}^i \otimes \mathbf{g}_j \ .$$

3.2. Evaluate the spherical and deviatoric part of the KIRCHHOFF stress tensor $\boldsymbol{\tau}$ defined by

$$\boldsymbol{\tau} = \kappa \ln J\, \mathbf{g} + \mu \operatorname{dev} \bar{\mathbf{b}}_e \ ,$$

where κ and μ are material constants and $\bar{\mathbf{b}}_e$ is a symmetric second-order tensor.

3.3. Using the expression

$$\rho_o \Psi = \frac{1}{2} \kappa \ln J + \frac{1}{2} \mu\, (\operatorname{tr} \bar{\mathbf{b}} - 3)\ , \quad \bar{\mathbf{b}} = J^{-2/3}\, \mathbf{b}\ ,$$

where κ and μ are constants and $J = \det \mathbf{F}$, evaluate from $\boldsymbol{\tau} = 2\rho_o \dfrac{\partial \Psi}{\partial \mathbf{b}} \mathbf{b}$ the KIRCHHOFF stress tensor $\boldsymbol{\tau}$ and show that $\Psi_{,b}$ and $\mathbf{b}$ are coaxial.

4 Time derivative

In this section the notion of material time derivative is introduced which is then used to define the velocity and the acceleration vector. Finally the material time derivatives of some geometrical variables such as those of volume, surface and line elements are given in spatial formulation.

4.1 Definitions

It is suitable to start with a short repetition. As has been shown in section 2.1 the motion of a body can be described by using LAGRANGIAN coordinates $\mathbf{X}$ or, alternatively, EULER coordinates $\mathbf{x}$ as independent variables. If the JACOBIAN $J = \det \mathbf{F} > 0$, then the correspondance between the position vectors $\mathbf{X}$ and $\mathbf{x}$ is one to one and we may write

$$\mathbf{x} = \boldsymbol{\Phi}(\mathbf{X}, t) \ , \quad \mathbf{X} = \boldsymbol{\Phi}^{-1}(\mathbf{x}, t) \ , \tag{4.1.1}$$

where $\boldsymbol{\Phi}$ indicates $\mathbf{x}$ to be a function of $\mathbf{X}$ and time t. Similar holds for the notation $\boldsymbol{\Phi}^{-1}$. In the following, we will make use only of the first expression.

The material time derivative plays an important role to describe the rate of change of variables during a deformation process. Special care is to be taken to calculate the material derivative, if a variable in spatial description is considered. To show this, we first consider the vector function $\mathbf{f}$ representing e.g. the acceleration vector and suppose that it is given in the following forms

$$\textit{material description:}\ \ \mathbf{f} = \hat{\mathbf{f}}(\mathbf{X}, t, t_o) \ ,$$

$$\textit{spatial description:}\ \ \mathbf{f} = \hat{\mathbf{f}}^*(\mathbf{x}, t) \ . \tag{4.1.2}$$

Herein the notation $\hat{\mathbf{f}}(\mathbf{X}, t, t_o)$ means that $\mathbf{f}$ is a function of the arguments $\mathbf{X}$, t and t_o, while $\hat{\mathbf{f}}^*(\mathbf{x}, t)$ indicates $\mathbf{f}$ to be a function of $\mathbf{x}$ and t. The notation t_o is used for the reference time occuring explicitly in the material description. Note that in (4.1.2) the material coordinates $\mathbf{X}$ are independent of t.

The possibility to present any variable according to (4.1.2) in two different forms makes it suitable to distinguish between two types of time derivative. The first one is the *partial time derivative* to be constructed in its usual form. On the contrary, the *material time derivative* is a differentiation with respect to time t holding the material coordinates $\mathbf{X}$ constant. If this rule is applied to a variable in spatial formulation $\mathbf{x}$ and t, it should be remembered that $\mathbf{x} = \hat{\mathbf{x}}(\mathbf{X}, t)$ is a function of t. Both types of time derivatives will be shown to be identical operations if applied to variables in material description. We first introduce the following notations

for partial time derivative: $\frac{\partial \mathbf{f}}{\partial t}$

for material time derivative: $\frac{D\mathbf{f}}{Dt} = \dot{\mathbf{f}}$

We remark that alternative notations can be found in literature: In Malvern 1969 e.g., $\dot{\mathbf{f}}_{\mathbf{X}}$ stands for the material time derivative, where index $\mathbf{X}$ identifies the material coordinates $\mathbf{X}$ to be held constant when forming the material time derivative.

If we use the material description (4.1.2) for $\mathbf{f}$, its material time derivative is given by:

$$\frac{D\mathbf{f}}{Dt} = \dot{\mathbf{f}} = \frac{\partial \hat{\mathbf{f}}(\mathbf{X}, t, t_o)}{\partial t} . \tag{4.1.3}$$

Starting, however, from the spatial description (4.1.2) we have to remember that $\mathbf{x} = \mathbf{x}(\mathbf{X}, t)$ is a function of t. Thus, by using the chain rule of differentiation and the components $\mathbf{x} = x^i \mathbf{i}_i$ defined with respect to the orthonormal basis $\mathbf{i}_i$, we obtain

$$\frac{D\mathbf{f}}{Dt} = \dot{\mathbf{f}} = \frac{\partial \hat{\mathbf{f}}^*(\mathbf{x}(t), t)}{\partial x^i}\frac{\partial x^i}{\partial t} + \frac{\partial \hat{\mathbf{f}}^*(\mathbf{x}(t), t)}{\partial t} . \tag{4.1.4}$$

If we use in accordance with (1.6.7) the notations

$$\mathbf{v} = \frac{\partial x^i}{\partial t}\mathbf{i}_i = \frac{\partial \mathbf{x}}{\partial t} , \quad \operatorname{grad} \mathbf{f} = \frac{\partial \mathbf{f}}{\partial x^i} \otimes \mathbf{i}^i , \tag{4.1.5}$$

where for convenience the notation $\hat{\mathbf{f}}^*$ has been replaced by $\mathbf{f}$, equation (4.1.4) takes the form:

$$\frac{D\mathbf{f}}{Dt} = \dot{\mathbf{f}} = (\operatorname{grad} \mathbf{f})\, \mathbf{v} + \frac{\partial \mathbf{f}}{\partial t} . \tag{4.1.6}$$

Here, $\mathbf{v}$ denotes the velocity vector. Equation (4.1.3) shows that for a function in material formulation $\mathbf{f} = \hat{\mathbf{f}}(\mathbf{X}, t, t_o)$ partial and material time derivatives are identical operations. From (4.1.6) we see that the material derivative of a function $\mathbf{f}$ in spatial formulation consists of two parts. The *local time derivative* $\partial \mathbf{f}/\partial t$ describes the change of $\mathbf{f}$ by considering the point $\mathbf{x}$ fixed, while the *convective time derivative* (grad $\mathbf{f}$) $\mathbf{v}$ is due to the dependence of the position $\mathbf{x}$ on time t. This is illustrated in Fig. 4.1.

The material time derivative of arbitrary tensor functions of the spatial position $\mathbf{x}$ and t can be constructed by a procedure similar to that applied above to the vector $\mathbf{f}$. In the following, useful results are summarized

for a scalar ψ:

$$\dot{\psi} = \frac{D\psi}{Dt} = \frac{\partial \psi}{\partial t} + (\operatorname{grad} \psi) \cdot \mathbf{v} , \tag{4.1.7}$$

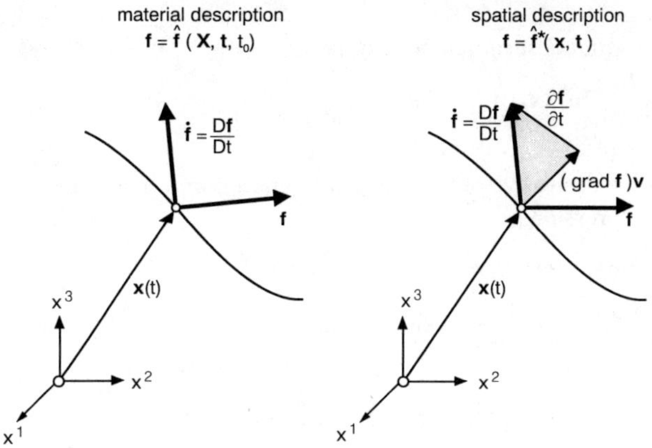

Fig. 4.1. Illustration of the material time derivative of a vector **f**

for a vector **f**: $$\dot{\mathbf{f}} = \frac{D\mathbf{f}}{Dt} = \frac{\partial \mathbf{f}}{\partial t} + (\text{grad}\,\mathbf{f})\,\mathbf{v} \,, \tag{4.1.8}$$

for a second-order tensor $\boldsymbol{\sigma}$: $$\dot{\boldsymbol{\sigma}} = \frac{D\boldsymbol{\sigma}}{Dt} = \frac{\partial \boldsymbol{\sigma}}{\partial t} + (\text{grad}\,\boldsymbol{\sigma})\,\mathbf{v} \,, \tag{4.1.9}$$

where ψ, **f** and $\boldsymbol{\sigma}$ are supposed to be given in spatial formulation. The gradient operator occuring in the above relations are defined according to (1.6.6) to (1.6.8) by

$$\text{grad}\,\psi = \frac{\partial \psi}{\partial x^i}\,\mathbf{i}^i \,, \quad \text{grad}\,\mathbf{f} = \frac{\partial \mathbf{f}}{\partial x^i} \otimes \mathbf{i}^i \,, \quad \text{grad}\,\boldsymbol{\sigma} = \frac{\partial \boldsymbol{\sigma}}{\partial x^i} \otimes \mathbf{i}^i \,. \tag{4.1.10}$$

4.2 Velocity and acceleration

To construct the velocity vector **v** and the acceleration vector **a**, we make use of the relation (4.1.3) or (4.1.6). Let $\mathbf{u} = \mathbf{x} - \mathbf{X}$ be the displacement vector of a characteristic point from its position **X** in the undeformed state to its position **x** in the deformed state. The *velocity vector* **v** is then defined by

$$\mathbf{v} = \frac{D\mathbf{x}}{Dt} = \frac{D\,(\mathbf{X} + \mathbf{u})}{Dt} = \frac{D\mathbf{X}}{Dt} + \frac{D\mathbf{u}}{Dt} = \frac{D\mathbf{u}}{Dt} \,, \tag{4.2.1}$$

where **u** may be given in material or spatial formulation. Accordingly, the material time derivative D**u**/Dt is to be calculated in two different ways (4.1.3) or (4.1.6) leading to

for material description $\mathbf{u} = \hat{\mathbf{u}}\,(\mathbf{X}, t, t_o)$:

$$\mathbf{v} = \frac{D\mathbf{u}}{Dt} = \frac{D\hat{\mathbf{u}}\,(\mathbf{X}, t, t_o)}{Dt} = \frac{\partial \mathbf{u}}{\partial t} \,, \tag{4.2.2}$$

and for spatial description $\mathbf{u} = \hat{\mathbf{u}}^*(\mathbf{x}, t)$:

$$\mathbf{v} = \frac{D\mathbf{u}}{Dt} = \frac{D\hat{\mathbf{u}}^*(\mathbf{x}, t)}{Dt} = \frac{\partial \mathbf{u}}{\partial t} + (\operatorname{grad}\mathbf{u})\,\mathbf{v} \ . \tag{4.2.3}$$

The material time derivative of $\mathbf{v}$ is the *acceleration vector* $\mathbf{a}$ described similarly to $\mathbf{v}$ by the following expressions

for material description $\mathbf{v} = \hat{\mathbf{v}}(\mathbf{X}, t, t_o)$:

$$\mathbf{a} = \frac{D\mathbf{v}}{Dt} = \frac{D\hat{\mathbf{v}}(\mathbf{X}, t, t_o)}{Dt} = \frac{\partial \mathbf{v}}{\partial t} \ , \tag{4.2.4}$$

and for spatial description $\mathbf{v} = \hat{\mathbf{v}}^*(\mathbf{x}, t)$:

$$\mathbf{a} = \frac{D\mathbf{v}}{Dt} = \frac{D\hat{\mathbf{v}}^*(\mathbf{x}, t)}{Dt} = \frac{\partial \mathbf{v}}{\partial t} + (\operatorname{grad}\mathbf{v})\,\mathbf{v} \ , \tag{4.2.5}$$

where grad $\mathbf{v}$ is to be constructed according to the second relation in (4.1.10).

4.3 Examples for material time derivative

In this section the material time derivatives of the geometrical elements $d\mathbf{x}$, dV, dA associated with the deformed body are calculated. An important deformation variable which will be used in this context is the *spatial velocity gradient* $\mathbf{l}$ introduced in (2.9.11) in the form

$$\mathbf{l} = \operatorname{grad}\dot{\mathbf{x}} = \dot{\mathbf{x}}_{,i} \otimes \mathbf{g}^i = \dot{\mathbf{F}}\,\mathbf{F}^{-1} \ . \tag{4.3.1}$$

As shown in (2.9.12) the tensor $\mathbf{l}$ permits to transform the basis $\mathbf{g}_i$ according to

$$\dot{\mathbf{g}}_i = \mathbf{l}\,\mathbf{g}_i \tag{4.3.2}$$

into the material time derivative $\dot{\mathbf{g}}_i$. The symmetrical part of $\mathbf{l}$ defined in (2.9.16)

$$\mathbf{d} = \frac{1}{2}(\mathbf{l} + \mathbf{l}^T) \tag{4.3.3}$$

is called the rate of deformation tensor. Using (1.6.10), (4.3.1) as well as the identity $\operatorname{tr}\mathbf{l} = \operatorname{tr}\mathbf{d}$ due to (4.3.3), $\operatorname{div}\dot{\mathbf{x}}$ can be expressed alternatively by:

$$\operatorname{div}\dot{\mathbf{x}} = \operatorname{grad}\dot{\mathbf{x}} : \mathbf{I} = \mathbf{l} : \mathbf{I} = \operatorname{tr}\mathbf{l} = \operatorname{tr}\mathbf{d} \ . \tag{4.3.4}$$

Note that $\operatorname{div}\dot{\mathbf{x}}$ is a scalar-valued function.

Line element dx. The material time derivative of the line element $d\mathbf{x}$ in the deformed configuration has already been derived in (2.9.14). The corresponding result is a linear transformation in terms of the spatial velocity gradient $\mathbf{l}$:

$$d\mathbf{v} := d\dot{\mathbf{x}} = \mathbf{l}\, d\mathbf{x} \ . \tag{4.3.5}$$

Determinant of F. Starting from (2.2.31) we first obtain

$$\dot{J} = \overline{\det \mathbf{F}}^{\boldsymbol{\cdot}} = \overline{\sqrt{\frac{g}{G}}}^{\boldsymbol{\cdot}} = \frac{\dot{\overline{\sqrt{g}}}}{\sqrt{G}} \ . \tag{4.3.6}$$

To evaluate $\dot{\overline{\sqrt{g}}}$ we use the identities

$$\sqrt{g} = [\mathbf{g}_1\ \mathbf{g}_2\ \mathbf{g}_3]$$

$$\mathbf{g}_1 \times \mathbf{g}_2 = \sqrt{g}\, \mathbf{g}^3 \ , \quad \mathbf{g}_2 \times \mathbf{g}_3 = \sqrt{g}\, \mathbf{g}^1 \ , \quad \mathbf{g}_3 \times \mathbf{g}_1 = \sqrt{g}\, \mathbf{g}^2$$

obtained from $\varepsilon_{rst} = [\mathbf{g}_r\ \mathbf{g}_s\ \mathbf{g}_t]$ and $\mathbf{g}_r \times \mathbf{g}_s = \varepsilon_{rst}\, \mathbf{g}^t$. By considering in addition (4.3.2) we may then write

$$\begin{aligned} \dot{\overline{\sqrt{g}}} &= \dot{\overline{[\mathbf{g}_1\ \mathbf{g}_2\ \mathbf{g}_3]}} \\ &= \dot{\mathbf{g}}_1 \cdot (\mathbf{g}_2 \times \mathbf{g}_3) + \dot{\mathbf{g}}_2 \cdot (\mathbf{g}_3 \times \mathbf{g}_1) + \dot{\mathbf{g}}_3 \cdot (\mathbf{g}_1 \times \mathbf{g}_2) \\ &= \sqrt{g}\, (\dot{\mathbf{g}}_1 \cdot \mathbf{g}^1 + \dot{\mathbf{g}}_2 \cdot \mathbf{g}^2 + \dot{\mathbf{g}}_3 \cdot \mathbf{g}^3] \\ &= \sqrt{g}\, (\dot{\mathbf{g}}_i \cdot \mathbf{g}^i) = \sqrt{g}\, \mathbf{g}_i\, \mathbf{l}^T \mathbf{g}^i = \sqrt{g}\, \mathbf{l}^T : \mathbf{g}_i \otimes \mathbf{g}^i = \sqrt{g}\, \mathbf{l} : \mathbf{I} \end{aligned} \tag{4.3.7}$$

leading in view of (4.3.4) to

$$\dot{\overline{\sqrt{g}}} = \sqrt{g}\, \mathbf{l} : \mathbf{I} = \sqrt{g}\, \operatorname{div} \dot{\mathbf{x}} = \sqrt{g}\, \operatorname{tr} \mathbf{l} = \sqrt{g}\, \operatorname{tr} \mathbf{d} \ . \tag{4.3.8}$$

Substitution of this result into (4.3.6) finally gives:

$$\dot{J} = \overline{\det \mathbf{F}}^{\boldsymbol{\cdot}} = J \operatorname{div} \dot{\mathbf{x}} = \det \mathbf{F} \operatorname{div} \dot{\mathbf{x}} = J \operatorname{tr} \mathbf{l} = J \operatorname{tr} \mathbf{d} \ , \tag{4.3.9}$$

where $\dot{\mathbf{x}} = \dot{\mathbf{u}} = \mathbf{v}$ in accordance with (4.2.1).

Volume element dV. Since from (2.2.31)

$$\dot{\overline{dV}} = \dot{J}\, dV_o = \dot{J}\, J^{-1}\, dV \tag{4.3.10}$$

the result is in view of (4.3.9)

$$\dot{\overline{dV}} = \operatorname{div} \dot{\mathbf{x}}\, dV = \operatorname{div} \mathbf{v}\, dV = \operatorname{tr} \mathbf{l}\, dV = \operatorname{tr} \mathbf{d}\, dV \ . \tag{4.3.11}$$

Surface element dA. The starting point of the derivation is equation (2.2.35) which using (4.3.6) gives

$$\dot{\mathbf{n}}\, dA + \mathbf{n}\, \dot{\overline{dA}} = \dot{J}\, \mathbf{F}^{-T} \mathbf{N}\, dA_o + J\, \dot{\mathbf{F}}^{-T} \mathbf{N}\, dA_o \ . \tag{4.3.12}$$

In view of the identities

$$\mathbf{n} \cdot \mathbf{n} = 1 \quad \text{and} \quad \dot{\mathbf{n}} \cdot \mathbf{n} = 0$$

relation (4.3.12) is transformed, after scalar multiplication with **n**, into:

$$\dot{\overline{dA}} = \dot{J}\, \mathbf{n}\, \mathbf{F}^{-T}\, \mathbf{N}\, dA_o + J\, \mathbf{n}\, \dot{\mathbf{F}}^{-T}\, \mathbf{N}\, dA_o \ . \qquad (4.3.13)$$

We now express $\mathbf{N}\, dA_o$ according to (2.2.35) by

$$\mathbf{N}\, dA_o = J^{-1}\, \mathbf{F}^T\, \mathbf{n}\, dA$$

and consider the relation

$$\dot{\mathbf{F}}^{-T}\, \mathbf{F}^T = -\mathbf{F}^{-T}\, \dot{\mathbf{F}}^T$$

deduced from the identity

$$\dot{\overline{\mathbf{F}^{-T}\, \mathbf{F}^T}} = \dot{\mathbf{F}}^{-T}\, \mathbf{F}^T + \mathbf{F}^{-T}\, \dot{\mathbf{F}}^T = \dot{\mathbf{I}} = \mathbf{0} \ .$$

Thus equation (4.3.13) takes the form:

$$\dot{\overline{dA}} = \dot{J}\, J^{-1}\, \mathbf{n}\, \mathbf{F}^{-T}\, \mathbf{F}^T\, \mathbf{n}\, dA + \mathbf{n}\, \dot{\mathbf{F}}^{-T}\, \mathbf{F}^T\, \mathbf{n}\, dA = (\dot{J}\, J^{-1} - \mathbf{n}\, \mathbf{F}^{-T}\, \dot{\mathbf{F}}^T\, \mathbf{n})\, dA \ . \qquad (4.3.14)$$

The consideration of (4.3.9) together with (4.3.1), (4.3.4) permits finally to present (4.3.14) in the following forms:

$$\dot{\overline{dA}} = (\operatorname{div} \dot{\mathbf{x}} - \mathbf{n}\, \mathbf{l}^T\, \mathbf{n})\, dA = (\operatorname{tr} \mathbf{d} - \mathbf{n}\, \mathbf{d}\, \mathbf{n})\, dA \ . \qquad (4.3.15)$$

In deriving the second expression the fact has been considered that $\mathbf{n}\, \mathbf{k}\, \mathbf{n} = \mathbf{n}\, (\operatorname{sym} \mathbf{k})\, \mathbf{n}$ for any second-order tensor **k**. Accordingly, $\mathbf{l}^T$ has been replaced by **d**.

Application. Form the material time derivative of the JACOBIAN J starting from (2.4.43)

$$J = \det \mathbf{F} \ .$$

By definition we have

$$\dot{J} = \frac{DJ}{Dt} = \frac{\partial (\det \mathbf{F})}{\partial F^i_{\cdot j}} \frac{\partial F^i_{\cdot j}}{\partial t} = \frac{\partial (\det \mathbf{F})}{\partial F^i_{\cdot j}}\, \mathbf{G}^i \otimes \mathbf{G}_j : \frac{\partial F^m_{\cdot n}}{\partial t}\, \mathbf{G}_m \otimes \mathbf{G}^n = \frac{\partial (\det \mathbf{F})}{\partial \mathbf{F}} : \dot{\mathbf{F}} \ . \qquad (4.3.16)$$

Since $III_F = \det \mathbf{F}$ we may use the rule (1.8.10) for forming $(\det \mathbf{F})_{,\mathbf{F}}$. Thus, by considering (1.3.51) and (4.3.1) we obtain

$$\dot{J} = J\, \mathbf{F}^{-T} : \dot{\mathbf{F}} = J \operatorname{tr} (\dot{\mathbf{F}}\, \mathbf{F}^{-1}) = J \operatorname{tr} \mathbf{l} = J \operatorname{tr} \mathbf{d} \ , \qquad (4.3.17)$$

in accordance with (4.3.9).

Application. Consider a scalar-valued function ψ depending on the invariants of the left CAUCHY-GREEN tensor **b**. Such a function is expressible in terms of the mixed components of **b**.

$$\psi = \psi\, (I_b, II_b, III_b) = \psi\, (b^j_i) \ . \qquad (4.3.18)$$

We emphazise that ψ contains only the components b^j_i as time-dependent variables. In this case the material time derivative $\dot{\psi}$ can be constructed as follows:

$$\dot{\psi} = \frac{D\psi}{Dt} = \frac{\partial\psi}{\partial b^j_i}\dot{b}^j_i = \psi,_\mathbf{b} : L_v\,\mathbf{b} \ , \tag{4.3.19}$$

where the partial derivative $\psi,_\mathbf{b}$ and the LIE-derivative $L_v\,\mathbf{b}$

$$\psi,_\mathbf{b} = \frac{\partial\psi}{\partial b^j_i}\mathbf{g}_i \otimes \mathbf{g}^j \ , \quad L_v\,\mathbf{b} = L_v\,(b^n_m\,\mathbf{g}^m \otimes \mathbf{g}_n) = \dot{b}^n_m\,\mathbf{g}^m \otimes \mathbf{g}_n \tag{4.3.20}$$

have been used as abbreviations. Our aim is to show that the expression given in (4.3.19) for $\dot{\psi}$ can be replaced by $\dot{\psi} = \psi,_\mathbf{b} : \dot{\mathbf{b}}$ under the assumption (4.3.18).

To this end we express in (4.3.19) the LIE-derivative $L_v\,\mathbf{b}$ according to (2.10.27). Then, the corresponding result can be, by means of (1.3.24), transformed as follows:

$$\begin{aligned}\dot{\psi} &= \psi,_\mathbf{b} : L_v\,\mathbf{b} = \psi,_\mathbf{b} : (\dot{\mathbf{b}} + \mathbf{l}^T\,\mathbf{b} - \mathbf{b}\,\mathbf{l}^T) \\ &= \psi,_\mathbf{b} : \dot{\mathbf{b}} + \psi,_\mathbf{b} : (\mathbf{l}^T\,\mathbf{b}) - \psi,_\mathbf{b} : ((\mathbf{b}\,\mathbf{l}^T) \\ &= \psi,_\mathbf{b} : \dot{\mathbf{b}} + (\psi,_\mathbf{b}\,\mathbf{b}) : \mathbf{l}^T - (\mathbf{b}\,\psi,_\mathbf{b}) : \mathbf{l}^T \ .\end{aligned} \tag{4.3.21}$$

As is shown in (2.4.54), the assumption (4.3.18) implies that the tensors $\psi,_\mathbf{b}$ and $\mathbf{b}$ are coaxial:

$$\psi,_\mathbf{b}\,\mathbf{b} = \mathbf{b}\,\psi,_\mathbf{b} \ . \tag{4.3.22}$$

Thus, (4.3.21) is given in terms of $\dot{\mathbf{b}}$:

$$\dot{\psi} = \psi,_\mathbf{b} : L_v\,\mathbf{b} = \psi,_\mathbf{b} : \dot{\mathbf{b}} \tag{4.3.23}$$

demonstrating the initial statement.

Application. Prove that the LIE-derivative $L_v\,\mathbf{b} = \dot{b}^j_i\,\mathbf{g}^i \otimes \mathbf{g}_j$ of the left CAUCHY-GREEN tensor $\mathbf{b}$ is coaxial with $\mathbf{b}$.

From (2.10.27) we have

$$L_v\,\mathbf{b} = \dot{b}^j_i\,\mathbf{g}^i \otimes \mathbf{g}_j = \dot{\mathbf{b}} + \mathbf{b}\,\mathbf{l} - \mathbf{l}\,\mathbf{b} \ , \tag{4.3.24}$$

where in view of (2.6.30)

$$\mathbf{b} = \sum_{i=1}^{3} \lambda_i^2\,\mathbf{n}_i \otimes \mathbf{n}_i = \sum_{i=1}^{3} \Lambda_i\,\mathbf{n}_i \otimes \mathbf{n}_i \tag{4.3.25}$$

$$\dot{\mathbf{b}} = \sum_{i=1}^{3}\left(2\,\dot{\lambda}_i\,\lambda_i\,\mathbf{n}_i \otimes \mathbf{n}_i + \lambda_i^2\,\dot{\mathbf{n}}_i \otimes \mathbf{n}_i + \lambda_i^2\,\mathbf{n}_i \otimes \dot{\mathbf{n}}_i\right) . \tag{4.3.26}$$

If we identify the base vectors $\mathbf{G}_i$ and $\mathbf{G}^i$ of the undeformed configuration with the (unit) eigenvectors $\mathbf{N}_i$ of $\mathbf{C}$ then, their counterparts in the deformed state are given by $\mathbf{g}_i = \lambda_i\,\mathbf{n}_i$ and $\mathbf{g}^i = \frac{1}{\lambda_i}\mathbf{n}_i$, respectively. It follows according to (2.9.11)

$$\mathbf{l} = \sum_{i=1}^{3}\left(\frac{1}{\lambda_i}(\lambda_i\,\mathbf{n}_i)^{\cdot} \otimes \mathbf{n}_i\right) = \sum_{i=1}^{3}\left(\frac{\dot{\lambda}_i}{\lambda_i}\mathbf{n}_i \otimes \mathbf{n}_i + \dot{\mathbf{n}}_i \otimes \mathbf{n}_i\right) . \tag{4.3.27}$$

which in view of the identity

$$\dot{\mathbf{I}} = \overline{\left(\sum_{i=1}^{3} \mathbf{n}_i \otimes \mathbf{n}_i\right)}^{\boldsymbol{\cdot}} = \mathbf{0} \quad \rightarrow \quad \sum_{i=1}^{3} \dot{\mathbf{n}}_i \otimes \mathbf{n}_i = -\sum_{i=1}^{3} \mathbf{n}_i \otimes \dot{\mathbf{n}}_i \tag{4.3.28}$$

can be also expressed in the form

$$\mathbf{l} = \sum_{i=1}^{3} \left(\frac{\dot{\lambda}_i}{\lambda_i} \mathbf{n}_i \otimes \mathbf{n}_i - \mathbf{n}_i \otimes \dot{\mathbf{n}}_i \right) . \tag{4.3.29}$$

To construct $\mathbf{l}\,\mathbf{b}$ we use (4.3.25) and (4.3.27) which, by considering the identity $\mathbf{n}_i \cdot \mathbf{n}_j = \delta_{ij}$, lead to

$$\begin{aligned} \mathbf{l}\,\mathbf{b} &= \left(\sum_{i=1}^{3} \left(\frac{\dot{\lambda}_i}{\lambda_i} \mathbf{n}_i \otimes \mathbf{n}_i + \dot{\mathbf{n}}_i \otimes \mathbf{n}_i \right)\right)\left(\sum_{j=1}^{3} \lambda_j^2\, \mathbf{n}_j \otimes \mathbf{n}_j \right) \\ &= \sum_{i=1}^{3} \left(\lambda_i \dot{\lambda}_i\, \mathbf{n}_i \otimes \mathbf{n}_i + \lambda_i^2\, \dot{\mathbf{n}}_i \otimes \mathbf{n}_i \right) . \end{aligned} \tag{4.3.30}$$

Similarly, we find from (4.3.25) and (4.3.29)

$$\begin{aligned} \mathbf{b}\,\mathbf{l} &= \left(\sum_{i=1}^{3} \lambda_i^2\, \mathbf{n}_i \otimes \mathbf{n}_i \right)\left(\sum_{j=1}^{3} \left(\frac{\dot{\lambda}_j}{\lambda_j} \mathbf{n}_j \otimes \mathbf{n}_j - \mathbf{n}_j \otimes \dot{\mathbf{n}}_j \right)\right) \\ &= \sum_{i=1}^{3} \left(\lambda_i \dot{\lambda}_i\, \mathbf{n}_i \otimes \mathbf{n}_i - \lambda_i^2\, \mathbf{n}_i \otimes \dot{\mathbf{n}}_i \right) . \end{aligned} \tag{4.3.31}$$

With (4.3.25), (4.3.30) and (4.3.31) and $\Lambda_i = \lambda_i^2$, equation (4.3.24) takes the form

$$\mathrm{L_v}\, \mathbf{b} = \sum_{i=1}^{3} 2\, \lambda_i \dot{\lambda}_i\, \mathbf{n}_i \otimes \mathbf{n}_i = \sum_{i=1}^{3} \overline{(\lambda_i)^2}^{\boldsymbol{\cdot}}\, \mathbf{n}_i \otimes \mathbf{n}_i = \sum_{i=1}^{3} \dot{\Lambda}_i\, \mathbf{n}_i \otimes \mathbf{n}_i \tag{4.3.32}$$

showing that the LIE-derivative $\mathrm{L_v}\, \mathbf{b} = \dot{b}^i_j\, \mathbf{g}^i \otimes \mathbf{g}_j$ and $\mathbf{b}$ are coaxial.

Exercises

4.1. Prove the following equality:

$$\dot{u}_i\, \mathbf{g}^i + u_i\, \dot{\mathbf{g}}^i = \dot{u}^i\, \mathbf{g}_i + u^i\, \dot{\mathbf{g}}_i \; .$$

4.2. Remembering that, for the scalar-valued function $\psi = \psi\,(\mathrm{I_a}, \mathrm{II_a}, \mathrm{III_a})$, the tensors $\psi_{,\mathbf{a}}$ and $\mathbf{a}$ are coaxial prove the following equalities:

$$\dot{\psi} = \psi_{,\mathbf{a}} : (\mathbf{l}\,\mathbf{a} + \mathbf{a}\,\mathbf{l}^{\mathrm{T}} + \mathrm{L_v}\,\mathbf{a}) = 2\,\psi_{,\mathbf{a}}\,\mathbf{a} : (\mathbf{d} + \tfrac{1}{2}\,\mathrm{L_v}\,(\mathbf{a})\,\mathbf{a}^{-1})$$

where $\mathrm{L_v}\,\mathbf{a} = \dot{a}^{ij}\, \mathbf{g}_i \otimes \mathbf{g}_j$ and $\mathbf{a} = \mathbf{a}^{\mathrm{T}}$.

4.3. Let $\psi = \psi(\mathbf{C})$ be an arbitrary scalar-valued function of the right CAUCHY-GREEN tensor $\mathbf{C}$. Construct the material time derivative $\dot{\psi}$.

4.4. Construct the material time derivatives of the invariants of $\mathbf{C}$ and $\mathbf{b}$ and present the results in component and symbolic notation.

4.5 Establish the relation between $\dot{\mathbf{E}} = \dot{E}_{ij}\, \mathbf{G}^i \otimes \mathbf{G}^j$ and $\mathrm{L_v}\,\mathbf{e} = \dot{e}_{ij}\, \mathbf{g}^i \otimes \mathbf{g}^j$, where $\mathbf{E}$ is the GREEN-LAGRANGE strain tensor and $\mathbf{e}$ is the ALMANSI strain tensor.

5 Balance laws

This chapter introduces balance laws: conservation of mass, balance of momentum, balance of moment of momentum, balance of kinetic energy and conservation of energy being the most important general principles of continuum mechanics. The equations of motion are derived in material and spatial formulation. Furthermore the symmetry of the CAUCHY *tensor is shown as a consequence of balance of moment of momentum. Finally the principle of virtual work is derived coupling kinematically admissible virtual deformations with real forces and stresses as weak formulation of the equations of motion.*

5.1 Conservation of mass

We consider the reference configuration B_o of a body at time t_o taking after deformation the configuration B at time t. If ρ is the mass density of the body in the configuration B then its total mass in the considered current state is

$$m(t) = \iiint_V \rho \, dV \ , \tag{5.1.1}$$

where, as usual, dV denotes the volume element of the body in the state B. Note that the mass density $\rho = \rho(\mathbf{x},t)$ is a function of spatial coordinates $\mathbf{x}$ and time t.

The law of conservation of mass states that the mass of a body remains constant during the deformation process. This permits to evaluate the mass density ρ_o of the body in its initial configuration B_o in terms of the current value ρ. If we write equation (5.1.1) for the initial state B_o and express the undeformed volume element dV_o according to (2.2.31)

$$dV_o = \frac{1}{\det \mathbf{F}} dV \tag{5.1.2}$$

we find

$$\iiint_V \left(\rho - \frac{\rho_o}{\det \mathbf{F}} \right) dV = 0 \ . \tag{5.1.3}$$

Since this relation holds for arbitrary subdomains of the volume V, the integrand must vanish in any point of the actual configuration B leading to

$$\rho_o = \rho \det \mathbf{F} \ , \tag{5.1.4}$$

as local formulation of *conservation of mass*. It should be noticed that this statement only holds for deformation processes where no mass is created or destroyed and there is no inflow of mass through the boundary surface of the body.

Since $\rho_o = \rho_o(\mathbf{X}, t_o)$ is a fixed function of $\mathbf{X}$ at the reference time t_o and the material coordinates $\mathbf{X}$ are independent of t, equation (5.1.4) implies that $\rho \det \mathbf{F}$ remains unaltered during deformation. Thus by considering (4.3.9) we receive with $\dot{\mathbf{x}} = \mathbf{v}$ as an alternative formulation to (5.1.4)

$$\frac{D}{Dt}(\rho \det \mathbf{F}) = 0 \quad \rightarrow \quad \dot{\rho} + \rho \operatorname{div} \mathbf{v} = 0 \ , \tag{5.1.5}$$

which is known as *continuity condition*. As usual, $\dot{\rho} = D\rho/Dt$ denotes the material time derivative of ρ, which presents a differentiation with respect to time holding $\mathbf{X}$ constant.* Expressing $\det \mathbf{F}$ according to (5.1.2), equation (5.1.4) may be also given in the form

$$\frac{D}{Dt}(\rho \, dV) = 0 \quad \text{or} \quad \rho \, dV = \rho_o \, dV_o \ . \tag{5.1.6}$$

5.2 Balance of momentum

The body in its actual configuration B at time t is supposed to be subjected to the velocity field $\dot{\mathbf{u}} = \dot{\mathbf{x}}$, where $\mathbf{u} = \mathbf{x} - \mathbf{X}$ is the displacement vector. The *total momentum* of the body at an arbitrary time t of the motion is defined by

$$\mathbf{J} := \iiint\limits_V \rho \, \dot{\mathbf{x}} \, dV \tag{5.2.1}$$

with the density ρ referring to the actual deformed configuration B. As usual the notation $(\dot{...})$ stands for material time derivative. According to Fig. 5.1 we suppose that the body is subjected to volume forces $\mathbf{b}$ measured per unit mass and the surface forces $\mathbf{t}$ acting upon the boundary surface A in its actual state B. The forces $\mathbf{t}$ are supposed to refer to the unit area of A. Thus the expression

$$\mathbf{K} := \mathbf{K}_t + \mathbf{K}_b = \iint\limits_A \mathbf{t} \, dA + \iiint\limits_V \rho \, \mathbf{b} \, dV \tag{5.2.2}$$

describes the resultant external force $\mathbf{K}$ exerted on the body in its actual position B, $\mathbf{K}_t$ and $\mathbf{K}_b$ being its constituent parts due to $\mathbf{t}$ and $\mathbf{b}$.

The *law of balance of momentum* states that the rate of change of the total momentum $\dot{\mathbf{J}}$ is equal to the resultant external force $\mathbf{K}$ consisting of surface forces $\mathbf{K}_t$ and volume forces $\mathbf{K}_b$. Thus, by considering (5.2.1) and (5.2.2) this postulate is expressed by

* For a more detailed definition of the material time derivative we refer to section 4.1.

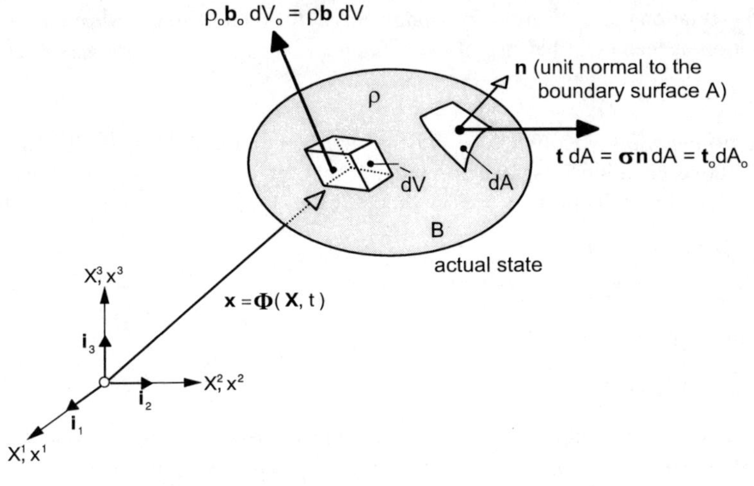

Fig. 5.1. Momentum balance

$$\dot{\mathbf{J}} := \frac{D\mathbf{J}}{Dt} = \mathbf{K} \quad \rightarrow \quad \frac{D}{Dt}\iiint_V \rho\, \dot{\mathbf{x}}\, dV = \iint_A \mathbf{t}\, dA + \iiint_V \rho\, \mathbf{b}\, dV \ , \qquad (5.2.3)$$

which is called the law of *balance of momentum*.

Since, according to (5.1.6), $\overline{\rho\, dV}^{\cdot} = 0$ and, therefore, the identity

$$\frac{D}{Dt}\iiint_V \rho\, \dot{\mathbf{x}}\, dV = \iiint_V \rho\, \ddot{\mathbf{x}}\, dV \qquad (5.2.4)$$

holds, equation (5.2.3) is also expressible alternatively in the form

$$\dot{\mathbf{J}} := \frac{D\mathbf{J}}{Dt} = \mathbf{K} \quad \rightarrow \quad \iiint_V \rho\, \ddot{\mathbf{x}}\, dV = \iint_A \mathbf{t}\, dA + \iiint_V \rho\, \mathbf{b}\, dV \ . \qquad (5.2.5)$$

Now, our aim is to transform the above global law of balance of momentum into a local formulation. To this end we transform the surface integral, by using the GAUSS-GREEN theorem (1.6.16) and the CAUCHY theorem (3.1.17), into a volume integral

$$\iint_A \mathbf{t}\, dA = \iint_A \boldsymbol{\sigma}\, \mathbf{n}\, dA = \iiint_V \operatorname{div} \boldsymbol{\sigma}\, dV \ . \qquad (5.2.6)$$

Thus, (5.2.5) becomes

$$\iiint_V \rho\, \ddot{\mathbf{x}}\, dV = \iiint_V \operatorname{div} \boldsymbol{\sigma}\, dV + \iiint_V \rho\, \mathbf{b}\, dV \ . \qquad (5.2.7)$$

Since this relation holds for arbitrary subdomains of the total volume V we obtain as local formulation of the law of balance of momentum

$$\rho\, \ddot{\mathbf{x}} = \operatorname{div} \boldsymbol{\sigma} + \rho\, \mathbf{b} \;, \tag{5.2.8}$$

the *CAUCHY equation of motion* to be satisfied at each point **x** of the actual configuration B. The above equation in spatial formulation may be also expressed in terms of the contravariant stress vector $\mathbf{t}^i$. Since, in view of (1.6.11), (3.1.6) as well as $\mathbf{g}_i\|_j = \mathbf{0}^*$, the relation

$$\operatorname{div} \boldsymbol{\sigma} = \sigma^{ij}\|_j\, \mathbf{g}_i = \mathbf{t}^j\|_j \tag{5.2.9}$$

holds, the result is of the form

$$\rho\, \ddot{\mathbf{x}} = \mathbf{t}^j\|_j + \rho\, \mathbf{b} \tag{5.2.10}$$

with the covariant derivative $(\quad)\|_j$ referring to the deformed basis $\mathbf{g}_i$. If the acceleration vector $\ddot{\mathbf{x}}$ vanishes, equation (5.2.8) reduces to the *equilibrium equation*

$$\operatorname{div} \boldsymbol{\sigma} + \rho\, \mathbf{b} = \mathbf{0} \;, \tag{5.2.11}$$

which, by using (5.2.9) and the decomposition $\mathbf{b} = b^i\, \mathbf{g}_i$, takes the form

$$\sigma^{ij}\|_j + \rho\, b^i = 0 \tag{5.2.12}$$

where, according to (3.2.3), $\boldsymbol{\sigma} = \sigma^{ij}\, \mathbf{g}_i \otimes \mathbf{g}_j$.

The law of balance of momentum (5.2.5) and its local form (5.2.8) are also expressible in material formulation. For this purpose we introduce surface forces $\mathbf{t}_o = dA/dA_o\, \mathbf{t}$ per unit area of the undeformed boundary surface A_o and denote the volume forces **b** per unit mass by

$$\mathbf{b}_o(\mathbf{X}) = \mathbf{b}(\boldsymbol{\Phi}(\mathbf{X}, t)) \tag{5.2.13}$$

which indicates that $\mathbf{b}_o$ is a function of the material coordinates **X**. In this context we recall that the notation **b** expresses the dependence of volume forces on spatial coordinates **x**. If we finally consider (5.1.6), equation (5.2.5) is equivalently replaced by

$$\iiint_{V_o} \rho_o\, \ddot{\mathbf{x}}\, dV_o = \iint_{A_o} \mathbf{t}_o\, dA_o + \iiint_{V_o} \rho_o\, \mathbf{b}_o\, dV_o \;. \tag{5.2.14}$$

The local form of this statement can be derived by a procedure similar to that applied to the spatial formulation (5.2.5). First, the surface integral is transformed by means of (1.6.16) and (3.2.14) into a volume integral

$$\iint_{A_o} \mathbf{t}_o\, dA_o = \iint_{A_o} \mathbf{P}\, \mathbf{N}\, dA_o = \iiint_{V_o} \mathrm{DIV}\, \mathbf{P}\, dV_o \;. \tag{5.2.15}$$

---

* see equation (1.2.24).

Inserting this expression into (5.2.14) delivers then in view of the arbitrariness of the considered subdomain

$$\rho_o \ddot{\mathbf{x}} = \mathrm{DIV}\,\mathbf{P} + \rho_o \mathbf{b}_o \ , \tag{5.2.16}$$

where $\mathbf{P} = P^{ij}\,\mathbf{G}_i \otimes \mathbf{G}_j$ denotes the first PIOLA-KIRCHHOFF stress tensor. Using in accordance with (1.6.11) the expression

$$\mathrm{DIV}\,\mathbf{P} = P^{ij}|_j\,\mathbf{G}_i \ , \tag{5.2.17}$$

as well as the decompositions

$$\ddot{\mathbf{x}} = \ddot{U}^i\,\mathbf{G}_i \ , \quad \mathbf{b}_o = b^i_o\,\mathbf{G}_i \ , \tag{5.2.18}$$

equation (5.2.16) is finally transformed into a component relation

$$P^{ij}|_j + \rho_o\,(b^i_o - \ddot{U}^i) = 0 \tag{5.2.19}$$

with the covariant derivative $(\ )|_j$ referring to the basis $\mathbf{G}_i$. Equations (5.2.16) and (5.2.19) given in material formulation are called *LAGRANGE equations of motion.*

5.3 Balance of moment of momentum

In this section the balance law of moment of momentum is derived. For bodies which are not subjected to couple loads, the so-called BOLTZMANN-continua, this principle has as consequence the symmetry of the CAUCHY stress tensor $\boldsymbol{\sigma} = \boldsymbol{\sigma}^T$.

We consider again a body subjected to the velocity field $\dot{\mathbf{x}} = \dot{\mathbf{u}}$ at time t and adopt the notations $\mathbf{t}$ and $\mathbf{b}$ introduced in section 5.2 for surface and volume forces, respectively. According to (5.2.1), the *moment of the momentum* $\mathbf{J}$ with respect to the origin 0 of a fixed orthogonal Cartesian reference frame is given by

$$\mathbf{L} := \iiint_V \mathbf{x} \times \rho\,\dot{\mathbf{x}}\,dV \ , \tag{5.3.1}$$

while the moment of the resultant external force $\mathbf{K}$ has by virtue of (5.2.2) the form

$$\mathbf{M} := \mathbf{M}_t + \mathbf{M}_b = \iint_A \mathbf{x} \times \mathbf{t}\,dA + \iiint_V \mathbf{x} \times \rho\,\mathbf{b}\,dV \ . \tag{5.3.2}$$

The *balance law of moment of momentum* states that the rate of change of the moment of momentum is equal to the resultant of moments of external forces acting on the deformed body. By means of the above definitions we then have

$$\dot{\mathbf{L}} := \frac{D\mathbf{L}}{Dt} = \mathbf{M} \quad \rightarrow \quad \frac{D}{Dt}\iiint_V \mathbf{x} \times \rho\,\dot{\mathbf{x}}\,dV = \iint_A \mathbf{x} \times \mathbf{t}\,dA + \iiint_V \mathbf{x} \times \rho\,\mathbf{b}\,dV \ . \tag{5.3.3}$$

For the simplification of this relation we first use the continuity condition (5.1.6) to transform the left-hand side term into

$$\dot{\mathbf{L}} := \frac{D}{Dt} \iiint_V \mathbf{x} \times \rho\, \dot{\mathbf{x}}\, dV = \iiint_V \mathbf{x} \times \rho\, \ddot{\mathbf{x}}\, dV \ . \tag{5.3.4}$$

The next step to be accomplished is the transformation of the surface integral

$$\mathbf{M}_t := \iint_A \mathbf{x} \times \mathbf{t}\, dA = \iint_A \mathbf{x} \times (\boldsymbol{\sigma}\, \mathbf{n})\, dA \tag{5.3.5}$$

introduced in (5.3.2) into a volume integral. For this purpose we use the following identity obtained by means of (5.2.9) and the equality $\mathbf{g}_i = \mathbf{x},_i = \mathbf{x}\|_i$

$$\begin{aligned}\iiint_V (\mathbf{x} \times \mathbf{t}^i)\|_i\, dV &= \iiint_V (\mathbf{x} \times \mathbf{t}^i\|_i)\, dV + \iiint_V (\mathbf{x},_i \times \mathbf{t}^i)\, dV \\ &= \iiint_V (\mathbf{x} \times \operatorname{div} \boldsymbol{\sigma})\, dV + \iiint_V (\mathbf{g}_i \times \mathbf{t}^i)\, dV \ .\end{aligned} \tag{5.3.6}$$

Since the vector product $\mathbf{A}^i = \mathbf{x} \times \mathbf{t}^i$ defines contravariant vectors $\mathbf{A}^i$, the GAUSS-GREEN theorem (1.6.17) can be applied to the integral on the left-hand side leading in view of the relation $\mathbf{t} = \mathbf{t}^i\, n_i$ by (3.1.13) to

$$\iiint_V (\mathbf{x} \times \mathbf{t}^i)\|_i\, dV = \iint_A \mathbf{x} \times \mathbf{t}^i\, n_i\, dA = \iint_A \mathbf{x} \times \mathbf{t}\, dA \ . \tag{5.3.7}$$

Thus (5.3.6) becomes

$$\iint_A \mathbf{x} \times \mathbf{t}\, dA = \iiint_V (\mathbf{x} \times \operatorname{div} \boldsymbol{\sigma} + \mathbf{g}_i \times \mathbf{t}^i)\, dV \ . \tag{5.3.8}$$

If we insert this result together with (5.3.4) into (5.3.3)

$$\iiint_V \left[\mathbf{x} \times (\operatorname{div} \boldsymbol{\sigma} + \rho\, \mathbf{b} - \rho\, \ddot{\mathbf{x}}) + \mathbf{g}_i \times \mathbf{t}^i\right] dV = \mathbf{0} \ , \tag{5.3.9}$$

and consider that, in view of the equation of motion (5.2.8), the expression in round brackets vanishes, we end up with

$$\iiint_V \mathbf{g}_i \times \mathbf{t}^i\, dV = \mathbf{0} \tag{5.3.10}$$

as simplified form of (5.3.3). Since this result holds for an arbitrary subdomain we deduce that the relation

$$\mathbf{g}_i \times \mathbf{t}^i = \mathbf{0} \tag{5.3.11}$$

must be satisfied in each point of the body. Therefore, (5.3.11) is the local form of the statement (5.3.10) or (5.3.3).

To show the symmetry of the CAUCHY stress tensor $\boldsymbol{\sigma} = \boldsymbol{\sigma}^T$, equation (5.3.11) has to be transformed into component form. We substitute the expression (3.1.6), $\mathbf{t}^i = \sigma^{ji}\,\mathbf{g}_j$, into (5.3.11) and then use the well-known identity $\mathbf{g}_i \times \mathbf{g}_j = \varepsilon_{kij}\,\mathbf{g}^k$ in terms of the permutation tensor ε_{kij}. This procedure delivers

$$\varepsilon_{kij}\,\sigma^{ji}\,\mathbf{g}^k = \sqrt{g}\left[(\sigma^{32}-\sigma^{23})\,\mathbf{g}^1 + (\sigma^{13}-\sigma^{31})\,\mathbf{g}^2 + (\sigma^{21}-\sigma^{12})\,\mathbf{g}^3\right] = \mathbf{0} \tag{5.3.12}$$

establishing, since all the vectors $\mathbf{g}_i$ (i = 1, 2, 3) are independent, the symmetry of the CAUCHY stress tensor

$$\sigma^{ij} = \sigma^{ji} \quad \text{or} \quad \boldsymbol{\sigma} = \boldsymbol{\sigma}^T \ . \tag{5.3.13}$$

We finally complete the statement (5.3.13) by further symmetry relations

$$\boldsymbol{\tau} = \boldsymbol{\tau}^T \ , \quad \mathbf{S} = \mathbf{S}^T \ , \quad \mathbf{P}\,\mathbf{F}^T = \mathbf{F}\,\mathbf{P}^T \tag{5.3.14}$$

immediately resulting from (3.2.29). Note that the first PIOLA-KIRCHHOFF stress tensor $\mathbf{P}$ itself is not symmetric, which is also true for the BIOT stress tensor $\mathbf{T}$ defined as $\mathbf{T} = \mathbf{R}^T\,\mathbf{P}$ in terms of $\mathbf{P}$ and the rotation tensor $\mathbf{R}$.

5.4 Balance of kinetic energy

The law of balance of kinetic energy is not an independent postulate, but can be directly obtained from the CAUCHY equation of motion (5.2.8)

$$\rho\,\ddot{\mathbf{x}} = \operatorname{div}\boldsymbol{\sigma} + \rho\,\mathbf{b} \tag{5.4.1}$$

corresponding to the local form of the balance of momentum. For this we form the scalar product of (5.4.1) with the velocity vector $\mathbf{v} = \dot{\mathbf{x}}$

$$\mathbf{v}\cdot\rho\,\dot{\mathbf{v}} = \mathbf{v}\cdot\operatorname{div}\boldsymbol{\sigma} + \mathbf{v}\cdot\rho\,\mathbf{b} \ , \tag{5.4.2}$$

and transform the above result by using the identities

$$\operatorname{div}(\mathbf{v}\,\boldsymbol{\sigma}) = \boldsymbol{\sigma} : \operatorname{grad}\mathbf{v} + \mathbf{v}\cdot\operatorname{div}\boldsymbol{\sigma} \tag{5.4.3}$$

$$\boldsymbol{\sigma} : \operatorname{grad}\mathbf{v} = \boldsymbol{\sigma} : \mathbf{l} = \boldsymbol{\sigma} : \mathbf{d} \tag{5.4.4}$$

deduced from (1.6.14), (2.9.9), (2.9.16) and the symmetry condition $\boldsymbol{\sigma} = \boldsymbol{\sigma}^T$. Thus we obtain

$$\rho\,\frac{D}{Dt}\left(\frac{1}{2}\,\mathbf{v}\cdot\mathbf{v}\right) = \operatorname{div}(\mathbf{v}\,\boldsymbol{\sigma}) + \mathbf{v}\cdot\rho\,\mathbf{b} - \boldsymbol{\sigma} : \mathbf{d} \tag{5.4.5}$$

as the local form of *balance of kinetic energy*. Herein, $\mathbf{d} = \frac{1}{2}(\mathbf{l} + \mathbf{l}^T)$ is the rate of deformation tensor and $(\dot{\ }) = D/Dt$ denotes as usual the material time derivative.

We now integrate (5.4.5) over the volume V. If we then consider that in view of (1.6.15) and (3.1.17) the identity

$$\iiint_V \operatorname{div}(\mathbf{v}\,\boldsymbol{\sigma})\,dV = \iint_A \mathbf{v}\cdot(\boldsymbol{\sigma}\,\mathbf{n})\,dA = \iint_A \mathbf{v}\cdot\mathbf{t}\,dA \tag{5.4.6}$$

holds and use, in addition, the continuity condition $\overline{\rho dV}^{\cdot} = 0$ we obtain

$$\frac{D}{Dt}\iiint_V \frac{1}{2}\rho\,\mathbf{v}\cdot\mathbf{v}\,dV = \iint_A \mathbf{t}\cdot\mathbf{v}\,dA + \iiint_V \rho\,\mathbf{b}\cdot\mathbf{v}\,dV - \iiint_V \boldsymbol{\sigma}:\mathbf{d}\,dV\ . \tag{5.4.7}$$

By using the following abbreviations:

kinetic energy:

$$K = \iiint_V \frac{1}{2}\rho\,\mathbf{v}\cdot\mathbf{v}\,dV \tag{5.4.8}$$

power of external forces:

$$L = \iint_A \mathbf{t}\cdot\mathbf{v}\,dA + \iiint_V \rho\,\mathbf{b}\cdot\mathbf{v}\,dV \tag{5.4.9}$$

stress power:

$$P = \iiint_V \boldsymbol{\sigma}:\mathbf{d}\,dV \tag{5.4.10}$$

equation (5.4.7) may be also given in short form

$$\dot{K} := \frac{D}{Dt}K = L - P \tag{5.4.11}$$

indicating that the rate of change of the kinetic energy is equal to the difference between the power of external forces **t**, **b** and the power of the stresses $\boldsymbol{\sigma}$. Note that, in the present spatial formulation, the stress power $\boldsymbol{\sigma}:\mathbf{d}$ (per unit deformed volume) is expressed in terms of the CAUCHY stress tensor $\boldsymbol{\sigma}$ and the rate of deformation tensor **d** which have been shown in (3.3.22) to be *energy conjugate*. Equation (5.4.11) expresses *balance of kinetic energy* and the expression $\boldsymbol{\sigma}:\mathbf{d}$ denotes the stress power per unit deformed volume.

The material formulation of the statement (5.4.7) can be derived in a similar form. In this case the derivation starts from the LAGRANGE equation of motion (5.2.16) and leads to

$$\frac{D}{Dt}\iiint_{V_o} \frac{1}{2}\rho_o\,\mathbf{v}\cdot\mathbf{v}\,dV_o = \iint_{A_o} \mathbf{t}_o\cdot\mathbf{v}\,dA_o + \iiint_{V_o} \rho_o\,\mathbf{b}_o\cdot\mathbf{v}\,dV_o - \iiint_{V_o} \mathbf{P}:\dot{\mathbf{F}}\,dV_o \tag{5.4.12}$$

with a stress power $\mathbf{P}:\dot{\mathbf{F}}$ given in terms of the first PIOLA-KIRCHHOFF tensor $\mathbf{P} = J\,\boldsymbol{\sigma}\,\mathbf{F}^{-T}$ and the deformation gradient **F**. The material variables **P** and $\dot{\mathbf{F}}$ have been shown in (3.3.22) to be energy conjugate which is also confirmed by the present derivation. Note that the stress power $\mathbf{P}:\dot{\mathbf{F}}$ refers to unit undeformed volume.

5.5 Conservation of energy

General remarks. In this section we present the postulate of energy balance and its local form, the energy equation corresponding to the first law of thermodynamics. The energy equation introduces an additional quantity, the internal energy. Consequently, it can be regarded as a useful addition to the equations of continuum mechanics only if one is able to relate it to the other state variables. In classical thermodynamics state equations deliver the required additional relations. Since *state equations* are part of the constitutive equations for a particular material model, the development of specific state equations will not be carried out in this chapter which is devoted to the presentation of principles applicable to arbitrary materials.

For our study we again consider the actual configuration B of a body occupying at time t a volume V bounded by a surface A (Fig. 5.1) under surface loads **t** per unit area of A and volume forces **b** per unit mass. As further external actions we consider here also thermal loads: a distributed *internal heat source* of value r per unit mass and a *heat transfer* through the boundary surface A described by the *heat flux vector* **q** per unit area of A. The thermal actions r and **q** are illustrated in Fig. 5.2. The first law of thermodynamics then relates the power of external forces and the heat introduced into the system to the rate of change of energy of the entire body. For convenience we start with definitions.

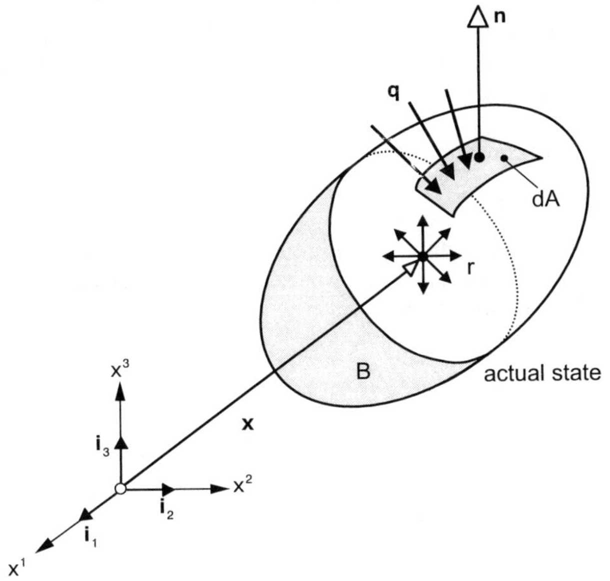

Fig. 5.2. Illustration to heat transfer

Definitions. The *power of external forces* has been already defined in (5.4.9) as

$$L = \iint_A \mathbf{t} \cdot \mathbf{v} \, dA + \iiint_V \rho \, \mathbf{b} \cdot \mathbf{v} \, dV , \tag{5.5.1}$$

where $\mathbf{v} = \dot{\mathbf{x}}$ is the velocity vector. The *heat input* Q due to the heat flux $\mathbf{q}$ and the internal heat source r is defined by:

$$Q = -\iint_A \mathbf{q} \cdot \mathbf{n} \, dA + \iiint_V \rho \, r \, dV , \tag{5.5.2}$$

$\mathbf{n}$ being the unit outward vector normal to the boundary surface A. The negative sign of the first integral means that $\mathbf{q} \cdot \mathbf{n}$ expresses the outward heat flux (per unit area).

We now confine our attention to the *total energy* E of the body. Let e be the *internal energy* per unit mass called also *specific internal energy* or *internal energy density*; then ρ e denotes the internal energy per unit deformed volume. The total energy E of the system is defined to consist of two parts, the *kinetic energy* K introduced in (5.4.8) and the internal energy corresponding to the volume integral of ρ e. Accordingly, E is given by

$$E = \iiint_V \frac{1}{2} \rho \, \mathbf{v} \cdot \mathbf{v} \, dV + \iiint_V \rho \, e \, dV . \tag{5.5.3}$$

First law of thermodynamics. The *conservation of energy* is a consequence of the first law of thermodynamics. It states that the rate of change of the total energy E is equal to the sum of the power of external forces L and the heat input Q so that

$$\dot{E} := \frac{DE}{Dt} = L + Q . \tag{5.5.4}$$

If we write the above equation in differential form

$$dE = L dt + Q dt = d^*L + d^*Q , \tag{5.5.5}$$

we see that the sum on the right-hand side corresponds to an exact differential: dE. This property does, however, not hold for the constituent parts d^*L and d^*Q denoted therefore by d^*. From (5.5.5), we also deduced that

$$\Delta E = (E)_2 - (E)_1 = \int_{t_1}^{t_2} (L + Q) \, dt . \tag{5.5.6}$$

Thus, in any change from state 1 to state 2 the change of the total energy E is determined by the initial and end values of E at time t_1 and t_2, respectively. Equations (5.5.4), (5.5.5) and (5.5.6) can be regarded as alternative formulations of the *first law of thermodynamics*.

Energy equation. By using (5.5.1) to (5.5.3) the relation (5.5.4) can be expressed as

$$\frac{D}{Dt}\iiint\limits_V \frac{1}{2}\rho\,\mathbf{v}\cdot\mathbf{v}\,dV + \frac{D}{Dt}\iiint\limits_V \rho\,e\,dV =$$

$$\iint\limits_A \mathbf{t}\cdot\mathbf{v}\,dA + \iiint\limits_V \rho\,\mathbf{b}\cdot\mathbf{v}\,dV - \iint\limits_A \mathbf{q}\cdot\mathbf{n}\,dA + \iiint\limits_V \rho\,r\,dV\ , \tag{5.5.7}$$

which will be used now to establish the energy equation as local form of the above statement. For this purpose we replace the first term on the left-hand side by the expression (5.4.7), consider the continuity condition $\overline{\rho\,\dot{dV}} = 0$ and use the transformation

$$\iint\limits_A \mathbf{q}\cdot\mathbf{n}\,dA = \iiint\limits_V \operatorname{div}\mathbf{q}\,dV \tag{5.5.8}$$

in accordance with GAUSS-GREEN theorem (1.6.15). This procedure leads to the following relation

$$\iiint\limits_V \left(\operatorname{div}\mathbf{q} + \rho\frac{De}{Dt} - \boldsymbol{\sigma}:\mathbf{d} - \rho\,r\right)dV = 0 \tag{5.5.9}$$

holding for arbitrary subdomains of the total volume V. Consequently, the integrand must vanish and the *energy equation*

$$\rho\frac{De}{Dt} := \rho\dot{e} = \boldsymbol{\sigma}:\mathbf{d} + \rho\,r - \operatorname{div}\mathbf{q} \tag{5.5.10}$$

is obtained as final result expressing the conservation of energy at each point. We recall that $\boldsymbol{\sigma}$ is the CAUCHY stress tensor (3.1.14) and its energy conjugate counterpart $\mathbf{d} = \frac{1}{2}(\mathbf{l} + \mathbf{l}^T)$ is the rate of deformation tensor (2.9.16). The expression $\boldsymbol{\sigma}:\mathbf{d}$ is called *stress power* per unit deformed volume.

Similarly, the energy equation in material formulation can be derived starting from the equivalent material formulation of the statement (5.5.7). In terms of the second PIOLA-KIRCHHOFF stress tensor $\mathbf{S}$ (3.2.22) and the GREEN-LAGRANGE strain tensor $\mathbf{E}$ (2.5.9) the final result reads as

$$\rho_o\frac{De}{Dt} := \rho_o\dot{e} = \mathbf{S}:\dot{\mathbf{E}} + \rho_o r - \operatorname{DIV}\mathbf{Q}\ , \tag{5.5.11}$$

where

$$\mathbf{Q} = J\,\mathbf{F}^{-1}\,\mathbf{q}\ . \tag{5.5.12}$$

Note that the above definition is due to the transformation

$$\iint_{A} \mathbf{q} \cdot \mathbf{n}\, dA = \iint_{A_o} J\, \mathbf{q} \cdot (\mathbf{N}\, \mathbf{F}^{-1})\, dA_o = \iint_{A_o} \mathbf{N} \cdot (J\, \mathbf{F}^{-1}\, \mathbf{q})\, dA_o$$

$$= \iiint_{V_o} \mathrm{DIV}\, (J\, \mathbf{F}^{-1}\, \mathbf{q})\, dV_o \,, \tag{5.5.13}$$

obtained with the help of (1.6.15), (2.2.31) and (2.2.35). In view of the equality

$$J = \frac{dV}{dV_o} = \frac{\rho_o}{\rho} \tag{5.5.14}$$

due to (2.2.31) and (5.1.6) it can be easily confirmed that, for vanishing thermal effects ($\mathbf{q} = 0$, $r = 0$), the results given in (5.5.10) and (5.5.11) are in coincidence with those presented in equation (3.3.22) in terms of the same stress and strain variables.

5.6 Principle of virtual work

As is illustrated in Fig. 5.3 we consider a body in its reference position B_o under volume forces $\rho_o\, \mathbf{b}_o$ per unit undeformed volume. For convenience we suppose that the boundary surface A_o of the body consists of two parts A_{ou} and $A_{o\sigma}$, where displacements $\mathbf{u}$ and forces $\mathbf{t}_o$ are prescribed, respectively. Considering according to (3.2.14) the equality $\mathbf{t}_o = \mathbf{P}\, \mathbf{N}$, the boundary conditions to be satisfied in this case can be formulated as

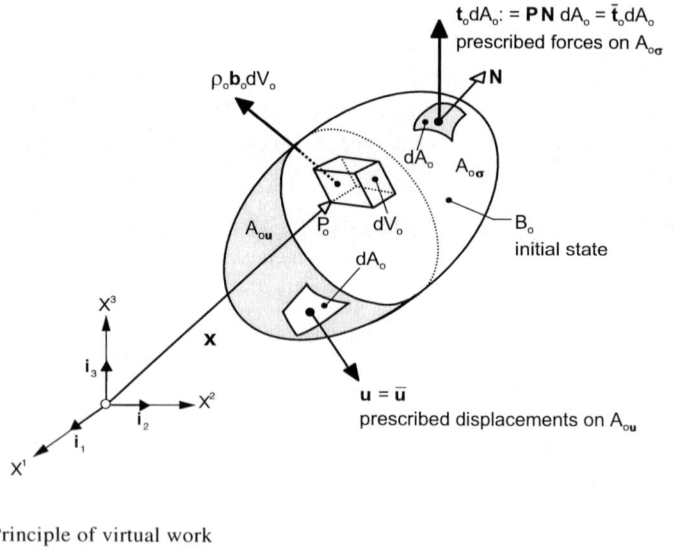

Fig. 5.3. Principle of virtual work

kinematic boundary conditions: $\mathbf{u} = \bar{\mathbf{u}}$ on A_{ou} , (5.6.1)

dynamic boundary conditions: $\mathbf{t}_o := \mathbf{P}\,\mathbf{N} = \bar{\mathbf{t}}_o$ on $A_{o\sigma}$, (5.6.2)

where the notation $(\bar{...})$ characterizes prescribed functions on the boundary surfaces.

The deformation of a body can be described by the displacement field $\mathbf{u} = \mathbf{x} - \mathbf{X}$ and the deformation gradient $\mathbf{F} = \mathbf{g}_i \otimes \mathbf{G}^i$. Note that other deformation variables e.g. the GREEN strain tensor $\mathbf{E}$ can be used instead of $\mathbf{F}$. The basic idea of the principle of virtual work is to couple *kinematically admissible virtual deformations* with force variables and stresses of the real deformation process. In the present case, where $\mathbf{u}$ and $\mathbf{F}$ are selected as deformation variables, kinematically admissible virtual deformations are understood to be the variations $\delta\mathbf{u}$ and $\delta\mathbf{F}$ which are, in accordance with $\mathbf{F} = \mathbf{g}_i \otimes \mathbf{G}^i$ and $\mathbf{u} = \bar{\mathbf{u}}$ on A_{ou}, subjected to the constraints

$$\delta\mathbf{F} = \delta\mathbf{g}_i \otimes \mathbf{G}^i = \delta\mathbf{u}_{,i} \otimes \mathbf{G}^i = \mathrm{GRAD}\,\delta\mathbf{u} \quad \text{in } V_o \tag{5.6.3}$$

$$\delta\mathbf{u} = \mathbf{0} \quad \text{on } A_{ou} \,. \tag{5.6.4}$$

Variations occuring in the above relations follow the well-known rules of the calculus of variation. According to LAGRANGE, the variation of an arbitrary tensor function $\mathbf{f} = \mathbf{f}(\mathbf{u})$ is defined by

$$\delta\mathbf{f} = \frac{d}{d\varepsilon}\mathbf{f}\,(\mathbf{u} + \varepsilon\,\delta\mathbf{u})\big|_{\varepsilon=0} \,, \tag{5.6.5}$$

where $\delta\mathbf{u}$ denotes kinematically admissible variations of the displacement vector $\mathbf{u}$ and ε is a scalar-valued parameter.

To derive the principle of virtual work we start from the equation of motion (5.2.16)

$$\mathrm{DIV}\,\mathbf{P} + \rho_o\mathbf{b}_o - \rho_o\,\ddot{\mathbf{x}} = \mathbf{0} \,, \tag{5.6.6}$$

the local form of momentum balance, which is now to be transformed into a *weak* formulation. We form the scalar product of (5.6.6) with the virtual displacements $\delta\mathbf{u}$ and integrate the result over the volume V_o to obtain

$$\iiint\limits_{V_o} \delta\mathbf{u} \cdot (\mathrm{DIV}\,\mathbf{P} + \rho_o\mathbf{b}_o - \rho_o\,\ddot{\mathbf{x}})\,dV_o = 0 \,. \tag{5.6.7}$$

By considering the product rule

$$\mathrm{DIV}\,(\delta\mathbf{u}\,\mathbf{P}) = \delta\mathbf{u} \cdot \mathrm{DIV}\,\mathbf{P} + \mathbf{P} : \mathrm{GRAD}\,\delta\mathbf{u} \tag{5.6.8}$$

according to (1.6.14) as well as the constraint (5.6.3) to be satisfied by $\delta\mathbf{F}$, we write

$$\iiint\limits_{V_o} \delta\mathbf{u} \cdot \mathrm{DIV}\,\mathbf{P}\,dV_o = \iiint\limits_{V_o} \mathrm{DIV}\,(\delta\mathbf{u}\,\mathbf{P})\,dV_o - \iiint\limits_{V_o} \mathbf{P} : \delta\mathbf{F}\,dV_o \,. \tag{5.6.9}$$

The first integral on the right-hand side can be transformed into a surface integral by means of GAUSS-GREEN theorem (1.6.15). By considering (5.6.2) and (5.6.4) the following result is obtained

$$\iiint_{V_o} \mathrm{DIV}\,(\delta\mathbf{u}\,\mathbf{P})\,dV_o = \iint_{A_o} (\delta\mathbf{u}\,\mathbf{P})\cdot\mathbf{N}\,dA_o = \iint_{A_{o\sigma}} \delta\mathbf{u}\cdot\bar{\mathbf{t}}_o\,dA_o\,. \tag{5.6.10}$$

Inserting (5.6.9) together with (5.6.10) into (5.6.7) delivers finally the *principle of virtual work*

$$\delta^* W = \delta^* W_{EXT} + \delta^* W_{INT} = 0$$

$$= \iiint_{V_o} \delta\mathbf{u}\cdot(\rho_o\,\mathbf{b}_o - \rho_o\,\ddot{\mathbf{x}})\,dV_o + \iint_{A_{o\sigma}} \delta\mathbf{u}\cdot\bar{\mathbf{t}}_o\,dA_o - \iiint_{V_o} \mathbf{P}:\delta\mathbf{F}\,dV_o\,, \tag{5.6.11}$$

where the variable fields $\delta\mathbf{u}$ and $\delta\mathbf{F}$ are subjected to the kinematic constraints (5.6.3) and (5.6.4). Provided that these constraints are fulfilled, the formulation (5.6.11) ensures the validity of (5.6.2) and (5.6.6). Thus the principle of virtual work (5.6.11) is a *weak* formulation of the equations of motion as well as the dynamic boundary conditions.

We conclude this section with an important remark. Equalities presented in (3.3.22) in terms of energy conjugate variables preserve their validity if the material time derivative $(\dot{...})$ is replaced by the variation symbol δ. Special attention is, however, to be confined to LIE-derivatives L_v which are to be changed into the LIE-variations δ_L. Similar to the definition (2.10.12), the LIE-variation to be denoted by δ_L is understood as a variation of *spatial* tensors holding the deformed base vectors $\mathbf{g}_i$ and $\mathbf{g}^i$ constant. Thus, we have

$$\delta_L\mathbf{e} = \delta e_{ij}\,\mathbf{g}^i\otimes\mathbf{g}^j\,, \tag{5.6.12}$$

which is analogous to $L_v\mathbf{e} = \dot{e}_{ij}\,\mathbf{g}^i\otimes\mathbf{g}^j$.

Consequently, equations (3.3.22) can be rewritten for virtual deformations as*

$$\rho_o\delta^* e = \mathbf{P}:\delta\mathbf{F} = \mathbf{S}:\delta\mathbf{E} = \mathbf{T}:\delta\mathbf{U}\,,$$

$$\rho_o\delta^* e = \boldsymbol{\tau}:\delta_L\mathbf{e} = \boldsymbol{\tau}:\frac{1}{2}\,\delta_L\,\mathbf{g} = J\,\boldsymbol{\sigma}:\delta_L\mathbf{e}\,,\quad \delta_L\,\mathbf{g} = \delta g_{ij}\,\mathbf{g}^i\otimes\mathbf{g}^j \tag{5.6.13}$$

permitting to express the internal virtual work $\delta^* W_{INT}$ in the following forms:

$$\delta^* W_{INT} = \iiint_{V_o} \mathbf{P}:\delta\mathbf{F}\,dV_o = \iiint_{V_o} \mathbf{S}:\delta\mathbf{E}\,dV_o = \iiint_{V_o} \boldsymbol{\tau}:\delta_L\mathbf{e}\,dV_o\,. \tag{5.6.14}$$

* The notation δ^* indicates an expression which need not to be a complete variation.

Here, the last expression in terms of the ALMANSI strain tensor **e** and the KIRCHHOFF stress tensor $\boldsymbol{\tau}$ is of significance for the spatial formulation, while the first two ones are given in material formulation.

Exercises

5.1. Transform the following relations into component form:

$$\rho\,\dot{e} = \boldsymbol{\sigma} : \mathbf{d} + \rho\, r - \operatorname{div} \mathbf{q} \ ,$$

$$\rho_o\,\dot{e} = \mathbf{S} : \dot{\mathbf{E}} + \rho_o\, r - \mathrm{DIV}\, \mathbf{Q} \ .$$

5.2. Transform the equation (5.2.16) into the form (5.4.12).

6 Constitutive modelling

This chapter is devoted to a systematical presentation of constitutive equations applicable to large strain analysis of arbitrary elastic bodies. The discussion starts with a brief survey of the principles which are relevant for the formulation of constitutive equations. Then, the definition of objective tensors is given serving in the sequel to assess the material objectivity of constitutive laws. Constitutive equations are presented for CAUCHY material and for hyperelasticity. Special attention is given to isotropic elasticity to enlighten particularly the application of the material symmetry principle. In this context relevant isotropic material models such as OGDEN, MOONEY-RIVLIN and NEO-HOOKE models are presented. ST. VENANT-KIRCHHOFF model is then derived through a linearization technique. The last cited model in turn is used to derive by a similar procedure the HOOKEAN law as the simplest model for elastic materials. Finally, useful connections between the HOOKEAN material law and some nonlinear material models are established again through linearization.

6.1 General principles

The relations which we have derived in the previous chapters may be classified essentially into two groups: *kinematic relations* describing the relation between strains (deformation measures) and the displacements $\mathbf{u} = \mathbf{x} - \mathbf{X}$ and *equations of motion* as connections between stresses and external loads. In view of the equality $\mathbf{g}_i = \mathbf{G}_i + \mathbf{u}_{,i}$ the equation $\mathbf{F} = \mathbf{g}_i \otimes \mathbf{G}^i$ given in (2.2.3) can be seen as the kinematic relation for the deformation gradient $\mathbf{F}$, while the equations of motion expressed in terms of the first PIOLA-KIRCHHOFF stress tensor $\mathbf{P}$ have been established in (5.2.16). From (3.3.22) or (3.3.23) we see that the material time derivative of the deformation gradient $\dot{\mathbf{F}}$ and the stress tensor $\mathbf{P}$ are energy conjugate tensors the double contraction of which $\rho_o\,\dot{e} = \mathbf{P}{:}\dot{\mathbf{F}} = P^{ij}\,\dot{F}_{ij}$ defines the so-called stress power and, for isothermal processes, the rate of change of internal energy $\rho_o\,\dot{e}$ (per unit undeformed volume). For the considered case we may therefore write by using the component relations (2.1.19), (2.2.16), (3.2.14) and (5.2.18):

$$\textit{kinematic relations:}\quad \mathbf{F} = \mathbf{G} + \mathbf{u}_{,i} \otimes \mathbf{G}^i \quad \rightarrow \quad F_{ij} = G_{ij} + U_{i|j}\ , \tag{6.1.1}$$

$$\textit{equations of motion:}\quad \mathrm{DIV}\,\mathbf{P} + \rho_o\,(\mathbf{b}_o - \ddot{\mathbf{x}}) = \mathbf{0} \quad \rightarrow \quad P^{ij}|_j + \rho_o\,(b^i_o - \ddot{U}^i) = 0\ , \tag{6.1.2}$$

$$\textit{stress power:}\quad \rho_o\,\dot{e} = \mathbf{P} : \dot{\mathbf{F}} = P^{ij}\,\dot{F}_{ij}\ . \tag{6.1.3}$$

Expressed in terms of the GREEN-LAGRANGE strain tensor $\mathbf{E}$ and the second PIOLA-KIRCHHOFF stress tensor $\mathbf{S}$ the above relations read as:

kinematic relations: $\mathbf{E} = \frac{1}{2}\,(\mathbf{F}^T\mathbf{F}-\mathbf{G}) \qquad \rightarrow \; E_{ij} = \frac{1}{2}\,(U_i|_j + U_j|_i + U_r|_i\, U^r|_j)$, (6.1.4)

equations of motion: $\mathrm{DIV}\,(\mathbf{F}\,\mathbf{S}) + \rho_o\,(\mathbf{b}_o-\ddot{\mathbf{x}}) = 0 \rightarrow (F^i_{\cdot m}\; S^{mj})|_j + \rho_o\,(b^i_o-\ddot{U}^i) = 0$, (6.1.5)

stress power: $\rho_o\,\dot{e} = \mathbf{S} : \dot{\mathbf{E}} = S^{ij}\,\dot{E}_{ij}$. (6.1.6)

The first equation has been directly adopted from (2.5.12) while the second one corresponds to the transformation of (6.1.2) by means of (3.2.22), $\mathbf{P} = \mathbf{F}\,\mathbf{S}$. It is clear that similar examples may be added in spatial formulation.

The above survey shows that it is essentially possible to formulate kinematic relations and equations of motion in terms of energy conjugate stress and strain variables. However, these equations alone are not sufficient to describe completely boundary value problems of solids.

Thus, the purpose of this chapter is to add to the kinematic relations and equations of motion a new set of equations, the constitutive equations also called material laws, in order to obtain a complete system of differential equations. Attention will be restricted here to elasticity by assuming purely mechanical phenomena.

Then, *constitutive equations* may be understood as one-to-one relations between appropriate stress and strain measures and it is possible to formulate them under certain conditions, which will be identified during derivation, in terms of energy conjugate variables summarized in a general form in (3.3.22) and (3.3.23).

In this context it is worthy to recall that equations of motion are the immediate consequence of the postulate of momentum balance, while kinematic relations are due to purely geometrical considerations. In this sense, both sets of equations are *exact*. It is also useful to remember that any set of kinematic relations introduced in the principle of virtual work leads, due to the GAUSS-GREEN theorem, to a well-defined set of equations of motion expressed in terms of that stress tensor which is energy conjugate to the adopted strain measure. The application of this procedure is shown in section 5.6. In contrast to the cited equation sets, constitutive equations are based on experimental observations and, thus, approximations by nature. Nevertheless there exists a number of conditions imposed on their formulation from the point of view of continuum mechanics. In the following we summarize some of them being particularly relevant for elasticity. For a more detailed discussion of this aspect we refer to MALVERN 1969, STEIN and BARTHOLD 1993.

For the sake of brevity we consider purely mechanical phenomena. Thus the fundamental postulates which are assumed to be valid in this case for any constitutive law can be summarized as follows:

- *Principle of determinism for stress.* The stress in a body is determined by the history of the motion of that body. This principle involves as special cases the classical history-independent Newtonian fluid and the classical laws of elasticity whose only history dependence consists in possessing a natural state to which the body will return upon unloading.

- *Principle of local action.* In determining the stress in a given material point M, the motion outside an arbitrarily small neighborhood of M may be disregarded.

- *Principle of material objectivity.* Constitutive equations must be invariant under changes of reference frame. This means that two observers, even if in a relative motion with respect to each other, observe the same stress in a given body. This principle which is also referred to as *principle of material frame-indifference* will be mathematically defined in section 6.3 using for this purpose the definition of objective tensors given in section 6.2.

- *Principle of material symmetry.* Mostly, the motion of a body is described by using material coordinates introduced in the reference configuration B_0. According to the principle of material symmetry, the constitutive equations must be invariant under any transformation of material coordinates belonging to a symmetry group S which is defined by the existing symmetry properties of the considered material. Thus, this principle expresses the influence of the transformation of the reference frame on the response of the stresses $\boldsymbol{\sigma}$. In section 6.4 the principle of material symmetry will be used to formulate constitutive equations for isotropic elastic materials. Its application combined with the principle of material objectivity will, moreover, show the important role of isotropic tensor functions for constitutive laws of isotropic materials.

To clearify the meaning of the principle of local action we consider a material point M defined by the position vector $\mathbf{x} = \boldsymbol{\chi}\,(\mathrm{M}, \mathrm{t})$. Let $\hat{\mathbf{x}} = \boldsymbol{\chi}\,(\mathrm{N}, \mathrm{t})$ denote the position vector of any other material point N. By means of the principle of local action, the stress $\boldsymbol{\sigma}\,(\mathrm{M}, \mathrm{t})$ at M is determined as follows. In a general form, the relative motion between N and M is described by

$$\hat{\mathbf{x}} - \mathbf{x} \;=\; \boldsymbol{\chi}\,(\mathrm{N}, \mathrm{t}) - \boldsymbol{\chi}\,(\mathrm{M}, \mathrm{t}) \;. \tag{6.1.7}$$

If we now assume that points N are in an infinitesimal neighborhood of M and suppose the differentiability of $\mathbf{x}$ the above relation can be replaced by

$$\boldsymbol{\chi}\,(\mathrm{N}, \mathrm{t}) - \boldsymbol{\chi}\,(\mathrm{M}, \mathrm{t}) \;=\; \frac{\partial \mathbf{x}}{\partial \mathbf{X}}\, \mathrm{d}\mathbf{X} \;=\; \mathbf{F}\, \mathrm{d}\mathbf{X} \;, \tag{6.1.8}$$

where $\mathbf{F} = \mathbf{F}\,(\mathrm{M}, \mathrm{t})$ is the deformation gradient at M at time t and $\mathrm{d}\mathbf{X}$ is the vector from M to N in the neighborhood considered. Hence, the distortion of the material in the vicinity of M is determined by the value $\mathbf{F}\,(\mathrm{M}, \mathrm{t})$ of the deformation gradient. Consequently, the stress $\boldsymbol{\sigma}\,(\mathrm{M}, \mathrm{t})$ which was supposed to be determined by the local configuration is completely determined by $\mathbf{F}\,(\mathrm{M}, \mathrm{t})$. This implies that the most general constitutive equation of an elastic material is of the form

$$\boldsymbol{\sigma} \;=\; \boldsymbol{\sigma}\,(\mathbf{F}) \;, \tag{6.1.9}$$

where $\boldsymbol{\sigma}(\mathbf{F})$ denotes a tensor-valued function of the single argument $\mathbf{F}$. This relation will be used in section 6.3 to define the so-called CAUCHY-elastic material.

6.2 Objective tensors

In this section we deal with objective tensors which will be used in the next section to define objective material laws. The starting point of our considerations is the deformed configuration B of a body at time t. Let $\mathbf{x} = \mathbf{x}(\mathbf{X}, t)$ be the position vector of a characteristic point P in B. If an orthogonal tensor $\mathbf{Q} = \mathbf{Q}(t)$ is applied to the position vector $\mathbf{x}$ the configuration B rotates into a new configuration $\overset{*}{\mathrm{B}}$. Then, the vector

$$\overset{*}{\mathbf{x}} = \mathbf{Q}(t)\,\mathbf{x} \tag{6.2.1}$$

determines the position of P in $\overset{*}{\mathrm{B}}$. Starting from the above relation, characteristic deformation tensors related to the new configuration $\overset{*}{\mathrm{B}}$ can be evaluated by means of the well-known definitions. We emphasize that $\mathbf{Q} = \mathbf{Q}(t)$ is a function only of time t since the transformation of B into $\overset{*}{\mathrm{B}}$ is supposed to be a rigid body rotation. We denote tensors associated with B by usual notations and use the asterisk ($\overset{*}{}$) for those obtained from the previous ones by replacing $\mathbf{x}$ by $\overset{*}{\mathbf{x}}$ or, alternatively, $\mathbf{F}$ by $\overset{*}{\mathbf{F}}$. In the following some useful results are summarized:

base vectors: $$\overset{*}{\mathbf{g}}_i = \overset{*}{\mathbf{x}}_{,i} = \mathbf{Q}\,\mathbf{g}_i\ , \quad \overset{*}{\mathbf{g}}{}^i = \mathbf{Q}\,\mathbf{g}^i\ , \tag{6.2.2}$$

deformation gradient: $$\overset{*}{\mathbf{F}} = \overset{*}{\mathbf{g}}_i \otimes \mathbf{G}^i = \mathbf{Q}\,\mathbf{F}\ , \tag{6.2.3}$$

left CAUCHY-GREEN tensor: $$\overset{*}{\mathbf{b}} = \overset{*}{\mathbf{F}}\,\overset{*}{\mathbf{F}}{}^{\mathrm{T}} = \mathbf{Q}\,\mathbf{b}\,\mathbf{Q}^{\mathrm{T}}\ , \tag{6.2.4}$$

determinant: $$\overset{*}{J} = \det\overset{*}{\mathbf{F}} = \det(\mathbf{Q}\,\mathbf{F}) = \det\mathbf{F} = J\ . \tag{6.2.5}$$

In the last relation the identity $\det\mathbf{Q} = 1$ has been used due to the orthogonality condition $\mathbf{Q}\,\mathbf{Q}^{\mathrm{T}} = \mathbf{I}$. If we consider a tensor-valued function e.g. of $\mathbf{b}$, $\mathbf{f} = \mathbf{f}(\mathbf{b})$, its value in the configuration $\overset{*}{\mathrm{B}}$ is obtained simply by replacing the argument $\mathbf{b}$ by $\overset{*}{\mathbf{b}}$, thus

$$\overset{*}{\mathbf{f}} = \mathbf{f}(\mathbf{Q}\,\mathbf{b}\,\mathbf{Q}^{\mathrm{T}})\ . \tag{6.2.6}$$

Objective tensors will be defined in close connection with the rotation of the configuration B into $\overset{*}{\mathrm{B}}$.

- A scalar-valued function e.g. of $\mathbf{F}$, $\alpha = \alpha(\mathbf{F})$, is said to be *objective* if its value remains unchanged in the new configuration $\overset{*}{\mathrm{B}}$:

  $$\overset{*}{\alpha} := \alpha(\mathbf{Q}\,\mathbf{F}) = \alpha(\mathbf{F})\ . \tag{6.2.7}$$

  An example is the determinant $J = \det\mathbf{F}$ given in (6.2.5).

- A vector-valued function e.g. of $\mathbf{F}$, $\mathbf{v} = \mathbf{v}(\mathbf{F})$, is said to be objective if it transforms in the same way as the base vectors $\mathbf{g}_i$ in the transformation of B into $\overset{*}{\mathrm{B}}$:

$$\overset{*}{\mathbf{v}} := \mathbf{v}(\mathbf{Q}\,\mathbf{F}) = \mathbf{Q}\,\mathbf{v}(\mathbf{F}) \;. \tag{6.2.8}$$

An example is the unit normal vector **n** of the deformed boundary surface of B which takes in the rotated configuration $\overset{*}{\mathrm{B}}$ the value $\overset{*}{\mathbf{n}} = \mathbf{Q}\,\mathbf{n}$.

- A second-order tensor depending e.g. on **F**, $\boldsymbol{\sigma} = \boldsymbol{\sigma}(\mathbf{F})$, is said to be *objective* if the equality

  $$\overset{*}{\boldsymbol{\sigma}} := \boldsymbol{\sigma}(\mathbf{Q}\,\mathbf{F}) = \mathbf{Q}\,\boldsymbol{\sigma}(\mathbf{F})\,\mathbf{Q}^{\mathrm{T}} \tag{6.2.9}$$

  holds for arbitrary orthogonal tensors **Q**. If we define the components of $\overset{*}{\boldsymbol{\sigma}}$ in a similar form as those of $\boldsymbol{\sigma}$:

  $$\overset{*}{\sigma}{}^{ij} = \overset{*}{\mathbf{g}}{}^{i}\,\overset{*}{\boldsymbol{\sigma}}\,\overset{*}{\mathbf{g}}{}^{j} \;, \qquad \sigma^{ij} = \mathbf{g}^{i}\,\boldsymbol{\sigma}\,\mathbf{g}^{j} \tag{6.2.10}$$

  we may deduce by considering (6.2.2) that the objectivity requirement (6.2.9) ensures the equality $\overset{*}{\sigma}{}^{ij} = \sigma^{ij}$:

  $$\overset{*}{\sigma}{}^{ij} = \mathbf{g}^{i}\,\mathbf{Q}^{\mathrm{T}}(\mathbf{Q}\,\boldsymbol{\sigma}\,\mathbf{Q}^{\mathrm{T}})\,\mathbf{Q}\,\mathbf{g}^{j} = \mathbf{g}^{i}\,\boldsymbol{\sigma}\,\mathbf{g}^{j} = \sigma^{ij} \;. \tag{6.2.11}$$

Similar holds for the components of any objective vector **v** as can be proved by a similar procedure. An example for an objective second-order tensor is the left CAUCHY-GREEN tensor **b** given in (6.2.4). A further consequence of (6.2.9) is that, if an objective second-order tensor $\boldsymbol{\sigma}$ is contracted by an objective vector **n**, then the result $\mathbf{t} = \boldsymbol{\sigma}\,\mathbf{n}$ is an objective vector **t**. To prove this we again consider (6.2.9) which by using the transformation $\overset{*}{\mathbf{n}} = \mathbf{Q}\,\mathbf{n}$ leads to

$$\overset{*}{\mathbf{t}} = \overset{*}{\boldsymbol{\sigma}}\,\overset{*}{\mathbf{n}} = (\mathbf{Q}\,\boldsymbol{\sigma}\,\mathbf{Q}^{\mathrm{T}})\,\mathbf{Q}\,\mathbf{n} = \mathbf{Q}\,\boldsymbol{\sigma}\,\mathbf{n} = \mathbf{Q}\,\mathbf{t} \;. \tag{6.2.12}$$

If we consider a tensor-valued function of **b**, $\boldsymbol{\sigma} = \boldsymbol{\sigma}(\mathbf{b})$, then the objectivity condition (6.2.9) is to be replaced in accordance with (6.2.4) by:

$$\overset{*}{\boldsymbol{\sigma}} := \boldsymbol{\sigma}(\mathbf{Q}\,\mathbf{b}\,\mathbf{Q}^{\mathrm{T}}) = \mathbf{Q}\,\boldsymbol{\sigma}(\mathbf{b})\,\mathbf{Q}^{\mathrm{T}} \;. \tag{6.2.13}$$

As has been discussed in section 2.11 such a relation indicates that $\boldsymbol{\sigma}(\mathbf{b})$ is an isotropic tensor function of **b**. Remember that *isotropic* tensor functions of **b** have in both configurations B and $\overset{*}{\mathrm{B}}$ the same components (see (2.11.9)). Thus we may state:

Remark. Isotropic tensor functions of **b** or $\mathbf{v} = \mathbf{b}^{1/2}$ are *objective* and characterized by the equality of their components in both configurations B and $\overset{*}{\mathrm{B}}$. This fact is illustrated in Fig. 6.1 on the example of simple tension.

By considering (6.2.3) and the definitions given in Table 2.2 it can be confirmed that the following *material* deformation tensors remain unchanged in the transformation of B into $\overset{*}{\mathrm{B}}$:

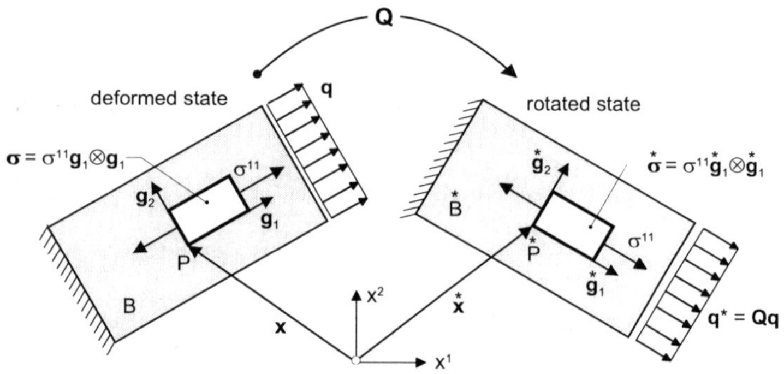

Fig. 6.1. Illustration of the frame-indifference of the material law on the example of simple tension

right CAUCHY-GREEN *tensor:* $\overset{*}{\mathbf{C}} := \mathbf{C}\,(\mathbf{Q}\,\mathbf{F}) = \mathbf{C}\,(\mathbf{F})$,

GREEN-LAGRANGE *strain tensor:* $\overset{*}{\mathbf{E}} := \mathbf{E}\,(\mathbf{Q}\,\mathbf{F}) = \mathbf{E}\,(\mathbf{F})$,

right stretch tensor: $\overset{*}{\mathbf{U}} := \mathbf{U}\,(\mathbf{Q}\,\mathbf{F}) = \mathbf{U}\,(\mathbf{F})$. (6.2.14)

The deformation gradient $\mathbf{F}$ transforms in the same manner as the base vectors $\mathbf{g}_i$ and is therefore not objective. Finally, we refer to the polar decomposition theorem $\mathbf{F} = \mathbf{R}\,\mathbf{U}$ to find the transformation rule for the rotation tensor $\mathbf{R}$. Since $\mathbf{U} = \mathbf{C}^{1/2}$ and $\overset{*}{\mathbf{C}} = \mathbf{C}$, we have the equality $\overset{*}{\mathbf{U}} = \mathbf{U}$ leading to

$$\overset{*}{\mathbf{R}} = \overset{*}{\mathbf{F}}\,\overset{*}{\mathbf{U}}^{-1} = \mathbf{Q}\,\mathbf{F}\,\mathbf{U}^{-1} = \mathbf{Q}\,\mathbf{R} \; . \tag{6.2.15}$$

Thus, $\mathbf{R}$ transforms in the same manner as $\mathbf{F}$ and is, therefore, not objective.

6.3 Elastic material

The starting point of the derivation is the CAUCHY stress tensor $\boldsymbol{\sigma} = \sigma^{ij}\,\mathbf{g}_i \otimes \mathbf{g}_j$ which defines, as has been shown in section 3.1, the real stresses in the deformed configuration B of a body. We recall that the actual deformed configuration B is determined with respect to the initial configuration B_o by the deformation gradient $\mathbf{F} = \mathbf{g}_i \otimes \mathbf{G}^i$ which has been used in chapter 2 to introduce various strain measures. *Constitutive equations* for elastic materials are then relations between stress tensors and corresponding deformation measures.

A material is said to be *elastic*, if the CAUCHY stress tensor $\boldsymbol{\sigma}$ can be expressed by

$$\boldsymbol{\sigma} = \boldsymbol{\sigma}\,(\mathbf{F}) \; , \tag{6.3.1}$$

that means by a tensor-valued function of a single tensor argument, the deformation gradient $\mathbf{F}$. If the considered material is not homogeneous, the function $\boldsymbol{\sigma}\,(\mathbf{F})$ may have for different material points different dependences on the argument $\mathbf{F}$. For the sake of brevity we exclude in the sequel an explicit dependence of σ on the coordinates $\mathbf{X}$ and time t. Equation (6.3.1) describes the so-called *CAUCHY-elastic material.*

For a further important definition we make use of the definitions from section 6.2. Thus we suppose that the deformed body is rotated into a new position $\overset{*}{B}$ by means of an orthogonal tensor $\mathbf{Q} = \mathbf{Q}\,(t)$ depending solely on time t. If we replace in (6.3.1) $\mathbf{F}$ by $\mathbf{Q\,F}$ we obtain the value of the stress tensor referring to the new configuration $\overset{*}{B}$: $\overset{*}{\sigma} = \sigma\,(\mathbf{Q\,F})$. As usual the components of $\overset{*}{\sigma}$ are defined by $\overset{*}{\sigma} = \overset{*}{\sigma}{}^{ij}\, \overset{*}{\mathbf{g}}_i \otimes \overset{*}{\mathbf{g}}_j$, where $\overset{*}{\mathbf{g}}_i = \mathbf{Q}\,\mathbf{g}_i$ are the base vectors of the configuration $\overset{*}{B}$. Note that the asterisk (*) indicates again the value of a tensor evaluated in the position $\overset{*}{B}$ in the same way as in the actual state B.

The material law (6.3.1) is said to be *objective* or *frame-indifferent* if the relation

$$\overset{*}{\sigma} := \sigma\,(\mathbf{Q\,F}) = \mathbf{Q}\,\sigma\,(\mathbf{F})\,\mathbf{Q}^T \tag{6.3.2}$$

holds for arbitrary orthogonal tensors $\mathbf{Q} = \mathbf{Q}\,(t)$. An important consequence of this condition is that the stress components σ^{ij} and $\overset{*}{\sigma}{}^{ij}$ are equal in both configurations B and $\overset{*}{B}$ as can be verified by considering (6.2.2) and (6.3.2):

$$\overset{*}{\sigma}{}^{ij} = \overset{*}{\mathbf{g}}{}^i\, \overset{*}{\sigma}\, \overset{*}{\mathbf{g}}{}^j = \mathbf{g}^i\, \mathbf{Q}^T\, (\mathbf{Q}\,\sigma\,\mathbf{Q}^T)\,\mathbf{Q}\,\mathbf{g}^j = \mathbf{g}^i\,\sigma\,\mathbf{g}^j = \sigma^{ij}\ . \tag{6.3.3}$$

Thus we see that an objective material law leaves the stress components σ^{ij} unchanged if the actual state B of a body is subjected to a rigid body rotation. This fact is illustrated in Fig. 6.1 on the example of simple tension.

Furthermore it can proved that, if the condition (6.3.2) holds, the constitutive law (6.3.1) can be expressed in terms of energy conjugate stress and strain variables defined in (3.3.22). To this end we require that equation (6.3.2) holds for the special selection $\mathbf{Q} = \mathbf{R}^T$, where $\mathbf{R}$ denotes as usual the rotation tensor used in the polar decomposition theorem (2.4.4)

$$\mathbf{F} = \mathbf{R\,U} \quad \to \quad \mathbf{U} = \mathbf{R}^T\,\mathbf{F}\ . \tag{6.3.4}$$

Thus, equation (6.3.2) takes the form

$$\mathbf{R}^T\,\sigma\,(\mathbf{F})\,\mathbf{R} = \sigma\,(\mathbf{R}^T\,\mathbf{F}) = \sigma\,(\mathbf{U}) = \hat{\sigma}\,(\mathbf{C})\ , \tag{6.3.5}$$

where, in accordance with the definition $\mathbf{U} = \mathbf{C}^{1/2}$, the notation

$$\hat{\sigma}\,(\mathbf{C}) = \sigma\,(\mathbf{C}^{1/2}) = \sigma\,(\mathbf{U}) \tag{6.3.6}$$

has been introduced. If we now consider equations (6.3.4) and (6.3.5) together with the identity

$$\det \mathbf{F} = \det\,(\mathbf{R\,U}) = \det \mathbf{R}\, \det \mathbf{U} = \det \mathbf{U} = \det \mathbf{C}^{1/2} \tag{6.3.7}$$

due to $\det \mathbf{R} = 1$ we obtain from (3.2.27) the following constitutive equation for the BIOT stress tensor $\mathbf{T}$:

$$\mathbf{T} = \det \mathbf{F}\, \mathbf{R}^{\mathrm{T}} \boldsymbol{\sigma}(\mathbf{F})\, \mathbf{F}^{-\mathrm{T}} = \det \mathbf{F}\, \mathbf{R}^{\mathrm{T}} \boldsymbol{\sigma}(\mathbf{F})\, \mathbf{R}\, \mathbf{U}^{-1}$$
$$= \det \mathbf{U}\, \boldsymbol{\sigma}(\mathbf{U})\, \mathbf{U}^{-1} = \mathbf{T}(\mathbf{U}) = \hat{\mathbf{T}}(\mathbf{H}) \,, \tag{6.3.8}$$

where by virtue of (2.4.30)

$$\hat{\mathbf{T}}(\mathbf{H}) = \mathbf{T}(\mathbf{H} + \mathbf{G}) = \mathbf{T}(\mathbf{U}) \,. \tag{6.3.9}$$

The relation (6.3.8) can also be used to express the second PIOLA-KIRCHHOFF stress tensor **S** in dependence on the GREEN-LAGRANGE strain tensor **E**. For this purpose we precontract it by $\mathbf{U}^{-1}$ to obtain by considering (3.2.29) and (6.3.6), (6.3.7):

$$\mathbf{S} = \mathbf{U}^{-1}\mathbf{T} = \det \mathbf{U}\, \mathbf{U}^{-1} \boldsymbol{\sigma}(\mathbf{U})\, \mathbf{U}^{-1}$$
$$= \det \mathbf{C}^{1/2}\, \mathbf{C}^{-1/2}\, \hat{\boldsymbol{\sigma}}(\mathbf{C})\, \mathbf{C}^{-1/2} = \mathbf{S}(\mathbf{C}) = \hat{\mathbf{S}}(\mathbf{E}) \tag{6.3.10}$$

with the notation

$$\hat{\mathbf{S}}(\mathbf{E}) = \mathbf{S}(2\mathbf{E} + \mathbf{G}) = \mathbf{S}(\mathbf{C}) \tag{6.3.11}$$

according to (2.5.9). Finally, by inserting (6.3.1) into (3.2.16) we find the constitutive law for the first PIOLA-KIRCHHOFF stress tensor **P**:

$$\mathbf{P} = \det \mathbf{F}\, \boldsymbol{\sigma}(\mathbf{F})\, \mathbf{F}^{-\mathrm{T}} = \mathbf{P}(\mathbf{F}) \,. \tag{6.3.12}$$

Equation (6.3.5) shows that any tensor-valued function $\boldsymbol{\sigma} = \boldsymbol{\sigma}(\mathbf{F})$ satisfying the objectivity condition (6.3.1) is transformable into a function $\boldsymbol{\sigma}(\mathbf{U})$ of **U** through the contraction $\mathbf{R}^{\mathrm{T}} \boldsymbol{\sigma}(\mathbf{F})\, \mathbf{R}$ in terms of the rotation tensor **R**. Evidently, the function $\boldsymbol{\sigma}(\mathbf{F})$ entering in (6.3.12) has to satisfy the condition (6.3.5) so that the corresponding constitutive law is objective. On the contrary, the objectivity of the constitutive laws (6.3.8) and (6.3.10) is assured for any selection of the functions $\boldsymbol{\sigma}(\mathbf{U})$ and $\hat{\boldsymbol{\sigma}}(\mathbf{C})$. Finally, we note that, in view of (6.3.5), the constitutive law (6.3.1) can be also given in the forms

$$\boldsymbol{\sigma} = \mathbf{R}\, \boldsymbol{\sigma}(\mathbf{U})\, \mathbf{R}^{\mathrm{T}} \,, \quad \boldsymbol{\sigma} = \mathbf{R}\, \hat{\boldsymbol{\sigma}}(\mathbf{C})\, \mathbf{R}^{\mathrm{T}} \,, \tag{6.3.13}$$

satisfying a priori the requirement of the frame-indifference.

6.4 Isotropic elastic material

In section 6.3 elastic materials have been considered in a general form. Now we restrict our attention to *isotropic* materials in order to find out the mathematical requirement which characterizes this property. We consider a particle* in the undeformed, unstressed configuration B_0 of the body (Fig. 6.2). If this particle is subjected to a deformation by means of the deformation gradient **F** then the CAUCHY stress tensor is given according to (6.3.1) by $\boldsymbol{\sigma} = \boldsymbol{\sigma}(\mathbf{F})$. If the considered particle is first rotated by an orthogonal tensor **Q** and then deformed by the same deformation gradient **F**, its final position is described by

* Particle is understood here as material point embedded in its infinitesimal vicinity.

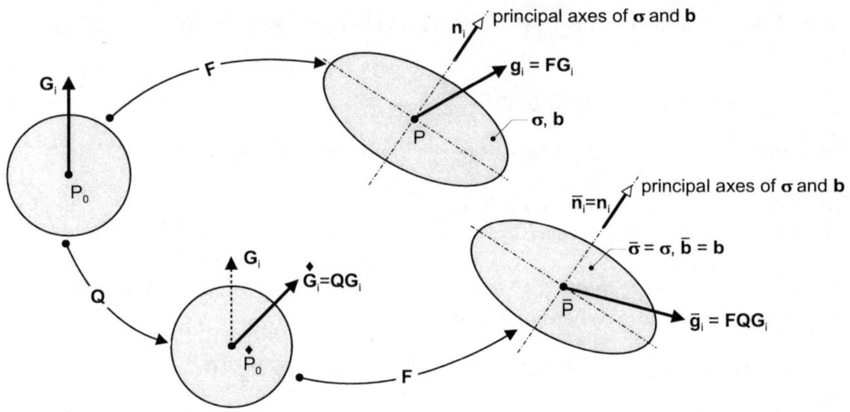

Fig. 6.2. Illustration of isotropic elasticity

$\bar{\mathbf{F}} = \mathbf{F\,Q}$ and the value of the stress tensor in this position is: $\bar{\boldsymbol{\sigma}} = \boldsymbol{\sigma}\,(\mathbf{F\,Q})$. A material is referred to as *isotropic* if the rotation of a particle has no influence on the stress tensor, that is, the condition

$$\bar{\sigma} := \sigma(\mathbf{F\,Q}) = \sigma(\mathbf{F}) \tag{6.4.1}$$

is satisfied for arbitrary orthogonal tensors $\mathbf{Q}$.

The above condition must be particularly fulfilled for the special selection $\mathbf{Q} = \mathbf{R}^{\mathrm{T}}$, where $\mathbf{R}$ is the rotation tensor used in (2.4.4):

$$\mathbf{F} = \mathbf{v\,R} \quad \rightarrow \quad \mathbf{v} = \mathbf{F\,R}^{\mathrm{T}} \tag{6.4.2}$$

Accordingly, the condition to be satisfied for *isotropic* materials is

$$\sigma(\mathbf{F}) = \sigma(\mathbf{F\,R}^{\mathrm{T}}) = \sigma(\mathbf{v}) = \sigma(\mathbf{b}^{1/2}) = \sigma(\mathbf{b}) \ . \tag{6.4.3}$$

Thus we see that, for isotropic materials, the function $\boldsymbol{\sigma}(\mathbf{F})$ depends on $\mathbf{F}$ only through its dependence on the left stretch tensor $\mathbf{v}$ or the left CAUCHY-GREEN tensor $\mathbf{b} = \mathbf{v}^2$. In other words, the isotropy requirement is automatically satisfied by arbitrary tensor-valued functions of $\mathbf{v}$ or $\mathbf{b}$: $\sigma(\mathbf{v}) = \sigma(\mathbf{b}^{1/2}) = \sigma(\mathbf{b})$. A suitable function $\sigma(\mathbf{F})$ for isotropy can therefore be obtained from $\sigma(\mathbf{v})$ and $\sigma(\mathbf{b})$ simply by setting $\mathbf{v} = (\mathbf{F\,F}^{\mathrm{T}})^{1/2}$ and $\mathbf{b} = \mathbf{F\,F}^{\mathrm{T}}$, respectively. From (6.4.3) we also deduce the following:

Remark. For isotropic elastic materials the CAUCHY stress tensor $\boldsymbol{\sigma}$ is determined by the six independent components of the symmetric tensor $\mathbf{v}$ or $\mathbf{b} = \mathbf{v}^2$.

Now we have to find out the requirement ensuring, in addition, the objectivity of the constitutive law $\sigma = \sigma(\mathbf{b})$. As usual, we suppose the deformed configuration B to be rotated into a new position $\overset{*}{\mathrm{B}}$ by means of an orthogonal tensor $\mathbf{Q} = \mathbf{Q}(t)$. Then the value of **b** in $\overset{*}{B}$ is, according to (6.2.4), given by $\mathbf{Q}\,\mathbf{b}\,\mathbf{Q}^{T}$ and the objectivity condition is of the form (6.2.13):

$$\overset{*}{\sigma} := \sigma(\mathbf{Q}\,\mathbf{b}\,\mathbf{Q}^{T}) = \mathbf{Q}\,\sigma(\mathbf{b})\,\mathbf{Q}^{T} \,, \tag{6.4.4}$$

which implies that $\sigma(\mathbf{b})$ is an isotropic tensor function. Thus, we may state:

Remark. Objective constitutive laws for *isotropic* materials are described by isotropic tensor functions $\boldsymbol{\sigma}(\mathbf{b})$ or, alternatively, $\sigma(\mathbf{v})$.

According to the *representation theorem* (2.11.18), any isotropic function $\boldsymbol{\sigma}(\mathbf{b})$ can be described by a quadratic polynomial in **b**

$$\boldsymbol{\sigma} = \boldsymbol{\sigma}(\mathbf{b}) = \varphi_0(i_\mathbf{b})\,\mathbf{g} + \varphi_1(i_\mathbf{b})\,\mathbf{b} + \varphi_2(i_\mathbf{b})\,\mathbf{b}^2 \,, \quad i_\mathbf{b} = I_\mathbf{b}, II_\mathbf{b}, III_\mathbf{b} \,, \tag{6.4.5}$$

where φ_0, φ_1 and φ_2 are functions of the invariants $i_\mathbf{b}$ of **b**. Note that $\sigma(\mathbf{v})$ can be described by a similar quadratic polynomial in **v** with coefficients depending on the invariants of **v**.

The relation (6.4.5) can be transformed into a material formulation in terms of the second PIOLA-KIRCHHOFF stress tensor **S** and the right CAUCHY-GREEN tensor **C**. If we replace in (3.2.22) $\boldsymbol{\sigma}$ by (6.4.5) and det **F** by J and consider (2.4.34), (2.4.35) we first find

$$\mathbf{S} = \mathbf{S}(\mathbf{C}) = J\,\varphi_0\,(i_\mathbf{C})\,\mathbf{C}^{-1} + J\,\varphi_1\,(i_\mathbf{C})\,\mathbf{G} + J\,\varphi_2\,(i_\mathbf{C})\,\mathbf{C} \,. \tag{6.4.6}$$

Herein all tensors on the right-hand side are isotropic tensor functions of **C** (see, section 2.11). Accordingly, by applying the representation theorem we have similar to (6.4.5)

$$\mathbf{S} = \mathbf{S}(\mathbf{C}) = \gamma_0\,(i_\mathbf{C})\,\mathbf{G} + \gamma_1\,(i_\mathbf{C})\,\mathbf{C} + \gamma_2\,(i_\mathbf{C})\,\mathbf{C}^2 \,, \quad i_\mathbf{C} = I_\mathbf{C}, II_\mathbf{C}, III_\mathbf{C} \tag{6.4.7}$$

with coefficients $\gamma_0, \gamma_1, \gamma_2$ being functions of the invariants $i_\mathbf{C}$ of **C**.

We close this section by summarizing the following important results:

Remark. An *objective* material law for isotropic materials is described by an isotropic tensor function $\boldsymbol{\sigma}(\mathbf{b})$ of **b** which can be represented by a quadratic polynomial of the form (6.4.5). From (6.4.5) we see that the tensors $\boldsymbol{\sigma} = \boldsymbol{\sigma}(\mathbf{b})$ and **b** are *coaxial.* From the mechanical point of view this means that, for isotropic materials, the stress tensor $\boldsymbol{\sigma}$ and the left CAUCHY-GREEN tensor **b** have the same principal axes $\mathbf{n}_i$. This property resulting from the conditions (6.4.3) and (6.4.4) is illustrated in Fig. 6.2 by considering the fact that the tensors $\boldsymbol{\sigma}$ and **b** have the same values at P and $\bar{P}$. From (6.4.7) we see that also the material tensors **S** and **C** are coaxial.

6.5 Derivatives of scalar-valued functions

This section is essentially devoted to mathematical considerations. The purpose is the derivation of some useful formulae for partial or time derivatives of scalar-valued functions. The corresponding results will serve in the subsequent sections to shorten the derivations.

First-order partial derivatives. Consider a scalar-valued function $\Psi = \Psi(\mathbf{C})$ of the right CAUCHY-GREEN tensor $\mathbf{C}$ related to the deformation gradient $\mathbf{F}$ by (2.4.34):

$$\mathbf{C} = \mathbf{F}^T \mathbf{F} \;\rightarrow\; C_{ij} = (F_{ik})^T F^k_{\cdot j} = F_{ki} F^k_{\cdot j} = F_{ki} F_{sj} G^{sk} \;. \tag{6.5.1}$$

Our aim is first to establish the relation between the partial derivatives $\Psi,_{\mathbf{F}}$ and $\Psi,_{\mathbf{C}}$. In a further step the derivative $\Psi,_{\mathbf{C}}$ will be transformed into $\Psi,_{\mathbf{U}}$ starting in this case from the equation (2.4.34):

$$\mathbf{C} = \mathbf{U}\,\mathbf{U} = \mathbf{U}^T \mathbf{U} \;. \tag{6.5.2}$$

Note that both tensors $\mathbf{C}$ and $\mathbf{U}$ are symmetric and this property will be used systematically in the subsequent derivation.

According to (1.7.4) and using the chain rule we form the partial derivative $\Psi,_{\mathbf{F}}$:

$$\Psi,_{\mathbf{F}} = \frac{\partial \Psi}{\partial F_{mn}} \mathbf{G}_m \otimes \mathbf{G}_n = \frac{\partial \Psi}{\partial C_{ij}} \frac{\partial C_{ij}}{\partial F_{mn}} \mathbf{G}_m \otimes \mathbf{G}_n \;, \tag{6.5.3}$$

where all tensor components refer to the initial basis $\mathbf{G}^i \otimes \mathbf{G}^j$. In view of (6.5.1) and (1.7.9) we may also write

$$\begin{aligned} \frac{\partial C_{ij}}{\partial F_{mn}} &= \frac{\partial F_{ki}}{\partial F_{mn}} F_{sj} G^{sk} + \frac{\partial F_{sj}}{\partial F_{mn}} F_{ki} G^{sk} \\ &= \delta^m_k \delta^n_i F_{sj} G^{sk} + \delta^m_s \delta^n_j F_{ki} G^{sk} \\ &= \delta^n_i F_{sj} G^{sm} \quad + \delta^n_j F_{ki} G^{mk} \end{aligned} \tag{6.5.4}$$

so that, with $C_{ij} = C_{ji}$, equation (6.5.3) becomes:

$$\Psi,_{\mathbf{F}} = F^m_{\cdot j} \frac{\partial \Psi}{\partial C_{ij}} \mathbf{G}_m \otimes \mathbf{G}_i + F^m_{\cdot i} \frac{\partial \Psi}{\partial C_{ji}} \mathbf{G}_m \otimes \mathbf{G}_j = 2\, F^m_{\cdot j} \frac{\partial \Psi}{\partial C_{ij}} \mathbf{G}_m \otimes \mathbf{G}_i \;. \tag{6.5.5}$$

This relation, in turn, is equivalent to

$$\Psi,_{\mathbf{F}} = 2\mathbf{F}\, \Psi,_{\mathbf{C}} \;. \tag{6.5.6}$$

In view of the similarity of the connections (6.5.1) and (6.5.2), the relation (6.5.6) holds also if we replace $\mathbf{F}$ by $\mathbf{U}$. Thus we have

$$\Psi,_{\mathbf{U}} = 2\mathbf{U}\, \Psi,_{\mathbf{C}} \;, \tag{6.5.7}$$

which is, due to the symmetry property $\Psi,_{\mathbf{U}} = (\Psi,_{\mathbf{U}})^T$, also expressible as

$$\Psi,_{\mathbf{U}} = \mathbf{U}\, \Psi,_{\mathbf{C}} + \Psi,_{\mathbf{C}}\, \mathbf{U} \;. \tag{6.5.8}$$

From (6.5.7) and (6.5.8) we furthermore see that

$$\mathbf{U}\,\Psi,_{\mathbf{C}} = \Psi,_{\mathbf{C}}\,\mathbf{U} \tag{6.5.9}$$

which, according to (1.9.19), indicates the symmetric tensors $\mathbf{U}$ and $\Psi,_{\mathbf{C}}$ to be *coaxial*.

The results obtained above can be generalized to arbitrary scalar-valued functions $\Psi = \Psi(\mathbf{A})$ with a *symmetric* argument $\mathbf{A} = \mathbf{A}^{\mathrm{T}}$ as follows

$$\Psi,_{\mathbf{B}} = 2\,\mathbf{B}\,\Psi,_{\mathbf{A}} \qquad \text{for} \quad \mathbf{A} = \mathbf{B}^{\mathrm{T}}\mathbf{B}\ , \tag{6.5.10}$$

$$\Psi,_{\mathbf{B}} = 2\,\Psi,_{\mathbf{A}}\,\mathbf{B} \qquad \text{for} \quad \mathbf{A} = \mathbf{B}\,\mathbf{B}^{\mathrm{T}}\ , \tag{6.5.11}$$

$$\begin{aligned}\Psi,_{\mathbf{B}} &= 2\,\mathbf{B}\,\Psi,_{\mathbf{A}} = 2\,\Psi,_{\mathbf{A}}\,\mathbf{B} \\ &= \mathbf{B}\,\Psi,_{\mathbf{A}} + \Psi,_{\mathbf{A}}\,\mathbf{B} \qquad \text{for} \quad \mathbf{A} = \mathbf{B}\,\mathbf{B} \quad \text{and} \quad \mathbf{B} = \mathbf{B}^{\mathrm{T}}\ .\end{aligned} \tag{6.5.12}$$

Now attention is restricted to functions depending on the invariants of $\mathbf{C}$:

$$\Psi = \Psi(\mathrm{I_C}, \mathrm{II_C}, \mathrm{III_C}) = \Psi(\mathrm{I_b}, \mathrm{II_b}, \mathrm{III_b})\ , \tag{6.5.13}$$

which may be expressed, due to (2.4.41), also in dependence of the invariants of the left CAUCHY-GREEN tensor $\mathbf{b}$. Then, we have according to (2.4.50) and (2.4.51):

$$\Psi,_{\mathbf{C}} = \left(\frac{\partial\Psi}{\partial \mathrm{I_C}} + \frac{\partial\Psi}{\partial \mathrm{II_C}}\,\mathrm{I_C}\right)\mathbf{G} - \frac{\partial\Psi}{\partial \mathrm{II_C}}\,\mathbf{C} + \frac{\partial\Psi}{\partial \mathrm{III_C}}\,\mathrm{III_C}\,\mathbf{C}^{-1}\ , \tag{6.5.14}$$

$$\Psi,_{\mathbf{b}} = \left(\frac{\partial\Psi}{\partial \mathrm{I_b}} + \frac{\partial\Psi}{\partial \mathrm{II_b}}\,\mathrm{I_b}\right)\mathbf{g} - \frac{\partial\Psi}{\partial \mathrm{II_b}}\,\mathbf{b} + \frac{\partial\Psi}{\partial \mathrm{III_b}}\,\mathrm{III_b}\,\mathbf{b}^{-1}\ . \tag{6.5.15}$$

It is also useful to recall equations (2.4.53) and (2.4.54)

$$\Psi,_{\mathbf{b}}\,\mathbf{b} = \mathbf{F}\,\Psi,_{\mathbf{C}}\,\mathbf{F}^{\mathrm{T}}\ , \tag{6.5.16}$$

$$\Psi,_{\mathbf{b}}\,\mathbf{b} = \mathbf{b}\,\Psi,_{\mathbf{b}}\ , \tag{6.5.17}$$

the last one indicating the coaxiality of $\mathbf{b}$ and $\Psi,_{\mathbf{b}}$.

Second partial derivatives. We again consider a scalar-valued function $\Psi = \Psi(\mathbf{C})$ depending on a material tensor, e.g. $\mathbf{C}$. The second partial derivative of Ψ with respect to $\mathbf{C}$ is given according to (1.7.5) by

$$\Psi,_{\mathbf{C}\otimes\mathbf{C}} = \frac{\partial^2\Psi}{\partial\mathbf{C}\,\partial\mathbf{C}} = \frac{\partial^2\Psi}{\partial C_{ij}\,\partial C_{mn}}\,\mathbf{G}_i\otimes\mathbf{G}_j\otimes\mathbf{G}_m\otimes\mathbf{G}_n\ . \tag{6.5.18}$$

This result can be also obtained, if we first form the partial derivative $\Psi,_{\mathbf{C}}$ and then apply the rule (1.7.6) holding for second-order tensors. Thus

$$\Psi,_{\mathbf{C}\otimes\mathbf{C}} = \frac{\partial}{\partial\mathbf{C}}\left(\frac{\partial\Psi}{\partial\mathbf{C}}\right) = \frac{\partial}{\partial C_{mn}}\left(\frac{\partial\Psi}{\partial C_{ij}}\right)\mathbf{G}_i \otimes \mathbf{G}_j \otimes \mathbf{G}_m \otimes \mathbf{G}_n \ . \tag{6.5.19}$$

Material time derivatives. For the calculation of the material time derivatives of scalar-valued functions it is important to distinguish between material and spatial formulation. This means that functions depending on material or spatial second-order tensors are to be treated differently. For a function $\Psi = \Psi(\mathbf{A})$ with an arbitrary material argument tensor $\mathbf{A} = A_{ij}\,\mathbf{G}^i \otimes \mathbf{G}^j$ the rule (1.7.8) can be applied to form the material time derivative $\dot{\Psi}$. This leads by considering (1.3.51) to

$$\dot{\Psi} := \frac{D\Psi}{Dt} = \frac{\partial\Psi}{\partial A_{ij}}\frac{\partial A_{ij}}{\partial t} = \Psi,_{\mathbf{A}} : \dot{\mathbf{A}} = \mathrm{tr}\,(\Psi,_{\mathbf{A}}\,\dot{\mathbf{A}}^T) = \mathrm{tr}\,[\dot{\mathbf{A}}\,(\Psi,_{\mathbf{A}})^T] \ , \tag{6.5.20}$$

where as usual

$$\Psi,_{\mathbf{A}} = \frac{\partial\Psi}{\partial A_{ij}}\mathbf{G}_i \otimes \mathbf{G}_j \ , \quad \dot{\mathbf{A}} = \dot{A}_{mn}\,\mathbf{G}^m \otimes \mathbf{G}^n \ . \tag{6.5.21}$$

If $\Psi = \Psi(\mathbf{a})$ depends on a second-order spatial tensor $\mathbf{a} = a_{ij}\,\mathbf{g}^i \otimes \mathbf{g}^j$ the above definition is to be replaced by

$$\dot{\Psi} = \frac{\partial\Psi}{\partial a_{ij}}\frac{\partial a_{ij}}{\partial t} = \left(\frac{\partial\Psi}{\partial a_{ij}}\mathbf{g}_i \otimes \mathbf{g}_j\right) : \left(\dot{a}_{mn}\,\mathbf{g}^m \otimes \mathbf{g}^n\right)$$

$$= \Psi,_{\mathbf{a}} : L_v\,(a_{mn}\,\mathbf{g}^m \otimes \mathbf{g}^n) = \Psi,_{\mathbf{a}} : L_v\,\mathbf{a} \ . \tag{6.5.22}$$

In contrast to (6.5.20) the above result depends with $L_v\,\mathbf{a}$ on the components of $\mathbf{a}$ used to express the function Ψ and therefore to form its partial derivative $\Psi,_{\mathbf{a}}$. The LIE-derivative can be replaced by the material time derivative, if $\Psi = \Psi(i_{\mathbf{a}}) = \Psi\,(a_i^k)$ depends on the invariants of a *symmetric* second-order tensor $\mathbf{a} = \mathbf{a}^T$. Then, according to (4.3.23)

$$\dot{\Psi} = \frac{D\Psi(i_{\mathbf{a}})}{Dt} = \Psi,_{\mathbf{a}} : L_v\,(a_i^k\,\mathbf{g}^i \otimes \mathbf{g}_k) = \Psi,_{\mathbf{a}} : \dot{\mathbf{a}} \ , \tag{6.5.23}$$

where the last expression holds for any arbitrary component relation of $\mathbf{a}$.

Application. Construct the partial derivatives of the functions J, ln J and tr $\mathbf{b}$ with respect to $\mathbf{b}$.

Since, in view of (2.4.41), (2.6.37), $\mathrm{tr}\,\mathbf{b} = I_{\mathbf{b}}$ and $J = \sqrt{III_{\mathbf{b}}}$, the problem can be solved by using the rules (1.8.7), (1.8.10) to form $(III_{\mathbf{b}}),_{\mathbf{b}}$ and $(I_{\mathbf{b}}),_{\mathbf{b}}$. The results are:

$$J,_{\mathbf{b}} = (III_{\mathbf{b}}^{1/2}),_{\mathbf{b}} = \frac{1}{2}III_{\mathbf{b}}^{-1/2}\,(III_{\mathbf{b}}),_{\mathbf{b}} = \frac{1}{2}\,III_{\mathbf{b}}^{-1/2}\,III_{\mathbf{b}}\,\mathbf{b}^{-1} = \frac{1}{2}\,J\,\mathbf{b}^{-1} \ ,$$

$$(\ln J),_{\mathbf{b}} = (\ln III_{\mathbf{b}}^{1/2}),_{\mathbf{b}} = III_{\mathbf{b}}^{-1/2}\,(III_{\mathbf{b}}^{1/2}),_{\mathbf{b}} = \frac{1}{2}\,\mathbf{b}^{-1} \ ,$$

$$(\mathrm{tr}\,\mathbf{b}),_{\mathbf{b}} = (I_{\mathbf{b}}),_{\mathbf{b}} = \mathbf{g} \ . \tag{6.5.24}$$

6.6 Hyperelastic material or GREEN-elastic materials

Stress power. For convenience we start with a short repetition of earlier results. In section 5.5 we have introduced the rate of the internal energy $\rho_o\dot{e}$ per unit undeformed volume. From (5.5.11) it is observable that, for isothermal processes r = 0 and $\mathbf{Q} = \mathbf{0}$, $\rho_o\dot{e}$ is equal to the stress power $\mathbf{S}:\dot{\mathbf{E}}$ (per unit undeformed volume). Alternative expressions for $\rho_o\dot{e}$ valid for this case are summarized in (3.3.22) and some of them are recorded below for the subsequent derivations:

$$\rho_o\dot{e} = \mathbf{P} : \dot{\mathbf{F}} = \mathbf{S} : \dot{\mathbf{E}} = \mathbf{S} : \frac{1}{2}\dot{\mathbf{C}} = \mathbf{T} : \dot{\mathbf{U}} = \text{sym}\,\mathbf{T} : \dot{\mathbf{U}} = \text{sym}\,\mathbf{T} : \dot{\mathbf{H}} \,. \tag{6.6.1}$$

Here, we have used the identity $\dot{\mathbf{C}} = 2\dot{\mathbf{E}}$ and considered that skew $\mathbf{T}:\dot{\mathbf{U}} = 0$ in accordance with the symmetry property $\dot{\mathbf{U}} = \dot{\mathbf{U}}^T$. In terms of spatial tensors, $\rho_o\dot{e}$ is given again according to (3.3.22) by

$$\rho_o\dot{e} = J\,\boldsymbol{\sigma} : L_v\,\mathbf{e} = \boldsymbol{\tau} : L_v\,\mathbf{e} = \boldsymbol{\tau} : \frac{1}{2}L_v\,\mathbf{g} \tag{6.6.2}$$

with the following LIE-derivatives

$$L_v\,\mathbf{e} = L_v\,(e_{ij}\,\mathbf{g}^i \otimes \mathbf{g}^j) = \dot{e}_{ij}\,\mathbf{g}^i \otimes \mathbf{g}^j \,, \quad L_v\,\mathbf{g} = L_v\,(g_{ij}\,\mathbf{g}^i \otimes \mathbf{g}^j) = \dot{g}_{ij}\,\mathbf{g}^i \otimes \mathbf{g}^j \,. \tag{6.6.3}$$

The definition of the strain and stress tensors appearing in the above relation are summarized in Table 2.2 and Table 3.1 while (2.10.27) involves the definition of the LIE-derivatives in general form. According to the above definitions the *total stress power* P of the body is given alternatively by:

$$P = \iiint_{V_o} \rho_o\dot{e}\,dV_o = \iiint_{V_o} \mathbf{S} : \dot{\mathbf{E}}\,dV_o = \iiint_{V_o} \boldsymbol{\tau} : L_v\,\mathbf{e}\,dV_o \,. \tag{6.6.4}$$

Definitions. A main characteristic of hyperelastic materials is the postulate of the existence of a strain energy $W = \rho_o\Psi$ depending on a deformation or strain tensor. The function W is called *elastic potential* and describes by definition a strain energy per unit undeformed volume so that the integral

$$\Pi = \iiint_{V_o} W\,dV_o = \iiint_{V_o} \rho_o\Psi\,dV_o \tag{6.6.5}$$

expresses the *total elastic potential* of the body. As usual ρ_o denotes the mass density of the undeformed body. Consequently, the notation Ψ in $W = \rho_o\Psi$ expresses a strain energy per unit mass and is called *specific strain energy* or *HELMHOLTZ free energy*.

We first suppose that W is a function of the GREEN-LAGRANGE strain tensor $\mathbf{E} = \frac{1}{2}(\mathbf{C} - \mathbf{G})$, where $\mathbf{C}$ is the right CAUCHY-GREEN tensor. Then, a material is said to be *hyperelastic* or *GREEN-elastic material* (Truesdell and Noll 1965) if there exists an elastic potential function $W = W(\mathbf{E})$ with material time derivative equal to the stress power $\rho_o\dot{e} = \mathbf{S} : \dot{\mathbf{E}}$ so that

$$\dot{W} := \frac{DW}{Dt} = \rho_o \dot{e} = \mathbf{S} : \dot{\mathbf{E}} = \mathbf{S} : \frac{1}{2}\dot{\mathbf{C}} \,. \tag{6.6.6}$$

Since, by virtue of (1.7.8), $\dot{W}$ can be expressed by

$$\dot{W} := \frac{DW}{Dt} = \frac{\partial W}{\partial E_{ij}} \frac{\partial E_{ij}}{\partial t} = \frac{\partial W}{\partial \mathbf{E}} : \dot{\mathbf{E}} \tag{6.6.7}$$

and as the relations (6.6.6) and (6.6.7) hold for arbitrary values of $\dot{\mathbf{E}}$ we obtain

$$\mathbf{S} = \frac{\partial W(\mathbf{E})}{\partial \mathbf{E}} = 2\frac{\partial W(\mathbf{C})}{\partial \mathbf{C}} \tag{6.6.8}$$

as constitutive equations of *hyperelastic* materials in material formulation.

We add further examples of hyperelastic material laws in material formulation:

first PIOLA-KIRCHHOFF stress tensor: $\mathbf{P} = \dfrac{\partial W(\mathbf{F})}{\partial \mathbf{F}}$, (6.6.9)

BIOT stress tensor: $\operatorname{sym}\mathbf{T} = \dfrac{\partial W(\mathbf{U})}{\partial \mathbf{U}} = \dfrac{\partial W(\mathbf{H})}{\partial \mathbf{H}}$, (6.6.10)

and in spatial formulation:

KIRCHHOFF stress tensor:

$$\boldsymbol{\tau} = J\boldsymbol{\sigma} = \frac{\partial W(e_{ij})}{\partial \mathbf{e}} = \frac{\partial W(e_{ij})}{\partial e_{ij}} \mathbf{g}_i \otimes \mathbf{g}_j$$

$$= 2\frac{\partial W(g_{ij})}{\partial \mathbf{g}} = 2\frac{\partial W(g_{ij})}{\partial g_{ij}} \mathbf{g}_i \otimes \mathbf{g}_j \,. \tag{6.6.11}$$

An essential characteristic of the above constitutive models is that they are expressed in terms of energy conjugate stress and strain tensors defined in (6.6.1) and (6.6.2). In contrast to the material formulations (6.6.8) to (6.6.10), where the differentiation can be carried out with respect to arbitrary tensor components, the relations given in (6.6.11) hold, if the partial derivatives are constructed with respect to the covariant components e_{ij} and g_{ij}. This fact has been emphasized by the notations $W(e_{ij})$ and $W(g_{ij})$.

Properties. Now, attention is focused on characteristic properties of the hyperelastic constitutive laws presented above. If we replace in (6.6.4) the stress tensor $\mathbf{S}$ by (6.6.8) the stress power P is transformed into the material time derivative $\dot{\Pi}$ of the total elastic potential (6.6.5):

$$P = \iiint_{V_o} \frac{\partial W}{\partial \mathbf{E}} : \dot{\mathbf{E}} \, dV_o = \iiint_{V_o} \frac{\partial W}{\partial E_{ij}} \dot{E}_{ij} \, dV_o = \iiint_{V_o} \dot{W} \, dV_o = \dot{\Pi} \,. \tag{6.6.12}$$

Thus the time integration between the limits t_1 and t_2 can be carried out leading to:

$$\int_{t_1}^{t_2} \mathrm{P}\,dt := \int_{t_1}^{t_2} \dot{\Pi}\,dt = \int_{t_1}^{t_2} d\Pi = \Pi(\mathbf{E}_2) - \Pi(\mathbf{E}_1)\ . \tag{6.6.13}$$

This result can be interpreted as follows. In any change from configuration 1 to configuration 2 the work performed by the stresses $\mathbf{S}$ along the path of the conjugate strains $\mathbf{E}$, the *internal work*, depends solely on the end values $\mathbf{E}_1 = \mathbf{E}(t_1)$ and $\mathbf{E}_2 = \mathbf{E}(t_2)$ of the strain tensor, but not on the particular deformation path relating the configurations 1 and 2.

Similar is valid for constitutive laws in spatial formulation. If we introduce (6.6.11) in (6.6.4) and consider again (6.6.5) we see that the stress power P becomes an exact material time derivative:

$$\mathrm{P} = \iiint_{V_o} \frac{\partial W}{\partial \mathbf{e}} : \mathrm{L}_v\,\mathbf{e}\,dV_o = \iiint_{V_o} \frac{\partial W}{\partial e_{ij}}\,\dot{e}_{ij}\,dV_o = \iiint_{V_o} \dot{W}\,dV_o = \dot{\Pi}\ . \tag{6.6.14}$$

Consequently, a common property of hyperelastic material models is the *path-independence* of the internal work performed by the stresses along the path of the conjugate strains.

Now, we refer to the constitutive law (6.6.8) in order to show that it satisfies the *principle of material objectivity* for any arbitrary selection of the function $W = W(\mathbf{C})$. To this end we introduce (6.6.8) into the definition (3.2.29) of the CAUCHY stress tensor $\boldsymbol{\sigma}$

$$\boldsymbol{\sigma} = J^{-1}\,\mathbf{F}\,\mathbf{S}\,\mathbf{F}^T = J^{-1}\,\mathbf{F}\left(2\frac{\partial W}{\partial \mathbf{C}}\right)\mathbf{F}^T\ . \tag{6.6.15}$$

Now the problem is to show that the above relation satisfies the objectivity requirement (6.3.2). From (6.2.14) we see that $\mathbf{C}$ remains unchanged, if the deformed configuration B is rotated into $\overset{*}{B}$ such that

$$\mathbf{C}^* = \mathbf{C} \;\rightarrow\; \frac{\partial W(\overset{*}{\mathbf{C}})}{\partial \overset{*}{\mathbf{C}}} = \frac{\partial W(\mathbf{C})}{\partial \mathbf{C}}\ . \tag{6.6.16}$$

By considering this result together with (6.2.3) and (6.2.5) the value of the stress tensor $\overset{*}{\boldsymbol{\sigma}}$ in $\overset{*}{B}$ is given according to (6.6.15) by:

$$\overset{*}{\boldsymbol{\sigma}} := \boldsymbol{\sigma}(\mathbf{Q}\,\mathbf{F}) = J^{-1}\,\mathbf{Q}\,\mathbf{F}\left(2\frac{\partial W}{\partial \mathbf{C}}\right)\mathbf{F}^T\,\mathbf{Q}^T = \mathbf{Q}\,\boldsymbol{\sigma}\,\mathbf{Q}^T \tag{6.6.17}$$

demonstrating the above statement. Evidently the same property holds, if W is expressed in terms of $\mathbf{E}$. On the contrary, an arbitrary selection of the function $W = W(\mathbf{F})$ in dependence of $\mathbf{F}$ does not ensure the objectivity of the constitutive law (6.6.9).

Our final goal is to show that the constitutive law (6.6.8) can be directly transformed into the forms (6.6.9) and (6.6.10). To this end we use for $\mathbf{P}$ and $\mathbf{T}$ the expressions given in

(3.2.29) in terms of **S**. By considering the differentiation rules (6.5.6) and (6.5.8) we then obtain

$$\mathbf{P} = \mathbf{F}\,\mathbf{S} = 2\,\mathbf{F}\,\frac{\partial W(\mathbf{C})}{\partial \mathbf{C}} = \frac{\partial W(\mathbf{F})}{\partial \mathbf{F}} , \tag{6.6.18}$$

$$\text{sym}\,\mathbf{T} = \frac{1}{2}(\mathbf{U}\,\mathbf{S} + \mathbf{S}\,\mathbf{U}) = \mathbf{U}\,\frac{\partial W(\mathbf{C})}{\partial \mathbf{C}} + \frac{\partial W(\mathbf{C})}{\partial \mathbf{C}}\,\mathbf{U} = \frac{\partial W(\mathbf{U})}{\partial \mathbf{U}} \tag{6.6.19}$$

confirming the above statement.

Material tensors. Material tensors $\mathbb{C}$ and $\mathbb{c}$ are fourth-order tensors which will be used to express the material time derivative $\dot{\mathbf{S}}$ and the LIE-derivative $L_v\,\boldsymbol{\tau}$ in terms of the corresponding energy conjugate strains, respectively. We first form the material time derivative of (6.6.8). This yields

$$\dot{\mathbf{S}} = \frac{\partial^2 W}{\partial E_{ij}\,\partial E_{mn}}\,\dot{E}_{mn}\,\mathbf{G}_i \otimes \mathbf{G}_j = 2\,\frac{\partial^2 W}{\partial C_{ij}\,\partial C_{mn}}\,\dot{C}_{mn}\,\mathbf{G}_i \otimes \mathbf{G}_j . \tag{6.6.20}$$

If we introduce under consideration of (2.5.9) and (6.6.8) the fourth-order tensor

$$\mathbb{C} := C^{ijmn}\,\mathbf{G}_i \otimes \mathbf{G}_j \otimes \mathbf{G}_m \otimes \mathbf{G}_n = \frac{\partial^2 W}{\partial \mathbf{E}\,\partial \mathbf{E}} = 4\,\frac{\partial^2 W}{\partial \mathbf{C}\,\partial \mathbf{C}} = 2\,\frac{\partial \mathbf{S}}{\partial \mathbf{C}} , \tag{6.6.21}$$

where similar to (6.5.18)

$$W_{,\mathbf{E}\otimes\mathbf{E}} := \frac{\partial^2 W}{\partial \mathbf{E}\,\partial \mathbf{E}} = \frac{\partial^2 W}{\partial E_{ij}\,\partial E_{mn}}\,\mathbf{G}_i \otimes \mathbf{G}_j \otimes \mathbf{G}_m \otimes \mathbf{G}_n , \tag{6.6.22}$$

equation (6.6.20) can be replaced by

$$\dot{\mathbf{S}} = \mathbb{C} : \dot{\mathbf{E}} = \frac{1}{2}\,\mathbb{C} : \dot{\mathbf{C}} . \tag{6.6.23}$$

We note that, in (6.6.22), arbitrary components of **E** may be used to form $W_{,\mathbf{E}\otimes\mathbf{E}}$.

In order to derive the spatial counterpart of the relation (6.6.23) we start from the component form of equation (6.6.11) the material time derivative of which delivers

$$\dot{\tau}^{ij} = \frac{\partial^2 W}{\partial e_{ij}\,\partial e_{mn}}\,\dot{e}_{mn} = 2\,\frac{\partial^2 W}{\partial g_{ij}\,\partial g_{mn}}\,\dot{g}_{mn} . \tag{6.6.24}$$

If we now introduce similar to (6.6.22) the fourth-order tensor

$$\mathbb{c} := c^{ijmn}\,\mathbf{g}_i \otimes \mathbf{g}_j \otimes \mathbf{g}_m \otimes \mathbf{g}_n = \frac{\partial^2 W(e_{ij})}{\partial \mathbf{e}\,\partial \mathbf{e}} = 4\,\frac{\partial^2 W(g_{ij})}{\partial \mathbf{g}\,\partial \mathbf{g}} = 2\,\frac{\partial \boldsymbol{\tau}}{\partial \mathbf{g}} \tag{6.6.25}$$

equation (6.6.24) becomes

$$L_v\,\boldsymbol{\tau} := L_v\,(\tau^{ij}\,\mathbf{g}_i \otimes \mathbf{g}_j) = \mathbb{c} : L_v\,(e_{mn}\,\mathbf{g}^m \otimes \mathbf{g}^n) = \frac{1}{2}\,\mathbb{c} : L_v\,(g_{mn}\,\mathbf{g}^m \otimes \mathbf{g}^n) . \tag{6.6.26}$$

We recall that in (6.6.25) the partial derivatives are to be carried out with respect to covariant components e_{ij} or g_{ij} as has been assumed in the derivation of (6.6.11).

Since $E_{ij} = e_{ij}$ and $C_{ij} = g_{ij}$, it can be deduced from (6.6.21) and (6.6.25) that $C^{ijkl} = c^{ijkl}$. This permits to present $\mathbb{c}$ as *push-forward* of $\mathbb{C}$ and vice versa $\mathbb{C}$ as *pull-back* of $\mathbb{c}$, thus

$$\mathbb{c} := c^{ijkl}\, \mathbf{g}_i \otimes \mathbf{g}_j \otimes \mathbf{g}_k \otimes \mathbf{g}_l = \Phi_* (\mathbb{C}) = \Phi_* (C^{ijkl}\, \mathbf{G}_i \otimes \mathbf{G}_j \otimes \mathbf{G}_k \otimes \mathbf{G}_l) \;,$$

$$= C^{ijkl}\, (\mathbf{F}\, \mathbf{G}_i) \otimes (\mathbf{G}_j\, \mathbf{F}^T) \otimes (\mathbf{F}\, \mathbf{G}_k) \otimes (\mathbf{G}_l\, \mathbf{F}^T) \;, \tag{6.6.27}$$

$$\mathbb{C} := C^{ijkl}\, \mathbf{G}_i \otimes \mathbf{G}_j \otimes \mathbf{G}_k \otimes \mathbf{G}_l = \Phi^* (\mathbb{c}) = \Phi^* (c^{ijkl}\, \mathbf{g}_i \otimes \mathbf{g}_j \otimes \mathbf{g}_k \otimes \mathbf{g}_l)$$

$$= c^{ijkl}\, (\mathbf{F}^{-1}\, \mathbf{g}_i) \otimes (\mathbf{g}_j\, \mathbf{F}^{-T}) \otimes (\mathbf{F}^{-1}\, \mathbf{g}_k) \otimes (\mathbf{g}_l\, \mathbf{F}^{-T}) \;. \tag{6.6.28}$$

The above relations are the generalisation of the rules presented in Fig. 2.11 to fourth-order tensors.

6.7 Isotropic hyperelastic material

Constitutive laws for compressible materials. Attention is now focused on *isotropic* hyperelastic materials. For convenience we recall the DOYLE-ERICKSEN formula given in (6.6.8)

$$\mathbf{S} = 2\,\rho_o \frac{\partial \psi (\mathbf{C})}{\partial \mathbf{C}} = 2 \frac{\partial W (\mathbf{C})}{\partial \mathbf{C}} \;, \tag{6.7.1}$$

where $W = \rho_o \psi$ is the elastic potential (per unit undeformed volume). In section 6.4 we have shown that objective constitutive laws for isotropic elastic materials are described by isotropic tensor functions $\boldsymbol{\sigma}(\mathbf{b})$ of the left CAUCHY-GREEN tensor $\mathbf{b}$, which can be, according to the representation theorem, expressed by a quadratic polynomial (6.4.5) with coefficients φ_0, φ_1 and φ_2 depending on the invariants of $\mathbf{b}$. A similar expression is given in (6.4.7) for the second PIOLA-KIRCHHOFF stress tensor $\mathbf{S}$ in terms of $\mathbf{C}$. Constitutive relations of the form (6.4.5) or (6.4.7) are ensured for *isotropic* elasticity if the elastic potential W is a function of the invariants of $\mathbf{b}$ or $\mathbf{C}$ (Ciarlet 1988):

$$W = W (I_C \,, II_C, III_C) = W (I_b, II_b, III_b) \;. \tag{6.7.2}$$

All possible elastic potentials for isotropic hyperelasticity are involved in the above formulation as special cases.

By making use of the definition $\boldsymbol{\tau} = \mathbf{F}\,\mathbf{S}\,\mathbf{F}^T$ of the KIRCHHOFF stress tensor $\boldsymbol{\tau}$ given by (3.2.29) and the identity (6.5.16)

$$W_{,b}\, \mathbf{b} = \mathbf{F}\, W_{,C}\, \mathbf{F}^T \tag{6.7.3}$$

holding for functions W of the form (6.7.2), the constitutive law (6.7.1) may be transformed into a *spatial* formulation. The result is

$$\boldsymbol{\tau} = 2\frac{\partial W(\mathbf{b})}{\partial \mathbf{b}}\mathbf{b} = 2\mathbf{b}\frac{\partial W(\mathbf{b})}{\partial \mathbf{b}}, \qquad (6.7.4)$$

where, in accordance with (6.5.17), the coaxiality of the tensors $W_{,\mathbf{b}}$ and $\mathbf{b}$ has been considered. In contrast to (6.7.1), the relation (6.7.4) holds only for isotropic materials since the validity of (6.7.3) is restricted to functions W of the form (6.7.2).

To specialize the general constitutive law (6.7.1) to elastic potentials of the form (6.7.2) we express the partial derivative $W_{,\mathbf{C}}$ according to (6.5.14) holding for this case. Thus the following constitutive law is obtained for *isotropic hyperelastic* materials

$$\mathbf{S} = 2\left[(a_I + a_{II}\, I_C)\,\mathbf{G} - a_{II}\,\mathbf{C} + a_{III}\, III_C\, \mathbf{C}^{-1}\right], \qquad (6.7.5)$$

where a_I, a_{II} and a_{III} are functions of the invariants of $\mathbf{C}$ or $\mathbf{b}$.

$$a_I = \frac{\partial W}{\partial I_C} = \frac{\partial W}{\partial I_b}, \quad a_{II} = \frac{\partial W}{\partial II_C} = \frac{\partial W}{\partial II_b}, \quad a_{III} = \frac{\partial W}{\partial III_C} = \frac{\partial W}{\partial III_b}. \qquad (6.7.6)$$

Similarly, we obtain from (6.5.15) and (6.7.4) the spatial counterpart of (6.7.5)

$$\boldsymbol{\tau} = J\boldsymbol{\sigma} = 2\left[(a_I + a_{II}\, I_b)\,\mathbf{b} - a_{II}\,\mathbf{b}^2 + a_{III}\, III_b\, \mathbf{g}\right], \qquad (6.7.7)$$

where $\boldsymbol{\sigma}$ is the CAUCHY stress tensor and $J = \det \mathbf{F} = \sqrt{III_b}$ by (2.4.43). Note that both results (6.7.5) and (6.7.7) are in agreement with the formulations given in (6.4.5) and (6.4.6) which demonstrates the suitability of the elastic potential (6.7.2) for modelling isotropic elastic materials. We finally transform both equations (6.7.5) and (6.7.7) under consideration of (2.4.45) and (2.4.46) into component relations. The comparison of the corresponding results shows the equality of the contravariant stress components S^{ij} and τ^{ij} in agreement with (3.2.24):

$$S^{ij} = \tau^{ij} = J\,\sigma^{ij} = 2\left[(a_I + a_{II}\, I_C)\, G^{ij} - a_{II}\, G^{im} G^{jn} g_{mn} + a_{III}\, III_C\, g^{ij}\right]. \qquad (6.7.8)$$

Herein, the invariants I_C and III_C of $\mathbf{C}$ may be replaced by those of $\mathbf{b}$.

Constitutive laws for incompressible materials. With slight modifications the above equations can be applied to *incompressible* isotropic materials. According to (2.4.43) incompressible deformations, which are also referred to as *isochoric*, are described by one of the following conditions:

$$J = \det \mathbf{F} = 1, \quad III_C = III_b = 1. \qquad (6.7.9)$$

Consequently, the elastic potential W is expressible in this case only in dependence on the first two invariants I_C and II_C

$$W = W(I_C, II_C) = W(I_b, II_b), \qquad (6.7.10)$$

and the constitutive equations (6.7.5) and (6.7.7) reduce therefore to

$$\mathbf{S} = 2\left[(a_I + a_{II}\, I_C)\, \mathbf{G} - a_{II}\, \mathbf{C} + a_{III}\, \mathbf{C}^{-1}\right], \tag{6.7.11}$$

$$\boldsymbol{\tau} = \boldsymbol{\sigma} = 2\left[(a_I + a_{II}\, I_b)\, \mathbf{b} - a_{II}\, \mathbf{b}^2 + a_{III}\, \mathbf{g}\right]. \tag{6.7.12}$$

The functions a_I and a_{II} can again be evaluated from (6.7.6). An essential distinction is, however, the equality $\boldsymbol{\tau} = \boldsymbol{\sigma}$ of the stress tensors defined in Table 3.1 due to (6.7.9). Furthermore, the function a_{III} can not be determined from (6.7.6) since the value of the derivative $\partial W/\partial III_C$ at $III_C = 1$ is unknown. The incompressibility condition (6.7.9) renders $a_{III} = p$ an unknown variable which corresponds to the hydrostatic pressure p. Its determination can be carried out through equilibrium and boundary conditions (Green and Zerna 1968, Başar and Ding 1997, Başar and Itskov 1998). Note that, by setting $J = 1$ and $III_C = 1$, the component relation (6.7.8) can be also used in the present case.

Series expansion of the elastic potential. We suppose that the function W introduced in (6.7.2) is continuously differentiable with respect to each invariant of $\mathbf{C}$. Then, W can be expanded in an infinite power series of the form (Ogden 1984, Barthold 1993):

$$W = W(I_C, II_C, III_C) = \sum_{p,q,r=0}^{\infty} c_{pqr}\, (I_C - 3)^p\, (II_C - 3)^q\, (III_C - 1)^r, \tag{6.7.13}$$

where the coefficients c_{pqr} are independent of the deformations. Note that in the undeformed state the elastic potential W vanishes since, in this case, $I_C = II_C = 3$ and $III_C = 1$. The relation (6.7.13) provides an exact representation of arbitrary objective elastic potentials of isotropic hyperelastic materials. In view of the relations

$$\begin{aligned} I_b &= I_C &&= (\lambda_1)^2 + (\lambda_2)^2 + (\lambda_3)^2, \\ II_b &= II_C &&= (\lambda_1)^2 (\lambda_2)^2 + (\lambda_2)^2 (\lambda_3)^2 + (\lambda_3)^2 (\lambda_1)^2, \\ III_b &= III_C &&= (\lambda_1)^2 (\lambda_2)^2 (\lambda_3)^2, \end{aligned} \tag{6.7.14}$$

which can be deduced with $\mathbf{C} = \mathbf{U}^2$ from (2.6.11) to (2.6.13), the series expansion (6.7.13) can be expressed in terms of the eigenvalues λ_i of $\mathbf{U}$

$$W = W(\lambda_1, \lambda_2, \lambda_3) = \sum_{p,q,r=0}^{\infty} a_{pqr}\Big(\Big[(\lambda_1)^p[(\lambda_2)^q + (\lambda_3)^q] + (\lambda_2)^p\,[(\lambda_3)^q + (\lambda_1)^q] + (\lambda_3)^p\,[(\lambda_1)^q + (\lambda_2)^q]\Big](\lambda_1\, \lambda_2\, \lambda_3)^r - 6\Big) \tag{6.7.15}$$

with coeffients a_{pqr} which are again independent of deformations. Both relations (6.7.13) and (6.7.15) hold for compressible materials. If an incompressible material with the condition $III_C = 1$ or $\lambda_1\, \lambda_2\, \lambda_3 = 1$ is considered they reduce to:

$$W = W(I_C, II_C) = \sum_{p,q=0}^{\infty} c_{pq}\, (I_C - 3)^p\, (II_C - 3)^q, \tag{6.7.16}$$

$$W = W(\lambda_1, \lambda_2, \lambda_3) = \sum_{p,q=0}^{\infty} a_{pq} \Big[(\lambda_1)^p [(\lambda_2)^q + (\lambda_3)^q] + (\lambda_2)^p [(\lambda_3)^q + (\lambda_1)^q]$$

$$+ (\lambda_3)^p [(\lambda_1)^q + (\lambda_2)^q] - 6 \Big] , \tag{6.7.17}$$

where c_{pq} and a_{pq} are the new material constants. We note that the above generally applicable relations (6.7.13), (6.7.15) for isotropic hyperelastic materials should deliver for small strains at the point $(I_C, II_C, III_C) = (3, 3, 1)$ or $(\lambda_1, \lambda_2, \lambda_3) = (1, 1, 1)$ the constitutive law and the elastic potential of the so-called ST-VENANT-KIRCHHOFF material model as will be shown in section 6.9.

Material tensor. We conclude this section by adding a short remark concerning the construction of the material tensor $\mathbb{c}$ in the case of isotropic hyperelastic materials. We recall that $\mathbb{c}$ has been introduced in (6.6.26) to relate the LIE-derivatives $L_v\, \tau^{ij}\, \mathbf{g}_i \otimes \mathbf{g}_j$ and $L_v\, e_{ij}\, \mathbf{g}^i \otimes \mathbf{g}^j$. For isotropic materials the elastic potential is expressible in terms of the mixed components $b^s_t = G^{sp}\, g_{pt}$ of the left CAUCHY-GREEN tensor (2.4.46). By considering the identity

$$\frac{\partial b^s_t}{\partial g_{ij}} = \frac{\partial\, (G^{sp}\, g_{pt})}{\partial g_{ij}} = G^{sp}\, \delta^i_p\, \delta^j_t = G^{si}\, \delta^j_t \; . \tag{6.7.18}$$

as well as the symmetry property $\mathbf{g} = \mathbf{g}^T$ it can be shown that

$$\frac{\partial^2 W\,(g_{ij})}{\partial \mathbf{g}\, \partial \mathbf{g}} = \frac{\partial^2 W}{\partial g_{ij}\, \partial g_{mn}}\, \mathbf{g}_i \otimes \mathbf{g}_j \otimes \mathbf{g}_m \otimes \mathbf{g}_n = \mathbf{b}\, \frac{\partial^2 W}{\partial \mathbf{b}\, \partial \mathbf{b}}\, \mathbf{b} \; . \tag{6.7.19}$$

This permits to replace (6.6.25) by the following relation (Stein and Barthold 1993):

$$\mathbb{c} = 4\mathbf{b}\, \frac{\partial^2 W}{\partial \mathbf{b}\, \partial \mathbf{b}}\, \mathbf{b} \; . \tag{6.7.20}$$

6.8 Special constitutive models for isotropic hyperelasticity

Our purpose is to introduce specific constitutive models for isotropic hyperelastic materials. In this context it is suitable to distinguish between two cases: constitutive models expressed in terms of the principal stretches λ_i and those given in terms of the invariants of $\mathbf{C}$ or $\mathbf{b}$. We recall that the invariants of $\mathbf{C}$ and $\mathbf{b}$ are identical so that the same elastic potential can be used without modification for material and spatial formulation.

The most relevant models given in terms of the principal stretches λ_i, the eigenvalues of the right stretch tensor $\mathbf{U}$, are the so-called OGDEN models which play for practical applications an important role. *Compressible OGDEN materials* are characterized by an elastic potential W of the form

$$W = W(\lambda_1, \lambda_2, \lambda_3) = \sum_{p=1}^{N} \frac{\mu_p}{\alpha_p} [(\lambda_1)^{\alpha_p} + (\lambda_2)^{\alpha_p} + (\lambda_3)^{\alpha_p} - 3] + g(J) , \qquad (6.8.1)$$

where μ_p, α_p are experimentally determined material constants and g(J) is a function of the determinant $J = \det \mathbf{F} = \sqrt{III_C}$. Thus, g(J) is responsible for compressible deformations. For its suitable selection we refer to Ciarlet 1983, Simo and Pister 1984, Marsden and Hughes 1983, Ball 1977. Remember that the elastic potential W refers as usual to unit undeformed volume.

For incompressible materials with J = 1 the expression (6.8.1) does not involve g(J) and its remaining part

$$W = W(\lambda_1, \lambda_2, \lambda_3) = \sum_{p=1}^{N} \frac{\mu_p}{\alpha_p} [(\lambda_1)^{\alpha_p} + (\lambda_2)^{\alpha_p} + (\lambda_3)^{\alpha_p} - 3] \qquad (6.8.2)$$

describes the so-called *OGDEN materials* (Odgen 1972, Odgen 1984). By linearization and comparison with the linear material model it can be shown that the material constants μ_p and α_p have to satisfy the *consistency* condition (see section 6.11):

$$\sum_{p=1}^{N} \mu_p \alpha_p = 2\mu , \qquad (6.8.3)$$

where $\mu = G$ is the shear modulus. A main distinction of the OGDEN model from (6.7.17) is that products of two differents eigenvalues λ_i (i = 1, 2, 3) are omitted in the elastic potential (6.8.2). But in contrast to (6.7.17), the principle stretches λ_i and, consequently, the strains can be considered in the OGDEN model with non-integer powers α_p. Comparisons with experimental results have shown that the OGDEN model provides an accurate simulation of incompressible rubber-like materials (Ogden 1972).

The OGDEN model (6.8.2) itself can be seen as special case of the VALANIS-LANDEL hypothesis (Valanis and Landel 1967) which separates the elastic potential into three parts such that each of them depends solely on a single eigenvalue λ_i (i = 1, 2, 3)

$$W = W(\lambda_1, \lambda_2, \lambda_3) = \omega(\lambda_1) + \omega(\lambda_2) + \omega(\lambda_3) . \qquad (6.8.4)$$

Herein $\omega(\lambda_i)$ denotes a function of λ_i.

For the finite element implementation of the OGDEN model we refer to Başar and Itskov 1998 and Eberlein et al. 1993. Its application to rubber-like membrane shells can be found in Wriggers and Taylor 1990, Gruttmann and Taylor 1992.

Compressible NEO-HOOKE models. Now we deal with hyperelastic models which are special cases of the formulation (6.7.13) and applicable therefore to compressible materials. As a relevant example of this class of materials we first consider the compressible NEO-HOOKE model (Simo et al. 1985, Eckstein 1999). The elastic potential W associated with this model depends on the first invariant I_C and the third invariant III_C

of the left CAUCHY-GREEN tensor $\mathbf{b}$ and can be expressed, through the relations (2.4.41), (2.4.43)

$$I_b = \mathrm{tr}\,\mathbf{b}\ ,\quad J = (III_b)^{1/2} = (\det \mathbf{b})^{1/2}\ , \tag{6.8.5}$$

alternatively by

$$\begin{aligned} W = W(I_b, III_b) &= U(III_b) && + \bar{W}(I_b, III_b) \\ &= \frac{1}{2}\kappa (\ln J)^2 && + \frac{1}{2}\mu (J^{-2/3}\,\mathrm{tr}\,\mathbf{b} - 3) \\ &= \frac{1}{2}\kappa (\ln \sqrt{III_b})^2 && + \frac{1}{2}\mu (III_b^{-1/3}\, I_b - 3)\ . \end{aligned} \tag{6.8.6}$$

The material constants involved in this model are the *bulk modulus* κ and the *LAMÉ constant* μ related to *YOUNG-modulus* E and the *POISSON-ratio* ν by

$$\kappa = \frac{E}{3(1-2\nu)}\ ,\quad \mu = \frac{E}{2(1+\nu)}\ . \tag{6.8.7}$$

The advantage of the above formulation is that the elastic potential is separated into a volumetric part U and a deviatoric one $\bar{W}$. The use of the left CAUCHY-GREEN tensor as deformation variable ensures, in addition, an easy consideration of hyperelasto-plastic material laws (Simo and Miehe 1992, Eckstein 1999, Başar and Itskov 1999).

To derive the constitutive relations related to (6.8.6) we refer to (6.7.7), which by evaluating the material constants a_I, a_{II} and a_{III} according to (6.7.6) and (6.8.6)

$$\begin{aligned} a_I &= \frac{\partial W}{\partial I_b} = \frac{1}{2}\mu J^{-2/3}\ ,\quad a_{II} = \frac{\partial W}{\partial II_b} = 0\ , \\ a_{III} &= \frac{\partial W}{\partial III_b} = \frac{1}{2}\kappa \ln J \frac{1}{J^2} - \frac{1}{6}\mu J^{-8/3} I_b \end{aligned} \tag{6.8.8}$$

delivers:

$$\begin{aligned} \boldsymbol{\tau} &= \mathrm{sph}\,\boldsymbol{\tau} \quad + \mathrm{dev}\,\boldsymbol{\tau} \\ &= \kappa \ln J\, \mathbf{g} + \mu J^{-2/3}\left[\mathbf{b} - \frac{1}{3}(\mathrm{tr}\,\mathbf{b})\,\mathbf{g}\right] \\ &= \kappa \ln J\, \mathbf{g} + \mu J^{-2/3}\,\mathrm{dev}\,\mathbf{b}\ . \end{aligned} \tag{6.8.9}$$

As can be deduced from (1.5.2) to (1.5.4) the first term on the right-hand side corresponds to the spherical part and the second one to the deviatoric part of the KIRCHHOFF stress tensor $\boldsymbol{\tau}$. Finally, we note that the material tensor introduced in (6.7.70) has in the present case the form (Eckstein 1999):

Table 6.1. Compressible *Neo-Hooke* material models for hyperelasticity

compressible Neo-Hooke material (Simo et al. 1985)	$W = W(I_b, III_b) = U(III_b) + \bar{W}(I_b, III_b) = \frac{1}{2}\kappa(\ln J)^2 + \frac{1}{2}\mu\left(J^{-2/3}\,\mathrm{tr}\,\mathbf{b} - 3\right)$
	$\boldsymbol{\tau} = 2\frac{\partial W}{\partial \mathbf{b}}\mathbf{b} = \mathrm{sph}\,\boldsymbol{\tau} + \mathrm{dev}\,\boldsymbol{\tau} = \kappa \ln J\,\mathbf{g} + \mu J^{-2/3}\,\mathrm{dev}\,\mathbf{b}$
compressible Neo-Hooke material (Ciarlet 1988, Simo and Hughes 1989)	$W = W(I_b, III_b) = \lambda\frac{J^2-1}{4} - \left(\frac{\lambda}{2} + \mu\right)\ln J + \frac{1}{2}\mu(\mathrm{tr}\,\mathbf{b} - 3)$
	$\boldsymbol{\tau} = 2\frac{\partial W}{\partial \mathbf{b}}\mathbf{b} = \frac{\lambda}{4}(J^2 - 1)\,\mathbf{g} + \frac{1}{2}\mu(\mathbf{b} - \mathbf{g})$
compressible Neo-Hooke material (Simo and Pister 1984)	$W = W(I_b, III_b) = \frac{\lambda}{2}(\ln J)^2 - \mu \ln J + \frac{1}{2}\mu(\mathrm{tr}\,\mathbf{b} - 3)$
	$\boldsymbol{\tau} = 2\frac{\partial \pi}{\partial \mathbf{b}}\mathbf{b} = \lambda \ln J\,\mathbf{g} + \mu(\mathbf{b} - \mathbf{g})$
compressible Neo-Hooke material (Blatz and Ko 1963)	$W = W(I_b, III_b) = \frac{1}{2}\mu\left[\mathrm{tr}\,\mathbf{b} - 3 + \frac{2}{\beta}(J^{-\beta} - 1)\right],\ \beta = \frac{2\nu}{1-2\nu}$
	$\boldsymbol{\tau} = 2\frac{\partial W}{\partial \mathbf{b}}\mathbf{b} = \mu\,\mathbf{b} - 2J^{-\beta}\,\mathbf{g}$

$$
\begin{aligned}
\mathbb{c} &= \mathrm{sph}\,\mathbb{c} + \mathrm{dev}\,\mathbb{c} \\
&= \kappa\left[\mathbf{g}\otimes\mathbf{g} - 2(\ln J)\,\mathbb{I}\right] + \frac{2}{3}\mu J^{-2/3}\left[\mathrm{tr}\,\mathbf{b}\,(\mathbb{I} - \frac{1}{3}\mathbf{g}\otimes\mathbf{g})\right. \\
&\qquad \left. - (\mathrm{dev}\,\mathbf{b}\otimes\mathbf{g} + \mathbf{g}\otimes\mathrm{dev}\,\mathbf{b})\right]
\end{aligned}
\tag{6.8.10}
$$

with the following fourth-order identity tensor:

$$
\mathbb{I} = \mathbf{g}_i\otimes\mathbf{g}_j\otimes\mathbf{g}^i\otimes\mathbf{g}^j = g^{ik}g^{jl}\,\mathbf{g}_i\otimes\mathbf{g}_j\otimes\mathbf{g}_k\otimes\mathbf{g}_l\ . \tag{6.8.11}
$$

Neo-Hooke material models are characterized by the fact that the second invariant II_b does not occur in the elastic potential $W = W(I_b, III_b)$. Further examples for compressible Neo-Hooke materials are given with corresponding constitutive equations in Table 6.1.

Incompressible Mooney-Rivlin and Neo-Hooke models. Attention is now turned to special cases of the formulation (6.7.16) applicable therefore to incompressible materials. The Mooney-Rivlin model is described by an elastic potential of the form (Mooney 1940, Rivlin 1949, Rivlin and Saunders 1951):

$$W = W(I_C, II_C) = c_1 (I_C - 3) + c_2 (II_C - 3) \tag{6.8.12}$$

depending on the first two invariants of **C** and involving two materials constants c_1 and c_2. For the simulation of certain incompressible rubber-like materials this model seems to be very suitable and has therefore received a broad application. For $c_2 = 0$, expression (6.8.12) reduces to

$$W = W(I_C) = c_1 (I_C - 3) \tag{6.8.13}$$

characterizing the so-called NEO-HOOKE materials (Treloar 1943).

The constitutive equations associated with the above models are derivable from (6.7.6) and (6.7.11) and have for the MOONEY-RIVLIN model the form

$$\mathbf{S} = 2\left[(c_1 + c_2)\,\mathbf{G} - c_2\,\mathbf{C} + p\,\mathbf{C}^{-1}\right] . \tag{6.8.14}$$

As has been discussed in section 6.7, $a_{III} = p$ is in the present case an unknown quantity corresponding to the hydrostatic pressure p and is to be calculated through equilibrium and boundary conditions. We also note that the above models require an explicit consideration of the incompressibility condition $III_C = 1$ leading to strongly nonlinear procedures in the analysis. For a detailed discussion of this aspect we refer to Schieck et al. 1992, Başar and Ding 1996, Başar and Ding 1997.

6.9 ST. VENANT-KIRCHHOFF material

In this section we use as strain measure the GREEN-LAGRANGE strain tensor **E** related to the right CAUCHY-GREEN tensor **C** by (2.5.9):

$$\mathbf{E} = \frac{1}{2}(\mathbf{F}^T\mathbf{F} - \mathbf{G}) = \frac{1}{2}(\mathbf{C} - \mathbf{G}) . \tag{6.9.1}$$

We recall that, in the undeformed reference state B_o of the body, $\mathbf{C} = \mathbf{G}$ and $\mathbf{E} = \mathbf{0}$. Our aim is the linearisation of the isotropic hyperelastic constitutive law (6.4.7) with respect to **E** at the point **X**. In terms of **E**, equation (6.4.7) reads as:

$$\mathbf{S}(\mathbf{E}) = \gamma_0(i_C)\,\mathbf{G} + \gamma_1(i_C)\,(2\mathbf{E} + \mathbf{G}) + \gamma_2(i_C)\,(4\mathbf{E}^2 + 4\mathbf{E} + \mathbf{G}) , \tag{6.9.2}$$

where $i_C = (I_C, II_C, III_C)$ denotes the invariants of **C**.

Let **E** denote as usual the value of the GREEN-LAGRANGE strain tensor in the actual deformed configuration determined by the position vector **x**. The linearization of the tensor-valued function **S** (**E**) with respect to **E** at the point **x** is a procedure defined by

$$L\,\mathbf{S}(\mathbf{E}, \Delta\mathbf{E}) = \mathbf{S}(\mathbf{E}) + \Delta\mathbf{S}(\mathbf{E}, \Delta\mathbf{E}) , \tag{6.9.3}$$

where the second term on the right-hand side is the *GATEAUX-derivative* or the *directional derivative* defined by

$$\Delta \mathbf{S}\,(\mathbf{E}, \Delta\mathbf{E}) \;=\; \frac{d}{d\varepsilon}\,\mathbf{S}\,(\mathbf{E} + \varepsilon\,\Delta\mathbf{E})\big|_{\varepsilon=0} \;. \qquad (6.9.4)$$

This rule is similar to that used for the evaluation of the first variation of **S** with respect to **E**. In view of the fact that the operator Δ defined by (6.9.4) is linear the product and the chain rule of differentiation hold in its application. Thus we obtain from (6.9.2) by means of (6.9.3) and (6.9.4)

$$\begin{aligned} \mathrm{L}\,\mathbf{S}\,(\mathbf{E}, \Delta\mathbf{E}) \;=\; & (\gamma_0 + \gamma_1 + \gamma_2)\,\mathbf{G} + 2\,(\gamma_1 + 2\gamma_2)\,\mathbf{E} + 4\gamma_2\,\mathbf{E}^2 \\ & + (\Delta\gamma_0 + \Delta\gamma_1 + \Delta\gamma_2)\,\mathbf{G} + 2\,(\Delta\gamma_1 + 2\Delta\gamma_2)\,\mathbf{E} + 4\Delta\gamma_2\,\mathbf{E}^2 \\ & + 2\,(\gamma_1 + 2\gamma_2)\,\Delta\mathbf{E} + 4\gamma_2\,(\mathbf{E}\,\Delta\mathbf{E} + \Delta\mathbf{E}\,\mathbf{E}) \;, \end{aligned} \qquad (6.9.5)$$

where

$$\Delta\gamma_K\,(\mathrm{i_C}) \;=\; \frac{\partial\gamma_K}{\partial \mathrm{I_C}}\,\Delta\mathrm{I_C} + \frac{\partial\gamma_K}{\partial \mathrm{II_C}}\,\Delta\mathrm{II_C} + \frac{\partial\gamma_K}{\partial \mathrm{III_C}}\,\Delta\mathrm{III_C}\,, \quad K = 0, 1, 2 \qquad (6.9.6)$$

and

$$\Delta\mathrm{i_C}\,(\mathbf{C}) \;=\; \frac{\partial \mathrm{i_C}}{\partial \mathbf{E}} : \Delta\mathbf{E} \;=\; 2\,\frac{\partial \mathrm{i_C}}{\partial \mathbf{C}} : \Delta\mathbf{E} \;, \qquad \mathrm{i_C} = \mathrm{I_C}, \mathrm{II_C}, \mathrm{III_C} \;. \qquad (6.9.7)$$

We recall that, in (6.9.5), the dependence of the coefficients $\gamma_K = \gamma_K\,(\mathrm{i_C})$ ($K = 0, 1, 2$) on the invariants $\mathrm{i_C}$ is assumed.

Our next goal is to apply the linearization (6.9.5) at the point $\mathbf{x} = \mathbf{X}$. Since in this case the equalities

$$\mathbf{C} = \mathbf{G}\;, \quad \mathbf{E} = \mathbf{0} \;\;\rightarrow\;\; \mathrm{i_C} \;=\; (\mathrm{I_C}, \mathrm{II_C}, \mathrm{III_C}) \;=\; (3, 3, 1) \;=\; \mathrm{i_G} \qquad (6.9.8)$$

hold, equation (6.9.5) is simplified into:

$$\begin{aligned} \mathrm{L}\,\mathbf{S}\,(\mathbf{E} = \mathbf{0}, \Delta\mathbf{E}) \;=\; & \Big(\big[\gamma_0\,(\mathrm{i_G}) + \gamma_1\,(\mathrm{i_G}) + \gamma_2\,(\mathrm{i_G})\big] + \big[\Delta\gamma_0\,(\mathrm{i_G}) + \Delta\gamma_1\,(\mathrm{i_G}) + \Delta\gamma_2\,(\mathrm{i_G})\big]\Big)\,\mathbf{G} \\ & + 2\big[\gamma_1\,(\mathrm{i_G}) + 2\,\gamma_2\,(\mathrm{i_G})\big]\,\Delta\mathbf{E} \;. \end{aligned} \qquad (6.9.9)$$

Herein, $\Delta\gamma_K\,(\mathrm{i_G})$ denotes the values of the directional derivatives $\Delta\gamma_K$ at the point $\mathbf{C} = \mathbf{G}$. To evaluate them we refer to (6.9.7). Considering the identities

$$(\mathrm{I_C})_{,\mathbf{C}} \;=\; \mathbf{G}\,, \quad (\mathrm{II_C})_{,\mathbf{C}} \;=\; (\mathrm{tr}\,\mathbf{C})\,\mathbf{G} - \mathbf{C}\,, \quad (\mathrm{III_C})_{,\mathbf{C}} \;=\; \mathbf{C}^2 - \mathrm{I_C}\,\mathbf{C} + \mathrm{II_C}\,\mathbf{G} \qquad (6.9.10)$$

following from (1.8.7) to (1.8.9) as well as (6.9.8) we obtain

$$(\Delta\mathrm{I_C})_{\mathbf{C}=\mathbf{G}} \;=\; 2\,\mathrm{tr}\,\Delta\mathbf{E}\,, \quad (\Delta\mathrm{II_C})_{\mathbf{C}=\mathbf{G}} \;=\; 4\,\mathrm{tr}\,\Delta\mathbf{E}\,, \quad (\Delta\mathrm{III_C})_{\mathbf{C}=\mathbf{G}} \;=\; 2\,\mathrm{tr}\,\Delta\mathbf{E} \qquad (6.9.11)$$

and, by means of this result, similarly from (6.9.6)

$$\Delta\gamma_K(i_G) = \operatorname{tr}\Delta\mathbf{E}\left(2\frac{\partial\gamma_K}{\partial I_C} + 4\frac{\partial\gamma_K}{\partial II_C} + 2\frac{\partial\gamma_K}{\partial III_C}\right)_{C=G} = \overset{*}{\gamma}_K(i_G)\operatorname{tr}\Delta\mathbf{E}\ ,\quad K = 0, 1, 2 \quad (6.9.12)$$

with the following abbreviation

$$\overset{*}{\gamma}_K(i_G) = \left(2\frac{\partial\gamma_K}{\partial I_C} + 4\frac{\partial\gamma_K}{\partial II_C} + 2\frac{\partial\gamma_K}{\partial III_C}\right)_{C=G}\ ,\quad K = 0, 1, 2 \quad (6.9.13)$$

The introduction of the above results into (6.9.9) delivers finally the linearization of the second PIOLA-KIRCHHOFF stress tensor $\mathbf{S}$ ($\mathbf{E}$) at $\mathbf{x} = \mathbf{X}$, e.i. $\mathbf{E} = \mathbf{0}$:

$$\mathrm{L}\,\mathbf{S}(\mathbf{E}=\mathbf{0}, \Delta\mathbf{E}) = -\pi\,\mathbf{G} + \lambda\,(\operatorname{tr}\Delta\mathbf{E})\,\mathbf{G} + 2\mu\,\Delta\mathbf{E} \quad (6.9.14)$$

with the abbreviations

$$\pi = -[\gamma_0(i_G) + \gamma_1(i_G) + \gamma_2(i_G)]\ ,$$

$$\lambda = \overset{*}{\gamma}_0(i_G) + \overset{*}{\gamma}_1(i_G) + \overset{*}{\gamma}_2(i_G)\ ,$$

$$\mu = \gamma_1(i_G) + 2\gamma_2(i_G)\ . \quad (6.9.15)$$

If we now suppose the initial configuration to be unstressed ($\pi = 0$) and change furthermore the notation $\Delta\mathbf{E}$ into $\mathbf{E}$ then, for sufficiently small strains, equation (6.9.14) becomes

$$\mathbf{S} = \lambda\,(\operatorname{tr}\mathbf{E})\,\mathbf{G} + 2\mu\,\mathbf{E}\ , \quad (6.9.16)$$

which characterizes the so-called *ST. VENANT-KIRCHHOFF material*. Here, λ and μ denote the LAMÉ-constants.

By means of (6.6.8) it can be verified that the elastic potential W = W ($\mathbf{E}$) of the ST. VENANT-KIRCHHOFF material is of the form

$$W(\mathbf{E}) = \frac{\lambda}{2}(\operatorname{tr}\mathbf{E})^2 + \mu\operatorname{tr}\mathbf{E}^2\ . \quad (6.9.17)$$

Alternative expressions of the above formulation are presented in the next section in connection with the HOOKEAN material described by the same elastic potential.

As can be seen from the above derivation, the constitutive law (6.9.16) holds for small strain analysis of isotropic hyperelastic materials. It is nevertheless applicable to geometrically nonlinear analysis with arbitrarily large displacements provided the strain-displacement relations are described by nonlinear equations (2.5.12).

The constitutive law (6.9.16) can be also expressed in terms of the first PIOLA-KIRCHHOFF stress tensor $\mathbf{P}$ and the KIRCHHOFF stress tensor $\boldsymbol{\tau}$. In view of (3.2.29), the results are of the form

$$\mathbf{P} = \mathbf{F}\,\mathbf{S} = \lambda\,(\operatorname{tr}\mathbf{E})\,\mathbf{F} + 2\mu\,\mathbf{F}\,\mathbf{E}\ , \quad (6.9.18)$$

$$\boldsymbol{\tau} = \mathbf{F}\,\mathbf{S}\,\mathbf{F}^{\mathrm{T}} = \lambda\,(\mathrm{tr}\,\mathbf{E})\,\mathbf{F}\,\mathbf{F}^{\mathrm{T}} + 2\mu\,\mathbf{F}\,\mathbf{E}\,\mathbf{F}^{\mathrm{T}} \;, \tag{6.9.19}$$

which reveals the underlying geometric nonlinearity of the ST. VENANT-KIRCHHOFF model.

6.10 HOOKEAN material

Linearization. In section 6.9 the ST. VENANT-KIRCHHOFF material model has been derived through the linearization of the constitutive law (6.4.7) with respect to the GREEN-LAGRANGE strain tensor $\mathbf{E}$ at the point $\mathbf{x} = \mathbf{X}$. The HOOKEAN material model to be derived in this section corresponds to the linearization of the ST. VENANT-KIRCHHOFF model (6.9.16) with respect to the displacement field $\mathbf{u} = \mathbf{x} - \mathbf{X}$ again at the point $\mathbf{x} = \mathbf{X}$, e.i. $\mathbf{u} = \mathbf{0}$. The linearization of the tensor-valued function $\mathbf{S}(\mathbf{u})$ is similar to (6.9.3) described by

$$\mathrm{L}\,\mathbf{S}\,(\mathbf{u}, \Delta\mathbf{u}) = \mathbf{S}\,(\mathbf{u}) + \Delta\mathbf{S}\,(\mathbf{u}, \Delta\mathbf{u}) \;, \tag{6.10.1}$$

where

$$\Delta\mathbf{S}\,(\mathbf{u}, \Delta\mathbf{u}) = \frac{\mathrm{d}}{\mathrm{d}\varepsilon}\,\mathbf{S}\,(\mathbf{u} + \varepsilon\,\Delta\mathbf{u})\big|_{\varepsilon=0} \tag{6.10.2}$$

is the directional derivative of $\mathbf{S}$. In the present case the above rules are to be applied at the point $\mathbf{u} = \mathbf{0}$. Thus we obtain from (6.9.16)

$$\mathrm{L}\,\mathbf{S}\,(\mathbf{u}, \Delta\mathbf{u})\big|_{\mathbf{u}=\mathbf{0}} = [\lambda\,(\mathrm{tr}\,\Delta\mathbf{E})\,\mathbf{G} + 2\mu\,\Delta\mathbf{E}]\big|_{\mathbf{u}=\mathbf{0}} \;. \tag{6.10.3}$$

In view of the well-known connections (2.1.21) and (2.2.3)

$$\mathbf{g}_i = \mathbf{G}_i + \mathbf{u}_{,i} \;, \qquad \mathbf{F} = \mathbf{g}_i \otimes \mathbf{G}^i \tag{6.10.4}$$

we have

$$\Delta\mathbf{F} = \Delta\Big((\mathbf{G}_i + \mathbf{u}_{,i}) \otimes \mathbf{G}^i\Big) = \Delta\mathbf{u}_{,i} \otimes \mathbf{G}^i = \mathrm{GRAD}\,\Delta\mathbf{u} \tag{6.10.5}$$

holding for any value of the displacement vector $\mathbf{u}$. Thus the directional derivative $(\Delta\mathbf{E})_{\mathbf{u}=\mathbf{0}}$ occuring in (6.10.3) is given in terms of the displacements $\Delta\mathbf{u}$ by

$$\begin{aligned}
\boldsymbol{\varepsilon} := \Delta\mathbf{E}\big|_{\mathbf{u}=\mathbf{0}} &= \frac{1}{2}\,\Delta\,(\mathbf{F}^{\mathrm{T}}\,\mathbf{F} - \mathbf{G})\big|_{\mathbf{u}=\mathbf{0}} \;, \\
&= \frac{1}{2}\Big(\mathrm{GRAD}\,\Delta\mathbf{u} + (\mathrm{GRAD}\,\Delta\mathbf{u})^{\mathrm{T}}\Big) \\
&= \frac{1}{2}\Big(\Delta\mathbf{u}_{,i} \cdot \mathbf{G}_j + \Delta\mathbf{u}_{,j} \cdot \mathbf{G}_i\Big)\,\mathbf{G}^i \otimes \mathbf{G}^j \\
&= \frac{1}{2}\Big(\Delta U_i|_j + \Delta U_j|_i\Big)\,\mathbf{G}^i \otimes \mathbf{G}^j \;,
\end{aligned} \tag{6.10.6}$$

where $\Delta\mathbf{u} = \Delta U_i \, \mathbf{G}^i$. Consequently, the main distinction of the present model from the ST. VENANT-KIRCHHOFF model is that the strain-displacement relation (2.5.12) is replaced by a linear one in terms of the displacements $\Delta\mathbf{u}$. This fact has been pointed out in (6.10.6) changing the notation $(\Delta\mathbf{E})_{\mathbf{u}=\mathbf{0}}$ into $\boldsymbol{\varepsilon}$. We also note that, within a small-displacement theory, there is no distinction between $\mathbf{u}$ and $\Delta\mathbf{u}$.

By a similar linearization procedure of the relations given in (3.2.29) it can be deduced that the distinction between various stress tensors is irrelevant in the case of small-displacement theory:

$$\boldsymbol{\sigma} = \boldsymbol{\tau} = \mathbf{S} = \mathbf{P} \; . \tag{6.10.7}$$

Thus, denoting in (6.10.3) the left-hand side term by $\boldsymbol{\sigma}$ and $\Delta\mathbf{E}$ by $\boldsymbol{\varepsilon}$ we obtain the constitutive law for a HOOKEAN material

$$\boldsymbol{\sigma} = \lambda \, (\operatorname{tr}\boldsymbol{\varepsilon}) \, \mathbf{I} + 2\mu \, \boldsymbol{\varepsilon} \tag{6.10.8}$$

involving the LAMÉ constant λ and the shear modulus $\mu = G$ as material constants. The elastic potential (6.9.17) of the ST. VENANT-KIRCHHOFF material retains in the present case the same form, thus

$$W = W(\boldsymbol{\varepsilon}) = \frac{\lambda}{2} (\operatorname{tr}\boldsymbol{\varepsilon})^2 + \mu \operatorname{tr}\boldsymbol{\varepsilon}^2 \; . \tag{6.10.9}$$

Material tensors. To derive further useful relations we first summarize the material constants associated with the HOOKEAN material (Malvern 1996):

YOUNG modulus: $$E = \frac{\mu\,(3\lambda+2\mu)}{\lambda+\mu} = 3\kappa\,(1-2\nu) = 2\mu\,(1+\nu) \; , \tag{6.10.10}$$

POISSON ratio: $$\nu = \frac{\lambda}{2\,(\lambda+\mu)} = \frac{\lambda}{(3\kappa-\lambda)} = \frac{E}{2\mu} - 1 \; , \tag{6.10.11}$$

LAMÉ constant: $$\mu = G = \frac{E}{2\,(1+\nu)} = \frac{3}{2}(\kappa-\lambda) = \frac{\lambda\,(1-2\nu)}{2\nu} \; , \tag{6.10.12}$$

LAMÉ constant: $$\lambda = \frac{E\nu}{(1+\nu)\,(1-2\nu)} = \frac{2\nu\mu}{1-2\nu} = \kappa - \frac{2}{3}\mu \; , \tag{6.10.13}$$

bulk modulus: $$\kappa = \frac{1}{3}(3\lambda+2\mu) = \frac{E}{3\,(1-2\nu)} = \frac{\lambda\,(1+\nu)}{3\nu} \; . \tag{6.10.14}$$

It is apparent that only two constants among those introduced above can be specified independently. Note that G is called *shear modulus.*

By considering (6.10.12) to (6.10.14) we introduce the fourth-order *material tensor* $\mathbb{C}$:

$$\begin{aligned} \mathbb{C} &= C^{ijmn}\,\mathbf{G}_i \otimes \mathbf{G}_j \otimes \mathbf{G}_m \otimes \mathbf{G}_n = \lambda\,\mathbf{I} \otimes \mathbf{I} + 2\mu\,\mathrm{II} \\ &= 2\mu\left(\mathrm{II} + \frac{\nu}{1-2\nu}\,\mathbf{I} \otimes \mathbf{I}\right) , \end{aligned} \tag{6.10.15}$$

as well as its inverse $\mathbb{C}^{-1}$:

$$\mathbb{C}^{-1} = (C^{-1})_{ijmn}\, \mathbf{G}^i \otimes \mathbf{G}^j \otimes \mathbf{G}^m \otimes \mathbf{G}^n = \frac{1}{2\mu}\, \mathrm{II} - \frac{\lambda}{2\mu\,(3\lambda+2\mu)}\, \mathbf{I} \otimes \mathbf{I}$$

$$= \frac{1}{E}\,[(1+\nu)\,\mathrm{II} - \nu\, \mathbf{I} \otimes \mathbf{I}]\ , \qquad (6.10.16)$$

where, as usual, $\mathbf{I} = \mathbf{G}_i \otimes \mathbf{G}^i$ and

$$\mathrm{II} = \mathbf{G}_i \otimes \mathbf{G}_j \otimes \mathbf{G}^i \otimes \mathbf{G}^j = G^{im}\, G^{jn}\, \mathbf{G}_i \otimes \mathbf{G}_j \otimes \mathbf{G}_m \otimes \mathbf{G}_n \qquad (6.10.17)$$

denotes the *fourth-order identity* tensor already introduced in (6.8.11) in spatial formulation. By considering the identities

$$\mathbf{I} : \mathrm{II} = \mathrm{II} : \mathbf{I} = \mathbf{I}\ , \quad \mathrm{II} : \mathrm{II} = \mathrm{II}\ , \quad (\mathbf{I} \otimes \mathbf{I}) : (\mathbf{I} \otimes \mathbf{I}) = 3\, \mathbf{I} \otimes \mathbf{I} \qquad (6.10.18)$$

in accordance with (6.10.17) it can be easily verified that the tensors $\mathbb{C}$ and $\mathbb{C}^{-1}$ are related by

$$\mathbb{C} : \mathbb{C}^{-1} = \mathbb{C}^{-1} : \mathbb{C} = \mathrm{II}\ . \qquad (6.10.19)$$

It is also suitable to express $\mathbb{C}$ and $\mathbb{C}^{-1}$ in terms of the bulk modulus κ and the LAMÉ constant μ. By using (6.10.13), (6.10.14) and (6.10.15), (6.10.16), the corresponding results are of the form:

$$\mathbb{C} = \mathrm{sph}\, \mathbb{C} + \mathrm{dev}\, \mathbb{C} = \kappa\, \mathbf{I} \otimes \mathbf{I} + 2\mu \left(\mathrm{II} - \frac{1}{3}\, \mathbf{I} \otimes \mathbf{I} \right) , \qquad (6.10.20)$$

$$\mathbb{C}^{-1} = \mathrm{sph}\, \mathbb{C}^{-1} + \mathrm{dev}\, \mathbb{C}^{-1} = \frac{1}{9\kappa}\, \mathbf{I} \otimes \mathbf{I} + \frac{1}{2\mu} \left(\mathrm{II} - \frac{1}{3}\, \mathbf{I} \otimes \mathbf{I} \right) . \qquad (6.10.21)$$

We see that both tensors are split into a spherical and a deviatoric part.

Alternative formulation of the constitutive law. In view of (6.10.15), the HOOKEAN law (6.10.8) may be given alternatively by

$$\boldsymbol{\sigma} = \mathbb{C} : \boldsymbol{\varepsilon} = \lambda\, (\mathrm{tr}\, \boldsymbol{\varepsilon})\, \mathbf{I} + 2\mu\, \boldsymbol{\varepsilon} = 2\mu \left[\boldsymbol{\varepsilon} + \frac{\nu}{1-2\nu}\, (\mathrm{tr}\, \boldsymbol{\varepsilon})\, \mathbf{I} \right] , \qquad (6.10.22)$$

which, by considering (6.10.19), can be solved for $\boldsymbol{\varepsilon}$ to obtain with (6.10.16):

$$\boldsymbol{\varepsilon} = \mathbb{C}^{-1} : \boldsymbol{\sigma} = \frac{1}{2\mu}\, \boldsymbol{\sigma} - \frac{\lambda}{2\mu\,(3\lambda+2\mu)}\, (\mathrm{tr}\, \boldsymbol{\sigma})\, \mathbf{I} = \frac{1}{E}\, [(1+\nu)\, \boldsymbol{\sigma} - \nu\, (\mathrm{tr}\, \boldsymbol{\sigma})\, \mathbf{I}]\ . \qquad (6.10.23)$$

If we use in (6.10.22) the expression (6.10.20) for $\mathbf{C}$ we see that

$$\boldsymbol{\sigma} = (\mathrm{sph}\, \mathbb{C} + \mathrm{dev}\, \mathbb{C}) : \boldsymbol{\varepsilon} = \mathrm{sph}\, \boldsymbol{\sigma} + \mathrm{dev}\, \boldsymbol{\sigma}$$

$$= \kappa\, (\mathrm{tr}\, \boldsymbol{\varepsilon})\, \mathbf{I} + 2\mu\, \mathrm{dev}\, \boldsymbol{\varepsilon}\ . \qquad (6.10.24)$$

This formulation demonstrates the advantage of the definition (6.10.20) which permits to decompose the stress tensor $\boldsymbol{\sigma}$ into a spherical part sph $\boldsymbol{\sigma}$ and a deviatoric part dev $\boldsymbol{\sigma}$. Similarly, we have from (6.10.21) and (6.10.23):

$$\boldsymbol{\varepsilon} = (\text{sph}\,\mathbb{C}^{-1} + \text{dev}\,\mathbb{C}^{-1}) : \boldsymbol{\sigma} = \text{sph}\,\boldsymbol{\varepsilon} + \text{dev}\,\boldsymbol{\varepsilon}$$

$$= \frac{1}{9\kappa}(\text{tr}\,\sigma)\,\mathbf{I} + \frac{1}{2\mu}\,\text{dev}\,\boldsymbol{\sigma} \ . \qquad (6.10.25)$$

We finally refer to the elastic potential (6.10.9), which we express, by using (6.10.15) and (6.10.22), (6.10.23), in the alternative forms:

$$W = \frac{1}{2}\boldsymbol{\varepsilon} : \mathbb{C} : \boldsymbol{\varepsilon} = \frac{1}{2}\boldsymbol{\sigma} : \boldsymbol{\varepsilon} = \frac{1}{2}\boldsymbol{\sigma} : \mathbb{C}^{-1} : \boldsymbol{\sigma} \ . \qquad (6.10.26)$$

Component relations. We conclude this section by adding useful component relations referring to curvilinear coordinates Θ^i. We have from (6.10.12), (6.10.13), (6.10.22):

$$\sigma^{ij} = 2\mu\,\varepsilon^{ij} + \lambda\,\varepsilon^k_k\,G^{ij}$$

$$= 2\mu\left(\varepsilon^{ij} + \frac{\nu}{1-2\nu}\,\varepsilon^k_k\,G^{ij}\right)$$

$$= \frac{E}{1+\nu}\left(\varepsilon^{ij} + \frac{\nu}{1-2\nu}\,\varepsilon^k_k\,G^{ij}\right) , \qquad (6.10.27)$$

and from (6.10.23)

$$\varepsilon_{ij} = \frac{1}{2\mu}\,\sigma_{ij} - \frac{\lambda}{2\mu\,(3\lambda+2\mu)}\,\sigma^k_k\,G_{ij}$$

$$= \frac{1}{E}\Big[(1+\nu)\,\sigma_{ij} - \nu\,\sigma^k_k\,G_{ij}\Big] \ . \qquad (6.10.28)$$

According to (6.10.15) and (6.10.16), we also have

$$C^{ijmn} = \mu\left[G^{im}\,G^{jn} + G^{in}\,G^{jm} + \frac{2\nu}{1-2\nu}\,G^{ij}\,G^{mn}\right] ,$$

$$(C^{-1})_{ijmn} = \frac{1}{E}\left[\frac{1+\nu}{2}\,(G_{im}\,G_{jn} + G_{in}\,G_{jm}) - \nu\,G_{ij}\,G_{mn}\right] \qquad (6.10.29)$$

demonstrating the following symmetry properties

$$C^{ijmn} = C^{jimn} = C^{ijnm} = C^{mnij} , \qquad (6.10.31)$$

$$(C^{-1})^{ijmn} = (C^{-1})^{jimn} = (C^{-1})^{ijnm} = (C^{-1})^{mnij} ,$$

in accordance with the initial definition of $\mathbb{C}$ in (6.6.21).

If we use orthogonal Cartesian coordinates the position of the indices in component relations is immaterial. In this case, we obtain from the last expression in (6.10.27) for the stress components σ_{ij} in matrix notation:

$$\begin{bmatrix} \sigma_{11} \\ \sigma_{22} \\ \sigma_{33} \\ \sigma_{12} \\ \sigma_{23} \\ \sigma_{13} \end{bmatrix} = \frac{E}{(1+\nu)\,(1-2\nu)} \begin{bmatrix} 1-\nu & \nu & \nu & 0 & 0 & 0 \\ \nu & 1-\nu & \nu & 0 & 0 & 0 \\ \nu & \nu & 1-\nu & 0 & 0 & 0 \\ 0 & 0 & 0 & (1-2\nu)/2 & 0 & 0 \\ 0 & 0 & 0 & 0 & (1-2\nu)/2 & 0 \\ 0 & 0 & 0 & 0 & 0 & (1-2\nu)/2 \end{bmatrix} \cdot \begin{bmatrix} \varepsilon_{11} \\ \varepsilon_{22} \\ \varepsilon_{33} \\ 2\varepsilon_{12} \\ 2\varepsilon_{23} \\ 2\varepsilon_{13} \end{bmatrix} \tag{6.10.32}$$

6.11 Linearization and comparison of various material models

Linearization rules. The objective in this section is to compare the OGDEN model (6.8.2) and the MOONEY-RIVLIN model (6.8.12) with the HOOKEAN model (6.10.9) in order to establish useful relations between the material constants of these models. We recall that the cited first two models hold for incompressible materials while the last one applies in the form (6.10.9) to compressible materials.

Let $W = W(\lambda_1, \lambda_2, \lambda_3)$ be the elastic potential of an isotropic elastic material. In this case, W is expressible in terms of the eigenvalues λ_i $(i = 1, 2, 3)$ of the right stretch tensor $\mathbf{U}$. For the derivation we use the variation of W with respect to the spatial coordinates $\mathbf{x}$ defined by

$$\delta W = \delta W(\mathbf{x}, \delta\mathbf{x}) = \frac{d}{d\varepsilon} W(\mathbf{x} + \varepsilon\,\delta\mathbf{x})\big|_{\varepsilon=0} = \left(\frac{\partial W}{\partial \mathbf{x}}\right) \cdot \delta\mathbf{x} \; . \tag{6.11.1}$$

The linearization of δW at a given point $\mathbf{x}$ is described by

$$L\,\delta W(\mathbf{x}, \delta\mathbf{x}; \Delta\mathbf{x}) = \delta W(\mathbf{x}, \delta\mathbf{x}) + \Delta\,\delta W(\mathbf{x}, \delta\mathbf{x}; \Delta\mathbf{x}) \; , \tag{6.11.2}$$

where the directional derivative $\Delta\,\delta W$ is defined by a rule similar to (6.11.1) as

$$\Delta\,\delta W = \frac{d}{d\varepsilon}\,\delta W(\mathbf{x} + \varepsilon\,\Delta\mathbf{x}, \delta\mathbf{x})\big|_{\varepsilon=0} = \Delta\mathbf{x}\,\frac{\partial^2 W}{\partial\mathbf{x}\,\partial\mathbf{x}}\,\delta\mathbf{x} \; . \tag{6.11.3}$$

If we replace in the above expression Δ by δ the corresponding result is the second variation $\delta^2 W$. The above rules will be applied in the sequel at the point $\mathbf{x} = \mathbf{X}$ where

$$\mathbf{x} = \mathbf{X} : \quad \boldsymbol{\varepsilon} := \mathbf{E} = \mathbf{0} \; , \quad \lambda_1 = \lambda_2 = \lambda_3 = 1 \; .$$

As will be confirmed soon the first term on the right-hand side of (6.11.2) vanishes in this particular case and the linearization is then solely described by the value of the directional derivative $\Delta\,\delta W$ at $\mathbf{x} = \mathbf{X}$. Furthermore, the linearization is to be carried out for incompressible materials. Thus, according to (6.7.9) and (6.7.14) the constraint

$$J = \lambda_1\,\lambda_2\,\lambda_3 = 1 \quad \rightarrow \quad \lambda_3 = \frac{1}{\lambda_1\,\lambda_2} \tag{6.11.4}$$

is to be satisfied by the stretch component λ_3.

HOOKEAN model. We first consider the HOOKEAN model (6.10.9)

$$W(\boldsymbol{\varepsilon}) = \frac{\lambda}{2}(\operatorname{tr}\boldsymbol{\varepsilon})^2 + \mu \operatorname{tr}\boldsymbol{\varepsilon}^2 \tag{6.11.5}$$

with LAMÉ constants λ and μ obtained by the linearization of the ST. VENANT-KIRCHHOFF model (6.9.16) with respect to the displacement field $\mathbf{u}$. Our aim is to express (6.11.5) in terms of λ_i and then to linearize the corresponding result according to (6.11.2). Remembering that $\boldsymbol{\varepsilon}$ is the notation of the GREEN-LAGRANGE strain tensor $\mathbf{E}$ in the linear theory we find from (2.5.9) and (2.6.17)

$$\boldsymbol{\varepsilon} = \frac{1}{2}(\mathbf{C} - \mathbf{G}) = \frac{1}{2}\sum_{i=1}^{3}\left[(\lambda_i)^2 - 1\right]\mathbf{N}_i \otimes \mathbf{N}_i \ . \tag{6.11.6}$$

Thus, $\operatorname{tr}\boldsymbol{\varepsilon}$ and $\operatorname{tr}\boldsymbol{\varepsilon}^2$ occuring in (6.11.5) are given by

$$\operatorname{tr}\boldsymbol{\varepsilon} = \frac{1}{2}\sum_{i=1}^{3}\left[(\lambda_i)^2 - 1\right] , \tag{6.11.7}$$

$$\operatorname{tr}\boldsymbol{\varepsilon}^2 = \frac{1}{4}\sum_{i=1}^{3}\left[(\lambda_i)^2 - 1\right]^2 . \tag{6.11.8}$$

To illustrate the application of the rules (6.11.1) and (6.11.3) we consider as example the expression (6.11.7), which yields

$$\delta \operatorname{tr}\boldsymbol{\varepsilon} = \sum_{i=1}^{3} \lambda_i\, \delta\lambda_i \ , \qquad \Delta\,\delta \operatorname{tr}\boldsymbol{\varepsilon} = \sum_{i=1}^{3} (\Delta\lambda_i\, \delta\lambda_i + \lambda_i\, \Delta\,\delta\lambda_i) \ . \tag{6.11.9}$$

The values of the above expressions at $\mathbf{X}$ are

$$\delta \operatorname{tr}\boldsymbol{\varepsilon}\big|_{\mathbf{x}=\mathbf{X}} = \sum_{i=1}^{3} \delta\lambda_i \ , \qquad \Delta\delta \operatorname{tr}\boldsymbol{\varepsilon}\big|_{\mathbf{x}=\mathbf{X}} = \sum_{i=1}^{3} (\Delta\lambda_i\, \delta\lambda_i + \Delta\,\delta\lambda_i) \ . \tag{6.11.10}$$

Similarly, we find by considering (6.11.7) and (6.11.8)

$$\delta(\operatorname{tr}\boldsymbol{\varepsilon})^2\big|_{\mathbf{x}=\mathbf{X}} = 0 \ , \qquad \Delta\delta(\operatorname{tr}\boldsymbol{\varepsilon})^2\big|_{\mathbf{x}=\mathbf{X}} = 2\sum_{i=1}^{3}\sum_{j=1}^{3} \Delta\lambda_i\, \delta\lambda_j \ , \tag{6.11.11}$$

$$\delta \operatorname{tr}\boldsymbol{\varepsilon}^2\big|_{\mathbf{x}=\mathbf{X}} = 0 \ , \qquad \Delta\delta \operatorname{tr}\boldsymbol{\varepsilon}^2\big|_{\mathbf{x}=\mathbf{X}} = 2\sum_{i=1}^{3} \Delta\lambda_i\, \delta\lambda_i \ . \tag{6.11.12}$$

The application of a similar procedure to (6.11.4) delivers the constraints

$$\delta\lambda_3\big|_{\mathbf{x}=\mathbf{X}} = -(\delta\lambda_1 + \delta\lambda_2) \ , \tag{6.11.13}$$

$$\Delta\,\delta\lambda_3\big|_{\mathbf{x}=\mathbf{X}} = 2(\delta\lambda_1\, \Delta\lambda_1 + \delta\lambda_2\, \Delta\lambda_2) + \delta\lambda_1\, \Delta\lambda_2 + \Delta\lambda_1\, \delta\lambda_2 - (\Delta\,\delta\lambda_1 + \Delta\,\delta\lambda_2) \tag{6.11.14}$$

to be satisfied for incompressible materials. Note that (6.11.13) also holds if we replace δ by Δ. Inserting the above expressions into (6.11.11) and (6.11.12) delivers

$$\Delta\,\delta\,(\mathrm{tr}\,\boldsymbol{\varepsilon})^2|_{\mathbf{x}=\mathbf{X}} = 0\ ,$$

$$\Delta\delta\,\mathrm{tr}\,\boldsymbol{\varepsilon}^2|_{\mathbf{x}=\mathbf{X}} = 2\,[2\,\Delta\lambda_1\,\delta\lambda_1 + \Delta\lambda_1\,\delta\lambda_2 + \delta\lambda_1\,\Delta\lambda_2 + 2\,\Delta\lambda_2\,\delta\lambda_2]\ , \tag{6.11.15}$$

and hence in view of (6.11.2), (6.11.5) and (6.11.11), (6.11.12)

$$\mathrm{L}\,\delta\mathrm{W}|_{\mathbf{x}=\mathbf{X}} = \Delta\,\delta\mathrm{W}|_{\mathbf{x}=\mathbf{X}} = 2\mu\,[2\,(\Delta\lambda_1\,\delta\lambda_1 + \Delta\lambda_2\,\delta\lambda_2) + \delta\lambda_1\,\Delta\lambda_2 + \Delta\lambda_1\,\delta\lambda_2] \tag{6.11.16}$$

corresponding to the formulation of the HOOKEAN model in terms of the eigenvalues λ_1 and λ_2. The above derivation shows that:

Remark. The variation of the elastic potential δW vanishes at the point **X**. The term $(\mathrm{tr}\,\boldsymbol{\varepsilon})^2$ occuring in (6.11.5) has no influence to the linearization response if incompressible materials are considered. Due again to the linearization procedure at **X** the terms of the form $\Delta\,\delta\lambda_i$ do not appear in (6.11.11), (6.11.12) and can be essentially omitted in the derivation.

OGDEN model. Now attention is confined to the incompressible OGDEN model (6.8.2):

$$\mathrm{W}\,(\lambda_1, \lambda_2, \lambda_3) = \sum_{p=1}^{N} \frac{\mu_p}{\alpha_p}\Big[(\lambda_1)^{\alpha_p} + (\lambda_2)^{\alpha_p} + (\lambda_3)^{\alpha_p} - 3\Big]\ , \tag{6.11.17}$$

where μ_p and α_p are material constants. For convenience we apply the rules (6.11.1) and (6.11.3) to $(\lambda_i)^\beta$ representing for $\beta = \alpha_p$ the terms involved in the OGDEN model. The results are

$$\delta\,(\lambda_i)^\beta = \beta\,(\lambda_i)^{\beta-1}\,\delta\lambda_i\ , \tag{6.11.18}$$

$$\Delta\,\delta\,(\lambda_i)^\beta = \beta\,(\beta-1)\,(\lambda_i)^{\beta-2}\,\Delta\lambda_i\,\delta\lambda_i + \beta\,(\lambda_i)^{\beta-1}\,\Delta\,\delta\lambda_i\ , \tag{6.11.19}$$

which reduce at the point **X** to

$$\delta\,(\lambda_i)^\beta|_{\mathbf{x}=\mathbf{X}} = \beta\,\delta\lambda_i\ , \tag{6.11.20}$$

$$\Delta\,\delta\,(\lambda_i)^\beta|_{\mathbf{x}=\mathbf{X}} = \beta\,(\beta-1)\,\Delta\lambda_i\,\delta\lambda_i + \beta\,\Delta\,\delta\lambda_i\ . \tag{6.11.21}$$

If we form the first variation of W and consider (6.11.20) together with the incompressibility condition (6.11.13) we see that $\delta\mathrm{W} = 0$ at the point **X**. By application of the rule (6.11.2) to (6.11.17) we then find with (6.11.21) and (6.11.13), (6.11.14):

$$\mathrm{L}\,\delta\mathrm{W}|_{\mathbf{x}=\mathbf{X}} = \Delta\,\delta\mathrm{W}|_{\mathbf{x}=\mathbf{X}} = \sum_{p=1}^{N} \mu_p\,\alpha_p\Big[2\,(\delta\lambda_1\,\Delta\lambda_1 + \delta\lambda_2\,\Delta\lambda_2) + \delta\lambda_1\,\Delta\lambda_2 + \Delta\lambda_1\,\delta\lambda_2\Big]\ , \tag{6.11.22}$$

where again terms of the form $\Delta\, \delta\lambda_i$ do not occur due to the linearization performed at the point **X**.

MOONEY-RIVLIN model. Attention is finally focused on the MOONEY-RIVLIN model (6.8.12):

$$W\,(I_C, II_C) \;=\; c_1\,(I_C - 3) + c_2\,(II_C - 3) \qquad (6.11.23)$$

with the material constants c_1 and c_2 where, according to (6.7.14), the following transformations hold for the invariants I_C and II_C

$$I_C \;=\; (\lambda_1)^2 + (\lambda_2)^2 + (\lambda_3)^2\ , \qquad (6.11.24)$$

$$II_C \;=\; (\lambda_1)^2\,(\lambda_2)^2 + (\lambda_2)^2\,(\lambda_3)^2 + (\lambda_3)^2\,(\lambda_1)^2\ . \qquad (6.11.25)$$

We recall that this material model holds like the OGDEN model (6.11.17) for incompressible materials. Consequently, the incompressibility conditions (6.11.13) and (6.11.14) are to be considered in the linearization. By a procedure similar to that used above we find after some calculations:

$$\delta I_C|_{x=X} \;=\; 0\ , \quad \Delta\,\delta I_C|_{x=X} \;=\; 4\Big[2\,(\delta\lambda_1\,\Delta\lambda_1 + \delta\lambda_2\,\Delta\lambda_2) + \delta\lambda_1\,\Delta\lambda_2 + \Delta\lambda_1\,\delta\lambda_2\Big]\ , \qquad (6.11.26)$$

$$\delta II_C|_{x=X} \;=\; 0\ , \quad \Delta\,\delta II_C|_{x=X} \;=\; 4\Big[2\,(\delta\lambda_1\,\Delta\lambda_1 + \delta\lambda_2\,\Delta\lambda_2) + \delta\lambda_1\,\Delta\lambda_2 + \Delta\lambda_1\,\delta\lambda_2\Big] \qquad (6.11.27)$$

and, by using these results, according to (6.11.2):

$$L\,\delta W|_{x=X} \;=\; \Delta\,\delta W|_{x=X} \;=\; 4\,(c_1 + c_2)\Big[2\,(\delta\lambda_1\,\Delta\lambda_1 + \delta\lambda_2\,\Delta\lambda_2) + \delta\lambda_1\,\Delta\lambda_2 + \Delta\lambda_1\,\delta\lambda_2\Big]\ , \qquad (6.11.28)$$

which corresponds to the linearization of the MOOONEY-RIVLIN model at **X**.

After linearization at the point **X** the nonlinear MOONEY-RIVLIN and OGDEN models should deliver as special case the HOOKEAN model holding for small-displacement theory. As can be observed from (6.11.16), (6.11.22) and (6.11.28) this requirement is satisfied only if

$$\mu \;=\; \frac{1}{2}\sum_{p=1}^{N} \mu_p\,\alpha_p \;=\; 2\,(c_1 + c_2)\ , \qquad (6.11.29)$$

in accordance with (6.8.3). Note that, for incompressible materials, the HOOKEAN model involves only the shear modulus $G = \mu$ as material constant. The relation between μ and c_1, c_2 given in (6.11.29) can be found e.g. in Schieck 1989.

Exercises

6.1. Examine if the following expressions define objective tensors in the sense of the definitions given in section 6.2:

$$\Psi = \operatorname{tr}\mathbf{C}\,\operatorname{tr}\mathbf{b}\ , \qquad \Psi = \mathbf{b}:\mathbf{g}\ , \qquad \Psi = \mathbf{R}:(\mathbf{R}\,\mathbf{b})\ ,$$
$$\Psi = (\mathbf{C}\,\mathbf{R}):(\mathbf{C}\,\mathbf{R})\ , \qquad \Psi = \mathbf{U}^2:(\mathbf{F}^T\,\mathbf{F})\ , \qquad \Psi = (\mathbf{F}\,\mathbf{R}^T):\mathbf{v}\ ,$$
$$\boldsymbol{\tau} = \lambda \ln J\,\mathbf{g} + \mu\,(\mathbf{b}-\mathbf{g}) \quad \text{with} \quad \lambda,\ \mu\text{: constants}\ ,$$
$$\boldsymbol{\tau} = \kappa \ln J\,\mathbf{g} + \mu\,J^{-2/3}\,\operatorname{dev}\mathbf{b} \quad \text{with} \quad \kappa,\ \mu\text{: constants}\ .$$

6.2. Examine if the following expressions are expressible in terms of the eigenvalues λ_i of the right stretch tensor $\mathbf{U}$:

$$\Psi = \mathbf{R}:\mathbf{b}\ , \qquad \Psi = \mathbf{b}:\mathbf{g}\ , \qquad \Psi = (\mathbf{F}\,\mathbf{R}^T):(\mathbf{R}\,\mathbf{F}^T)\ ,$$
$$\Psi = \frac{1}{2}\kappa\,(\ln J)^2 + \frac{1}{2}\mu\,(J^{-2/3}\operatorname{tr}\mathbf{b} - 3) \quad \text{with} \quad \kappa,\ \mu\text{: constants}\ .$$

6.3. Starting from the elastic potential $W = W(I_b, J) = \frac{\lambda}{2}(\ln J)^2 - \mu \ln J + \frac{1}{2}\mu\,(\operatorname{tr}\mathbf{b} - 3)$ with the material constants λ, μ derive the constitutive law for $\boldsymbol{\tau}$.

Appendix 1

Appendix 1 summarizes the essential notations and formulae of tensor calculus which will be used systematically in the derivations of the first chapter. Their thorough understanding is highly recommended for an easy study of the book.

A1.1 Index notation

With exception of scalar valued quantities, variables defined in a three-dimensional Euclidean space E3 are presented with indices:

$$A_{ij}\,,\ C^{ijkl}\,,\ \delta^j_i\,,\ B^{i}_{\cdot m}\ ,$$

where Latin indices represent the number 1, 2, 3. Thus a symbol with n indices is a short representation of 3^n components. In general, it is not allowed to change the order or the position (subscript or superscript) of the indices such that

$$B^{ij} \neq B^{ji} \qquad B^{ij} \neq B^{i}_{\cdot j} \neq B_{i}^{\cdot j} \neq B_{ij}\ .$$

A useful rule for compact formulations is the *summation rule*: If an index is repeated twice in different positions, e.i. as subscript and superscript, a summation is defined in the following sense:

$$A^i B_{ij} = A^k B_{kj} = \ldots = A^1 B_{1j} + A^2 B_{2j} + A^3 B_{3j}\ .$$

$$C^{ij} B_{ij} = C^{mn} B_{mn} = \ldots = C^{11} B_{11} + C^{12} B_{12} + C^{13} B_{13} + \ldots + C^{33} B_{33}\ . \qquad \text{(A1.1.1)}$$

Consequently, the summation rule does not hold for expressions of the form:

$$A^i B_{ii} = (A^1 B_{11}\,, A^2 B_{22}\,, A^3 B_{33})\ , \qquad g_{ii} = (g_{11}\,, g_{22}\,, g_{33})\ , \qquad \text{(A1.1.2)}$$

where an index is repeated more than twice or repeated indices appear in the same position.

For later use we introduce the *Kronecker-delta*

$$\delta_{ij} = \delta^{ij} = \delta^i_j = 1\ , \quad \textit{if} \ \ i=j$$

$$= 0\ , \quad \textit{if} \ \ i \neq j \qquad \text{(A1.1.3)}$$

as well as the components of the *permutation tensor* associated with an orthogonal Cartesian coordinate system:

$$e_{ijk} = 0 \quad \textit{if any two of the indices are equal;}$$
$$= +1 \quad \textit{if i, j, k is an even permutation of the numbers 1, 2, 3;}$$
$$= -1 \quad \textit{if i, j, k is a odd permutation of the number 1, 2, 3.} \qquad \text{(A1.1.4)}$$

so that e.g.:

$$e_{112} = e_{122} = 0 \ , \quad e_{123} = e_{231} = 1 \ , \quad e_{213} = e_{132} = -1 \ .$$

A simple consequence of the above definitions (A1.1.1) and (A1.1.3) is the *rule of change of indices* as shown by the following examples:

$$A_{ij}\,\delta^j_k = A_{ik} \ , \quad A_{ij}\,\delta^i_m\,\delta^j_n = A_{mn} \ , \quad \delta^i_j\,\delta^j_i = \delta^i_i = \delta^j_j = 3 \ .$$

Orthogonal Cartesian coordinates are denoted by x^i and the unit base vectors along the x^i-axes by $\mathbf{i}_i = \mathbf{i}^i$, where, by definition, the position of the index i is irrelevant. By means of (A1.1.3) and (A1.1.4), the properties of the orthonormal basis $\mathbf{i}_j$ can be expressed as follows:

$$\mathbf{i}_i \cdot \mathbf{i}_j = \delta_{ij} \ , \qquad \mathbf{i}^i \cdot \mathbf{i}^j = \delta^{ij} \ , \quad \mathbf{i}_i \cdot \mathbf{i}^j = \delta^j_i \ , \qquad \text{(A1.1.5)}$$

$$\mathbf{i}_i \times \mathbf{i}_j = e_{ijk}\,\mathbf{i}^k \ , \qquad \mathbf{i}^i \times \mathbf{i}^j = e^{ijk}\,\mathbf{i}_k \ ,$$

$$[\mathbf{i}_i \ \mathbf{i}_j \ \mathbf{i}_k] = e_{ijk} \ , \quad [\mathbf{i}^i \ \mathbf{i}^j \ \mathbf{i}^k] = e^{ijk} \ . \qquad \text{(A1.1.6)}$$

For vectorial operations we use the following notations:

scalar product: $\mathbf{A} \cdot \mathbf{B} = \mathbf{B} \cdot \mathbf{A}$,

vector product: $\mathbf{A} \times \mathbf{B} = -\mathbf{B} \times \mathbf{A}$,

mixed product: $[\mathbf{A} \ \mathbf{B} \ \mathbf{C}] = \mathbf{A} \cdot (\mathbf{B} \times \mathbf{C})$. (A1.1.7)

We, finally, consider components $A_{ij} = A_{ij}(\Theta^1, \Theta^2, \Theta^3)$ depending on the parameters Θ^i (i = 1, 2, 3). The partial derivatives of A_{ij} with respect to Θ^k will be denoted by

$$A_{ij,k} = \frac{\partial A_{ij}}{\partial \Theta^k} \ . \qquad \text{(A1.1.8)}$$

A1.2 Metric tensor and geometrical properties

Any point P of a 3D-Euclidean space E3 is determined with respect to a fixed orthonormal basis $\mathbf{i}_j = \mathbf{i}^j$ by the position vector

$$\mathbf{x} = x^i(\Theta^1, \Theta^2, \Theta^3)\,\mathbf{i}_i = \mathbf{x}(\Theta^1, \Theta^2, \Theta^3) \ , \qquad \text{(A1.2.1)}$$

where Θ^i are *curvilinear* coordinates. The above relation defines in E3 three sets of coordinate lines, the so-called Θ^i-curves, each of them obtainable by holding two of the parameters Θ^i constant. The *covariant* $\mathbf{g}_i$ and the *contravariant* base vectors $\mathbf{g}^i$ are defined by

$$\mathbf{g}_i = \mathbf{x}_{,i} = \frac{\partial x^j}{\partial \Theta^i}\,\mathbf{i}_j \;, \quad \mathbf{g}^i = \frac{\partial \Theta^i}{\partial x_j}\,\mathbf{i}_j \;, \tag{A1.2.2}$$

where, evidently, the summation rule holds for the index j. We then have

$$\mathbf{i}_i = \mathbf{i}^i = \frac{\partial \Theta^j}{\partial x^i}\,\mathbf{g}_j = \frac{\partial x^i}{\partial \Theta^j}\,\mathbf{g}^j \;, \tag{A1.2.3}$$

and also

$$\mathbf{g}_i = g_{ir}\,\mathbf{g}^r \;, \quad \mathbf{g}^i = g^{ir}\,\mathbf{g}_r \;, \quad \mathbf{g}_i \cdot \mathbf{g}^j = \delta_i^j \;, \tag{A1.2.4}$$

where

$$g_{ij} = \mathbf{g}_i \cdot \mathbf{g}_j = \frac{\partial x^m}{\partial \Theta^i}\frac{\partial x^n}{\partial \Theta^j}\,\delta_{mn} = \frac{\partial x^m}{\partial \Theta^i}\frac{\partial x^m}{\partial \Theta^j} \;,$$

$$g^{ij} = \mathbf{g}^i \cdot \mathbf{g}^j = \frac{\partial \Theta^i}{\partial x^m}\frac{\partial \Theta^j}{\partial x^n}\,\delta^{mn} = \frac{\partial \Theta^i}{\partial x^m}\frac{\partial \Theta^j}{\partial x^m} \;,$$

$$g_i^j = g_{im}\,g^{mj} = \delta_i^j \;. \tag{A1.2.5}$$

The components g_{ij} and g^{ij} are called covariant and contravariant components of the so-called *metric tensor* (*identity tensor*) denoted in chapter 1 by **I**.

We also introduce by considering (A1.2.5) the determinant g and its inverse g^{-1}

$$g = \left|g_{ij}\right| = \left|\frac{\partial x^i}{\partial \Theta^j}\right|^2 , \quad \frac{1}{g} = \left|g^{ij}\right| = \left|\frac{\partial \Theta^i}{\partial x^j}\right|^2 , \tag{A1.2.6}$$

which can be also evaluated according to

$$\sqrt{g} = [\mathbf{g}_1\;\mathbf{g}_2\;\mathbf{g}_3] \;, \quad \frac{1}{\sqrt{g}} = [\mathbf{g}^1\;\mathbf{g}^2\;\mathbf{g}^3] \;. \tag{A1.2.7}$$

From (A1.2.2) it is apparent that the covariant base vectors $\mathbf{g}_i$ are tangential to the coordinate lines Θ^i at any point P of E3 (Fig. A1.1). The last equation in (A1.2.4) indicates that $\mathbf{g}^i$ is perpendicular to $\mathbf{g}_j$ ($j \neq i$): an important characteristic of the two sets of base vectors $\mathbf{g}_i$ and $\mathbf{g}^i$ associated with each point in E3.

It is also useful to introduce unit base vectors in direction of $\mathbf{g}_i$ and $\mathbf{g}^i$

$$\mathbf{g}_{<i>} = \frac{\mathbf{g}_i}{\sqrt{g_{ii}}} \;, \quad \mathbf{g}^{<i>} = \frac{\mathbf{g}^i}{\sqrt{g^{ii}}} \;, \tag{A1.2.8}$$

which will be used to define physical components of vectors. The bracket < > is used to point out that $\mathbf{g}_{<i>}$ and $\mathbf{g}^{<i>}$ are not covariant and contravariant vectors, respectively.

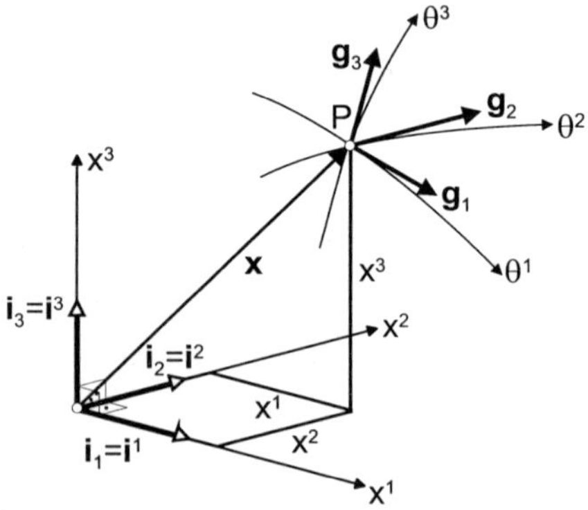

Fig. A1.1. Orthonormal reference frame $\mathbf{i}_i$, covariant base vectors $\mathbf{g}_i$

We close this section by adding the following definitions:

vectorial line element: $\mathbf{ds}_{<i>} = \mathbf{g}_i \, d\Theta^i$, (no summation over i) (A1.2.9)

magnitude of $\mathrm{ds}_{<i>}$: $\mathrm{ds}_{<i>} = \|\mathbf{ds}_{<i>}\| = \sqrt{g_{ii}}\, d\Theta^i$, (A1.2.10)

surface element: $dA_i = \sqrt{g\, g^{ii}}\, d\Theta^j\, d\Theta^k$, $(i \neq j \neq k)$ (A1.2.11)

volume element: $dV = [\mathbf{ds}_{<1>}\, \mathbf{ds}_{<2>}\, \mathbf{ds}_{<3>}] = \sqrt{g}\, d\Theta^1\, d\Theta^2\, d\Theta^3$. (A1.2.12)

Note that dV is the volume of the parallelepiped defined by the vectorial line elements $\mathbf{ds}_{<i>}$ (i = 1, 2, 3), while dA_i denotes the area of the parallelogram defined by $\mathbf{ds}_{<j>}$ and $\mathbf{ds}_{<k>}$ $(i \neq j \neq k)$. The unit vector normal to dA_i is $\mathbf{g}^{<i>} = \mathbf{g}^i/\sqrt{g^{ii}}$. The definition of dA_i and dV is illustrated in Fig. A1.2.

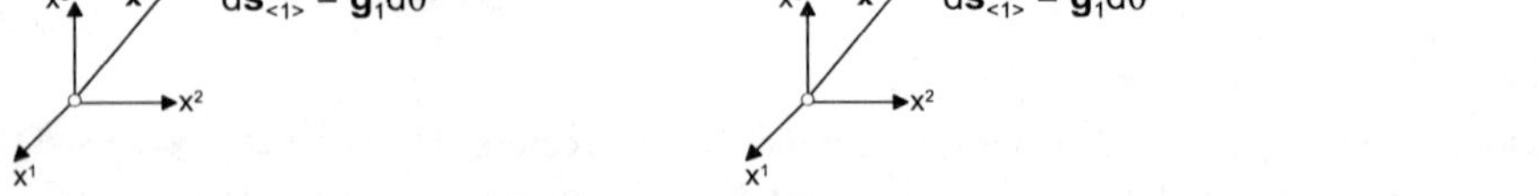

Fig. A1.2. Surface elements dA_i, volume element dV

A1.3 Vector decompositions, tensor components of first order

Let $\mathbf{u} = \mathbf{u}(\Theta^i)$ be a vector field in E3 representing e.g. a displacement field. Tensorial components of $\mathbf{u}$ are defined by

$$\mathbf{u} = u_i \mathbf{g}^i = u^i \mathbf{g}_i \,, \tag{A1.3.1}$$

where, in view of (A1.2.4), the following relations hold for the covariant and contravariant components u_i and u^i, respectively:

$$u_i = g_{ij} u^j \,, \quad u^i = g^{ij} u_j \,. \tag{A1.3.2}$$

It is apparent that the above transformations are similar to those given in (A1.2.4) for the base vectors $\mathbf{g}_i$ and $\mathbf{g}^i$ and define the rule of *lowering and raising of indices* which, in turn, can be extended to tensor components of arbitrary order. Examples are

$$A_{ij} g^{ik} = A^k_{\cdot j} \,, \quad A^{ij} g_{ik} = A_k^{\cdot j} \,, \quad A_{ij} g^{im} g^{jn} = A^{mn} \,. \tag{A1.3.3}$$

Equation (A1.3.1) involves two equivalent representations of the vector $\mathbf{u}$ in terms of the covariant and contravariant components. We also see that covariant components u_i refer to the contravariant basis $\mathbf{g}^i$ and vice versa. By using unit base vectors defined in (A1.2.8) we now introduce the physical components of $\mathbf{u}$ (notation with < >):

$$\mathbf{u} = u_{<i>} \mathbf{g}^{<i>} = u^{<i>} \mathbf{g}_{<i>} \,, \tag{A1.3.4}$$

which are, in view of (A1.2.8) and (A1.3.1), related to tensorial components by

$$u_{<i>} = u_i \sqrt{g^{ii}} \,, \quad u^{<i>} = u^i \sqrt{g_{ii}} \,. \tag{A1.3.5}$$

A1.4 Definition of higher-order tensor components

We suppose that curvilinear coordinates Θ^i are transformed into a new set of coordinates $\bar{\Theta}^i$ and that the correspondance between the coordinate systems Θ^i and $\bar{\Theta}^i$ is one to one with $|\partial\Theta^i/\partial\bar{\Theta}^j| > 0$:

$$\bar{\Theta}^i = \bar{\Theta}^i(\Theta^1, \Theta^2, \Theta^3) \,, \quad \Theta^i = \Theta^i(\bar{\Theta}^1, \bar{\Theta}^2, \bar{\Theta}^3) \,. \tag{A1.4.1}$$

If we evaluate the base vectors $\bar{\mathbf{g}}_i$ and $\bar{\mathbf{g}}^i$ associated with the new coordinates Θ^i according to (A1.2.2), we see by means of the chain rule of differentiation that

$$\bar{\mathbf{g}}_i = \frac{\partial\Theta^j}{\partial\bar{\Theta}^i} \mathbf{g}_j \,, \quad \mathbf{g}_i = \frac{\partial\bar{\Theta}^j}{\partial\Theta^i} \bar{\mathbf{g}}_j \,, \tag{A1.4.2}$$

$$\bar{\mathbf{g}}^i = \frac{\partial\bar{\Theta}^i}{\partial\Theta^j} \mathbf{g}^j \,, \quad \mathbf{g}^i = \frac{\partial\Theta^i}{\partial\bar{\Theta}^j} \bar{\mathbf{g}}^j \,. \tag{A1.4.3}$$

The first two equations define the *covariant transformation* and the last two ones the *contravariant transformation.* These rules will be used to define tensorial components of arbitrary order.

Let A_i be a set of three components defined in the coordinate system Θ^i and let $\bar{A}_i$ be the value of this set of components in a new coordinate system $\bar{\Theta}^i$. The components A_i are said to form *covariant tensor components* if the transformation

$$\bar{A}_i = \frac{\partial \Theta^j}{\partial \bar{\Theta}^i} A_j , \quad A_i = \frac{\partial \bar{\Theta}^j}{\partial \Theta^i} \bar{A}_j \tag{A1.4.4}$$

holds for arbitrary admissible coordinate transformation $\Theta^i \rightarrow \bar{\Theta}^i$. The above rules are similar to those holding for the covariant basis $\mathbf{g}_i$. Similarly, the A^i are said to form *contravariant tensor components* if the transformation rule is given, analogous to (A1.4.3) by:

$$\bar{A}^i = \frac{\partial \bar{\Theta}^i}{\partial \Theta^j} A^j , \quad A^i = \frac{\partial \Theta^i}{\partial \bar{\Theta}^j} \bar{A}^j . \tag{A1.4.5}$$

Note that, in the above transformations (A1.4.2) to (A1.4.5), the summation rule holds for repeated indices.

By using (A1.4.2) and (A1.4.3), it can easily be deduced from (A1.3.1) that the components of a vector are tensorial components of first order.

The above definitions can be extended to introduce tensor components of arbitrary order. We give some examples:

$$\bar{A}_{ij} = \frac{\partial \Theta^m}{\partial \bar{\Theta}^i} \frac{\partial \Theta^n}{\partial \bar{\Theta}^j} A_{mn} , \qquad A^{ij} = \frac{\partial \Theta^i}{\partial \bar{\Theta}^m} \frac{\partial \Theta^j}{\partial \bar{\Theta}^n} \bar{A}^{mn} ,$$

$$\bar{A}^{i}_{\cdot j} = \frac{\partial \bar{\Theta}^i}{\partial \Theta^m} \frac{\partial \Theta^n}{\partial \bar{\Theta}^j} A^{m}_{\cdot n} , \qquad A^{i}_{\cdot j} = \frac{\partial \Theta^i}{\partial \bar{\Theta}^m} \frac{\partial \bar{\Theta}^n}{\partial \Theta^j} \bar{A}^{m}_{\cdot n} ,$$

$$\bar{A}^{ij} = \frac{\partial \bar{\Theta}^i}{\partial \Theta^m} \frac{\partial \bar{\Theta}^j}{\partial \Theta^n} A^{mn} , \qquad A^{ij} = \frac{\partial \Theta^i}{\partial \bar{\Theta}^m} \frac{\partial \Theta^j}{\partial \bar{\Theta}^n} \bar{A}^{mn} , \tag{A1.4.6}$$

where

$$A^{m}_{\cdot n} = g^{mi} A_{in} , \qquad A^{mn} = g^{mi} g^{nj} A_{ij} . \tag{A1.4.7}$$

A scalar-valued variable is said to form a *zero-order tensor* or an *invariant scalar,* if it remains unchanged under all possible coordinate transformations: $A(\Theta^i) = A(\bar{\Theta}^i)$. An example is the temperature distribution $T = T(\Theta^i)$ in a body: evidently, the value of the temperature in any point of the body is independent of the selection of the coordinate system.

The following statement can be deduced from the above transformations which are useful to identify the tensorial property of an indexed quantity. If the tensor components are zero in a special coordinate system, they also vanish in any other coordinate system. This means that, the vanishing of tensor components is an *invariant* property.

Tensorial operations are understood as those rules which, starting from components identified as tensorial ones, lead always to new tensor components. In this sense, the following operations are allowed:

$$\text{addition:} \qquad A_{\alpha\beta} + B_{\alpha\beta} = C_{\alpha\beta} \;, \tag{A1.4.8}$$

$$\text{tensor product:} \qquad A_{\alpha\beta}\, B^{\rho\lambda} = C_{\alpha\beta}{}^{\cdot\cdot\rho\lambda} \;, \tag{A1.4.9}$$

$$\text{contraction:} \qquad A_{\alpha\rho}\, B^{\rho\lambda} = C_{\alpha}{}^{\cdot\lambda} \;, \qquad A_{\alpha\beta}\, B^{\alpha\beta} = C \;. \tag{A1.4.10}$$

The above operations mean that, if the participant components of the left-hand side are known to be tensor components, the components defined on the right-hand side are necessarily also tensor components.

A1.5 Permutation tensor

The components of the *permutation tensor* ε_{ijk} and ε^{ijk} referring to curvilinear coordinates Θ^i are defined by using the determinant g introduced in (A1.2.6) and the components e_{ijk} and e^{ijk} defined in (A1.1.4) as follows:

$$\varepsilon_{ijk} = \sqrt{g}\; e_{ijk} \;, \qquad \varepsilon^{ijk} = \frac{1}{\sqrt{g}}\, e^{ijk} \;, \tag{A1.5.1}$$

where both types of components are, as usual, related to each other by the process of lowering and raising of indices:

$$\varepsilon_{ijk} = g_{im}\, g_{jn}\, g_{kr}\, \varepsilon^{mnr} \;, \qquad \varepsilon^{ijk} = g^{im}\, g^{jn}\, g^{kr}\, \varepsilon_{mnr} \;. \tag{A1.5.2}$$

By considering (A1.2.6) or (A1.2.7) it can be deduced from (A1.5.1) that

$$\varepsilon_{ijk} = \frac{\partial x^m}{\partial \Theta^i}\frac{\partial x^n}{\partial \Theta^j}\frac{\partial x^r}{\partial \Theta^k}\, e_{mnr} \;, \qquad \varepsilon^{ijk} = \frac{\partial \Theta^i}{\partial x^m}\frac{\partial \Theta^j}{\partial x^n}\frac{\partial \Theta^k}{\partial x^r}\, e^{mnr} \;. \tag{A1.5.3}$$

Consequently, ε_{ijk} (ε^{ijk}) are tensor components and e_{ijk} (e^{ijk}) denote their values in an orthogonal Cartesian coordinate system (g = 1).

For later use we summarizes useful identities (Green and Zerna 1968)

$$\mathbf{g}_r \times \mathbf{g}_s = \varepsilon_{rst}\, \mathbf{g}^t \;, \qquad \mathbf{g}^r \times \mathbf{g}^s = \varepsilon^{rst}\, \mathbf{g}_t \;, \tag{A1.5.4}$$

$$\varepsilon_{rst} = [\mathbf{g}_r\, \mathbf{g}_s\, \mathbf{g}_t] \;, \qquad \varepsilon^{rst} = [\mathbf{g}^r\, \mathbf{g}^s\, \mathbf{g}^t] \;, \tag{A1.5.5}$$

$$\sqrt{g} = [\mathbf{g}_1\, \mathbf{g}_2\, \mathbf{g}_3] \;, \qquad \frac{1}{\sqrt{g}} = [\mathbf{g}^1\, \mathbf{g}^2\, \mathbf{g}^3] \;, \tag{A1.5.6}$$

$$\varepsilon^{rst}\,\varepsilon^{ijk}\,g_{ri}\,g_{sj}\,g_{tk} = 6\ , \qquad \varepsilon_{rst}\,\varepsilon_{ijk}\,g^{ri}\,g^{sj}\,g^{tk} = 6\ , \tag{A1.5.7}$$

$$\varepsilon^{rst}\,\varepsilon^{ijk}\,g_{sj}\,g_{tk} = 2g^{ri}\ , \qquad \varepsilon_{rst}\,\varepsilon_{ijk}\,g^{sj}\,g^{tk} = 2g_{ri}\ , \tag{A1.5.8}$$

which may be verified by observing that they are tensorial relations which are clearly satisfied in orthogonal Cartesian coordinate systems x^i. In particular, equations (A1.5.4) show that ε_{ijk} and ε^{ijk} are tensor components permitting to express vector products of the base vectors in tensorial formulation.

A1.6 Christoffel symbols, covariant differentiation

The partial derivatives of the base vectors $\mathbf{g}_i$ and $\mathbf{g}^i$ can be expressed as

$$\mathbf{g}_{i,j} = \Gamma_{ijk}\,\mathbf{g}^k = \Gamma^k_{ij}\,\mathbf{g}_k\ , \tag{A1.6.1}$$

$$\mathbf{g}^i{}_{,j} = -\Gamma^i_{jk}\,\mathbf{g}^k\ , \tag{A1.6.2}$$

where the symbols Γ_{ijk} and Γ^k_{ij} are called *Christoffel symbols of the first* and *second kind*, respectively. From the above definitions we see that

$$\Gamma_{ijk} = \Gamma_{jik} = \mathbf{g}_{i,j}\cdot\mathbf{g}_k = \mathbf{g}_{j,i}\cdot\mathbf{g}_k\ , \tag{A1.6.3}$$

$$\Gamma^k_{ij} = \Gamma^k_{ji} = \mathbf{g}_{i,j}\cdot\mathbf{g}^k = \mathbf{g}_{j,i}\cdot\mathbf{g}^k = -\,\mathbf{g}_i\cdot\mathbf{g}^k{}_{,j} = -\,\mathbf{g}_j\cdot\mathbf{g}^k{}_{,i}\ , \tag{A1.6.4}$$

and by considering these results, we may also write

$$\Gamma_{ijk} = \frac{1}{2}\,(g_{ik,j} + g_{kj,i} - g_{ij,k})\ , \tag{A1.6.5}$$

$$\Gamma^k_{ij} = \Gamma_{ijr}\,g^{rk} = \frac{1}{2}\,g^{rk}\,(g_{ir,j} + g_{rj,i} - g_{ij,r})\ , \tag{A1.6.6}$$

$$\Gamma^i_{ij} = \frac{1}{2}\,g^{ir}\,g_{ir,j} = \frac{1}{2g}\,\frac{\partial g}{\partial g_{is}}\,\frac{\partial g_{is}}{\partial \Theta^j} = \frac{1}{\sqrt{g}}\,\frac{\partial\sqrt{g}}{\partial\Theta^j}\ . \tag{A1.6.7}$$

Note that the last result is based on the identity

$$\frac{\partial g}{\partial g_{ij}} = g\,g^{ij}\ . \tag{A1.6.8}$$

Equations (A1.6.3) and (A1.6.4) indicate that both symbols Γ_{ijk} and Γ^k_{ij} are symmetric with respect to the indices i and j, and equation (A1.6.6) shows that they are transformable into each other by the process of lowering and raising of indices. We finally note that the Christoffel symbols are not tensor components. The transformation for Γ^k_{ij} reads as

$$\bar{\Gamma}^k_{ij} = \frac{\partial\Theta^r}{\partial\bar{\Theta}^i}\,\frac{\partial\Theta^s}{\partial\bar{\Theta}^j}\,\frac{\partial\bar{\Theta}^k}{\partial\Theta^t}\,\Gamma^t_{rs} + \frac{\partial^2\Theta^r}{\partial\bar{\Theta}^i\,\partial\bar{\Theta}^j}\,\frac{\partial\bar{\Theta}^k}{\partial\Theta^r} \tag{A1.6.9}$$

confirming clearly this statement (Sokolnikoff 1964).

Covariant differentiation to be denoted by a vertical line can be regarded as a generalization of the partial differentiation if curvilinear coordinates Θ^i are used. In the particular case of orthogonal Cartesian coordinates x^i both kinds of differentiation are identical. By definition, the covariant derivative of an invariant scalar Φ is identical with its partial derivative

$$\Phi|_k = \Phi,_k = \frac{\partial \Phi}{\partial \Theta^k} . \tag{A1.6.10}$$

For vector components u_i and u^i it is defined by

$$u_i|_j = u_{i,j} - \Gamma^k_{ij} u_k , \quad u^i|_j = u^i_{,j} + \Gamma^i_{kj} u^k , \tag{A1.6.11}$$

where terms with Christoffel symbols are due to the change of the base vectors $\mathbf{g}_i$ and $\mathbf{g}^i$ used in the decompositions $\mathbf{u} = u_i \mathbf{g}^i = u^i \mathbf{g}_i$. The above rules can be extended to arbitrary tensor components, e.g.:

$$A_{ij}|_r = A_{ij,r} - \Gamma^m_{ir} A_{mj} - \Gamma^m_{jr} A_{im} ,$$

$$A^i_{\cdot j}|_r = A^i_{\cdot j,r} + \Gamma^i_{rm} A^m_{\cdot j} - \Gamma^m_{jr} A^i_{\cdot m} ,$$

$$A^{ij}|_r = A^{ij}_{,r} + \Gamma^i_{rm} A^{mj} + \Gamma^j_{rm} A^{im} . \tag{A1.6.12}$$

The rules (A1.6.11) can be also used to define the covariant derivatives of covariant vectors $\mathbf{t}_i = g_{ir} \mathbf{t}^r$ and contravariant ones $\mathbf{t}^i = g^{ir} \mathbf{t}_r$, thus:

$$\mathbf{t}_i|_j = \mathbf{t}_{i,j} - \Gamma^k_{ij} \mathbf{t}_k , \quad \mathbf{t}^i|_j = \mathbf{t}^i_{,j} + \Gamma^i_{kj} \mathbf{t}^k . \tag{A1.6.13}$$

By means of these definitions and (A1.6.1), (A1.6.2) we see that

$$\mathbf{g}_i|_k = \mathbf{g}^i|_k = \mathbf{0} . \tag{A1.6.14}$$

Further variables which can be treated as constants in forming covariant derivatives are:

$$g_{ij}|_r = g^{ij}|_r = \delta^i_j|_r = 0 , \quad \varepsilon_{ijk}|_r = \varepsilon^{ijk}|_r = 0 . \tag{A1.6.14}$$

The above results known as *Ricci-lemma* can be easily confirmed specifying them to the orthogonal Cartesian coordinates x^i and remembering that the vanishing of tensor components is an invariant property.

The covariant differentiation is a rule which upon applying to tensor components conserves their tensorial properties. Consequently, the results obtained by application of the rules (A1.6.11) and (A1.6.12) are all tensorial components. This is, however, the case for the usual partial differentiation only for invariant scalar-valued functions which justifies the definition (A1.6.10). Finally, we notice that the order of covariant differentiation is irrelevant if the space is Euclidean, as in our case:

$$A_r|_{st} = A_r|_{ts} \ , \qquad A^r|_{ts} = A^r|_{st} \ . \tag{A1.6.15}$$

Similar property holds for tensor components of arbitrary order.

References

Abraham R, Marsden JE, Ratiu T (1988) Manifolds, Tensor Analysis and Applications. Springer-Verlag, New York, 2nd edition

Antman SS (1995) Nonlinear Problems of Elasticity. Applied Mathematical Sciences 107, Springer-Verlag, New York

Argyris J (1982) An Excursion into Large Rotations. Comp. Meth. Appl. Mech. Eng. 32, 85-155

Ball JM (1977) Convexity Conditions and Existence Theorems in Non-Linear Elasticity, Arch Rat. Mech. An. 63, 337-403

Barthold F-J (1993) Theorie und Numerik zur Berechnung und Optimierung von Strukturen aus isotropen, hyperelastischen Materialien. Forschungs- und Seminarberichte aus dem Bereich der Mechanik der Universität Hannover, Bericht-Nr. F 93/2

Başar Y, Itskov M (1999) Constitutive Model and Finite Element Formulation for Large Strain Elasto-Plastic Analysis of Shells. Computational Mechanics 23, 466-481

Başar Y, Itskov M (1998) Finite Element Formulation of the Odgen Material Model with Application to Rubber-Like Shells. Int. J. Numer. Meth. Engng. 42, 1279-1305

Başar Y, Ding Y (1997) Shear Deformation Models for Large-Strain Shell Analysis. Int. J. Solids Structures, Vol. 34, 1687-1708

Başar Y, Ding Y (1996) Finite-Element Analysis of Hyperelastic Thin Shells with Large Strains. Comp. Mech. 18, 200-214

Başar Y (1987) A Consistent Theory of Geometrically Non-Linear Shells with an Independent Rotation Vector. Int. J. Solids Structures 23, No. 10, 1401-1415

Başar Y (1986) Eine konsistente Theorie für Flächentragwerke endlicher Verformungen und deren Operatordarstellung auf variationstheoretischer Grundlage. Z. angew. Math. Mech. 66, 297-308

Başar Y (1986) Zur Struktur konsistenter inkrementeller Theorien für geometrisch nichtlineare Flächentragwerke und deren Operatordarstellung. Ingenieur-Archiv 56, 209-220

Başar Y, Krätzig WB (1985) Mechanik der Flächentragwerke. Vieweg Verlag, Braunschweig

Becker E, Bürger W (1975) Kontinuumsmechanik. B.G. Teubner, Stuttgart

Betten J (1993) Kontinuumsmechanik. Springer-Verlag, Berlin

Betten J (1987) Tensorrechnung für Ingenieure. B.G. Teubner Verlag

Betten J (1984) Interpolation Methods for Tensor Functions. In: Mathematical Modelling in Science and Technology. Eds. Avula, X.J.R., Pergamon Press, New York, 52-57

Blatz PJ, Ko WL (1962) Application of Finite Elasticity Theory to the Deformation of Rubbery Materials. Trans. Soc. Rheology 6, 223-251

Boer R de (1982) Vektor- und Tensorrechnung für Ingenieure. Springer-Verlag, Berlin

Ball JM (1977) Convexity Conditions and Existence Theorems in Nonlinear Elasticity. Archive for Rational Mechanics and Analysis 63, 337-403

Bonnet J, Wood RD (1998) Nonlinear Continuum Mechanics for Finite Element Analysis. Cambridge University Press

Bruhns OT (1992) Kontinuumsmechanik. Arbeitsheft zur Vorlesung, Ruhr-Universität Bochum

Büchter N (1992) Zusammenführung von Degenerationskonzept und Schalentheorie bei endlichen Rotationen. Bericht Nr. 14, Institut für Baustatik, Universität Stuttgart

Büchter N, Ramm E (1992) Shell Theory Versus Degeneration – A Comparison in Large Rotation Finite Element Analysis. Int. J. Num. Meth. Eng. 34, 39-59

Bufler H (1985) The Biot Stresses in Nonlinear Elasticity and the Associated Generalized Variational Principles. Ing. Archiv 55, 450-462

Chadwick P, Ogden RW (1971) A Theorem of Tensor Calculus and its Application to Isotropic Elasticity. Archive for Rational Mechanics and Analysis 44, 54-68

Chadwick P, Ogden RW (1971) On the Definition of Elastic Moduli. Archive for Rational Mechanics and Analysis 44, 41-53

Chang TY, Saleeb AF, Li G (1991) Large Strain Analysis of Rubber-Like Materials Based on a Pertubed Lagrangian Variational Principle, Computational Mechanics 8, 221-233

Ciarlet PG (1983) Lectures on Three Dimensional Elasticity. Springer-Verlag, Berlin

Ding Y (1989) Finite-Rotations-Elemente zur geometrisch nichtlinearen Analyse allgemeiner Flächentragwerke. Technisch Wissenschaftliche Mitteilungen Nr. 89-6. Institut für Konstruktiven Ingenieurbau der Ruhr-Universität Bochum

Doyle TC, Ericksen JL (1956) Nonlinear Elasticity. In: Advances in Appl. Mech. IV, Academic Press, New York, 53-115

Eberlein R, Wriggers P, Taylor RL (1993) A Fully Non-linear Axisymmetrical Quasi-Kirchhoff-Type Shell Element for Rubber-Like Materials. Int. J. Numer. Meth. Engng. 36, 4027-4043

Eckstein A (1999) Zur Theorie und Finite-Element-Simulation von Schalen mit großen inelastischen Dehnungen und duktilen Schädigungen. Dissertation, Technisch-wissenschaftliche Mitteilungen Nr. 99-3. Institut für konstruktiven Ingenieurbau, Ruhr-Universität Bochum

Eringen AC (1967) Mechanics of Continua, John Wiley & Sons, New York

Flory PJ (1961) Thermodynamic Relations for High Elastic Materials. Trans. Faraday. Soc. 57, 829-838

Fung YC (1965) Foundations of Solids Mechanics. Prentice Hall Inc., New Jersey

Green AE, Adkins JE (1970) Large Elastic Deformations. Oxford University Press, Second Edition

Green AE, Zerna W (1968) Theoretical Elasticity. Clarendon Press, Oxford, second edition

Green AE, Adkins JE (1960) Large Elastic Deformations and Non-Linear Continuum Mechanics. At the Clarendon Press

Gruttmann F (1988) Theorie und Numerik schubelastischer Schalen mit endlichen Drehungen unter Verwendung der Biot-Spannungen. Forschungs- und Seminarberichte aus dem Bereich der Mechanik der Universität Hannover, Bericht-Nr. F 88/1

Gruttmann F, Taylor RL (1992) Theory and Finite Element Formulation of Rubberlike Membrane Shells Using Principal Stretches. Int. J. Num. Meth. Engng. 35, 1111-1126

Hamel G (1967) Theoretische Mechanik. Springer-Verlag, Berlin, Heidelberg, New York

Hill R (1957) On Uniqueness and Stability in the Theory of Finite Elastic Strain. Journal of the Mechanics and Physics of Solids 5, 229-241

Hoger A, Carlson DE (1984) Determination of the Stretch and Rotation in the Polar Decomposition of the Deformation Gradient. Quart. Appl. Math. 42, 113-117

Hughes TJR, Pister KS (1978) Consistent Linearization in Mechanics of Solids and Structures. Computers & Structures 8, 391-397

Itskov M, Başar Y (1999) A generalised hyperelastic orthotropic constitutive model for reinforced rubber-like materials. First European Conference on Constitutive Models for Rubber, ECCMR '99, Vienna (Austria), 9-10 September 1999. Balkema Publishers (in press)

Itskov M, Başar Y (1999) On the theory of fourth-order tensors and their application in large strain elasticity. GAMM 99, Annual Meeting, Metz (France) 12-16 April 1999, Z. angew. Math. Mech. (in press)

Karasudhi P (1991) Foundations of Solids Mechanics. Kluver Academic Publishers, Dordrecht

Khan AS, Huang S (1995) Continuum Theory of Plasticity. John Wiley & Sons, New York

Krätzig, W.B (1993) "Best" Transverse Shearing and Stretching Shell Theory for Nonlinear Finite Element Simulations. Comp. Meth. Appl. Mech. Eng. 103, 135-160

Krätzig WB (1971) Thermodynamics of Deformations and Shell Theory. Technisch-wissenschaftliche Mitteilungen Nr. 71-3. Institut für Konstruktiven Ingenieurbau der Ruhr-Universität Bochum

Lammering R (1988) Beiträge zur Theorie und Numerik großer plastischer und kleiner elastischer Deformationen mit Schädigungseinfluß. Forschungs- und Seminarberichte aus dem Bereich der Mechanik der Universität Hannover, Bericht-Nr. F 88/3

Lemaitre J, Chaboche J-L (1990) Mechanics of Solid Materials. Cambridge University Press

Liu CH, Mang HA (1996) A Critical Assessment of Volumetric Strain Energy Functions for Hyperelasticity at Large Strains. Z. angew. Math. Mech. 76 (S5), 305-306

Lubliner J (1990) Plasticity Theory. Macmillan Publishing Company, New York

MacVean DB (1968) Die Elementarbeit in einem Kontinuum und die Zuordnung von Spannungs- und Verzerrungstensoren. ZAMP 19, 157-185

Malvern LE (1969) Introduction to the Mechanics of a Continuous Medium. Prentice Hall, Englewood Cliffs, New Jersey

Marsden JE, Hughes TJR (1983) Mathematical Foundations of Elasticity. Prentice Hall, Inc., Englewood Cliffs, New Jersey

Mason J (1980) Variational, Incremental and Energy Methods in Solid Mechanics and Shell Theory. Studies in Applied Mechanics 4, Elsevier Scientific Publishing Company, Amsterdam, Oxford, New York

Menzel W (1996) Gemischt-hybride Elementformulierungen für komplexe Schalenstrukturen unter endlichen Rotationen. Technisch-wissenschaftliche Mitteilungen Nr. 96-4. Institut für Konstruktiven Ingenieurbau der Ruhr-Universität Bochum

Meschke G, Liu WN (1997) A Study on the Significance of the Chosen Stress Measure in Finite Strain Plasticity. Z. angew. Math. Mech. 77, S219-S220

Miehe C (1997) Comparison of two Algorithms for the Computation of Fourth-Order Isotropic Tensor Functions. Computer and Structures 66, 37-43

Miehe C (1994) Aspects of the Formulation and Finite Element Implementation of Large Strain Isotropic Elasticity. International Journal for Numerical Methods in Engineering 37, 1981-2004

Miehe C (1993) Computation of Isotropic Tensor Functions. Communication in Numerical Methods in Engineering 9, 889-896

Mooney M (1940) A Theory of Elastic Deformations. Journal of Applied Physics 11, 582-598

Morman KN (1986) The Generalized Strain Measure with Application to Non Homogeneous Deformations in Rubber-Like Solids. J. Appl. Mech. 53, 726-728

Müller-Hoeppe N (1990) Beiträge zur Theorie und Numerik finiter inelastischer Deformationen. Forschungs- und Seminarberichte aus dem Bereich der Mechanik der Universität Hannover, Bericht-Nr. F 90/4

Naghdi PM (1972) The Theory of Shells and Plates. Handbuch der Physik. Band VI a/2: Festkörpermechanik II. Springer-Verlag Berlin, Heidelberg, New York, 425-640

Ogden RW (1984) Nonlinear Elastic Deformations. Ellis Horwood and John Wiley, Chichester

Ogden RW (1974) On Isotropic Tensors and Elastic Moduli, Proceedings of the Cambridge Philosophical Society 75, 427-436

Ogden RW (1972) Large Deformation Isotropic Elasticity: On the Correlation of Theory and Experiment for Incompressible Rubberlike Solids. Proceedings of the Royal Society of London, Series A 326, 565-584

Oldroyd JG (1950) Finite Strains in an Anisotropic Elastic Continuum. Proc. Roy. Soc. Lond., A 202, 407-419

Pietraszkiewicz W (1989) Geometrically Nonlinear Theories of Thin Elastic Shells. Advances in Mechanics 12, 52-130

Pietraszkiewicz W, Badur J (1983) Finite Rotations in the Description of Continuum Deformation. Int. J. Engng Sci. 21, No. 9, 1097-1115

Pietraszkiewicz W (1979) Finite Rotations and Lagrangian Description in the Non-Linear Theory of Shells, Polish Scientific Publishers, Warszawa-Poznan

Prager W (1961) Introduction of Mechanics of Continua. Ginn and Co, Boston

Reese S, Wriggers P (1997) A Material Model for Rubber-Like Polymers Exhibiting Plastic Deformation: Computational Aspects and a Comparison with Experimental Results. Computer Methods in Applied Mechanics and Engineering 148, 279-298

Rivlin RS, Sawyers KN (1976) The Strain-Energy Function for Elastomers.Transactions of the Society of Rheology 20(4), 545-557

Rivlin RS, Saunders DW (1951) Phil. Trans. Royal Soc. London A, 243-251ff

Rivlin, RS (1949) Proc. Camb. Phil. Soc., 45: 485ff

Roberts AJ (1994) A One-Dimensioned Introduction to Continuum Mechanics. World Scientific Publishing, Singapore

Saleeb AF, Chang TYP, Arnold SM (1992) On the Development of Explicit Robust Schemes for Implementation of a Class of Hyperelastic Models in Large-Strain Analysis of Rubbers. Int. J. Num. Meth. Engng. 33, 1237-1249

Sansour C (1992) Auf der polaren Zerlegung basierende Schalentheorie endlicher Rotationen und ihre Finite-Element Diskretisierung. Institut für Mechanik (Bauwesen) der Universität Stuttgart

Sansour C, Bufler H (1992) An Exact Finite Rotation Shell Theory, its Mixed Variational Formulation and its Finite Element Implementation. Int. J. Num. Meth. Eng. 34, 73-115

Schieck B, Pietraskiewicz W, Stumpf H (1992) Theory and Numerical Analysis of Shells Undergoing Large Elastic Strains. Int. J. Solids Structures 29, 689-709

Schieck B (1989) Große elastiche Dehnungen in Schalen aus hyperelastischen inkompressiblen Materialien. Mitteilungen aus dem Institut für Mechanik 69, Ruhr-Universität Bochum

Seth BR (1964) Generalized Strain Measure with Applications to Physical Problems. In: Second-Order-Effects in Elasticity, Plasticity and Fluid Dynamics, Eds. Reiner, M., Abir, D, Pergamon Press, Oxford, 162-172

Simo JC, Miehe C (1992) Associative Coupled Thermoplasticity at Finite Strains, Numerical Analysis and Implementation. Comp. Meth. Appl. Mech. Eng. 108, 319-339

Simo JC, Fox DD (1989) On a Stress Resultant Geometrically Exact Shell Model. Part I: Formulation and Optimal Parametrization. Comp. Meth. Appl. Mech. Eng. 72, 267-304

Simo JC, Fox DD, Rifai MS (1989) On a Stress Resultant Geometrically Exact Shell Model. Part II: The Linear Theory; Computational Aspects. Comp. Meth. Appl. Mech. Eng. 73, 53-92

Simo JC, Fox DD, Rifai MS (1990) On a Stress Resultant Geometrically Exact Shell Model. Part III: Computational Aspects of the Nonlinear Theory. Comp. Meth. Appl. Mech. Eng. 79, 21-70

Simo JC, Rifai MS, Fox DD (1990) On a Stress Resultant Geometrically Exact Shell Model. Part IV: Variable Thickness Shells with Through-the-thickness Stretching. Comp. Meth. Appl. Mech. Eng. 81, 91-126

Simo JC, Taylor RL (1991) Quasi-Incompressible Finite Elasticity in Principle Stretches. Continuum Basis and Numerical Algorithms. Comp. Meth. Appl. Mech. Eng. 85, 273-310

Simo JC, Taylor RL, Pister KS (1985) Variational and Projection Methods for the Volume Constraint in Finite Deformation Elasto-Plasticity. Comp. Meth. Appl. Mech. Engng. 51, 177-208

Simo JC, Pister KS (1984) Remarks on Rate Constitutive Equations for Finite Deformation Problems. Comp. Meth. Appl. Mech. Engng. 46, 201-215

Simo JC, Marsden JE (1984) On the Rotated Stress Tensor and the Material Version of the Doyle-Ericksen Formula, Archive for Rational Mechanics and Analysis 86, 213-231

Sokolnikoff IS (1964) Tensor Analysis, Theory and Applications of Continua. John Wiley & Sons, Second Edition

Spencer AJM (1984) Constitutive Theory for Strongly Anisotropic Solids. In: Continuum Theory of the Mechanics of Fibre-Reinforced Composites, ed. by A.J.M. Spencer. Springer Verlag Wien, 3-31

Spencer AJM (1980) Continuum Mechanics. Longman Group, Essex

Stein E, Barthold F-J (1993/94) Unterlagen zur Elastizitätstheorie. Institut für Baumechanik und Numerische Mechanik der Universität Hannover

Stumpf H, Schieck B (1994) Theory and Analysis of Shells Undergoing Finite Elasto-Plastic Strains and Rotations. Acta Mechanica 106, 1-21

Ting TCT (1996) Anisotropic Elasticity: Theory and Applications. Oxford University Press, New York

Ting TCT (1985) Determination of $\mathbf{C}^{1/2}$, $\mathbf{C}^{-1/2}$ and More General Isotropic Tensor Functions of $\mathbf{C}$. Journal of Elasticity 15, 319-323

Treloar LRG (1943) Trans. Faraday Soc., 39-241 ff

Truesdell C (1984) Rational Thermodynamics. Springer-Verlag, New York

Truesdell CA (1977) A First Course in Rational Continuum Mechanics. Academic Press, New York

Truesdell CA, Noll W (1965) The Nonlinear Field Theories. In: Handbuch der Physik, Bd. III/3. Eds.: Flügge, Springer-Verlag, Berlin

Valanis KS, Landel RF (1967) The Strain-Energy Function of a Hyperelastic Material in Terms of the Extension Ratios. J. Appl. Phys. 38, 2997-3002

Wriggers P (1988) Konsistente Linearisierung in der Kontinuumsmechanik und ihre Anwendung auf die Finite-Element-Methode. Forschungs- und Seminarberichte aus dem Bereich der Mechanik der Universität Hannover, Bericht-Nr. F 88/4

Wriggers P, Taylor RL (1990) A Fully Non-Linear Axisymmetrical Membrane Element for Rubber-Like Materials. Engng. Comput. 7, 303-310

Wriggers P, Stein E (1988) Die Verwendung von Lie-Ableitungen bei der Angabe von Spannungsflüssen für große Deformationen. Z. angew. Math. Mech. 68, T264-T267

Index